Security Issues and Solutions in 6G Communications and Beyond

Digvijay Pandey
Department of Technical Education, Government of Uttar Pradesh, India

Binay Kumar Pandey
Department of Information Technology, Govind Ballabh Pant University of Agriculture and Technology, India

Tanveer Ahmad
Tata Consultancy Services, UK

A volume in the Advances in Wireless Technologies and Telecommunication (AWTT) Book Series

Published in the United States of America by
IGI Global
Information Science Reference (an imprint of IGI Global)
701 E. Chocolate Avenue
Hershey PA, USA 17033
Tel: 717-533-8845
Fax: 717-533-8661
E-mail: cust@igi-global.com
Web site: http://www.igi-global.com

Library of Congress Cataloging-in-Publication Data

Nanotechnology Applications and Innovations for Improved Soil Health
Digvijay Pandey, Tanveer Ahmad, Binay Pandey
2024 Information Science Reference

ISBN: 979-8-3693-2931-3
eISBN: 979-8-3693-2932-0

British Cataloguing in Publication Data
A Cataloguing in Publication record for this book is available from the British Library.

The views expressed in this book are those of the authors, but not necessarily of the publisher.

For electronic access to this publication, please contact: eresources@igi-global.com.

Advances in Wireless Technologies and Telecommunication (AWTT) Book Series

Xiaoge Xu
University of Nottingham Ningbo China, China

ISSN:2327-3305
EISSN:2327-3313

Mission

The wireless computing industry is constantly evolving, redesigning the ways in which individuals share information. Wireless technology and telecommunication remain one of the most important technologies in business organizations. The utilization of these technologies has enhanced business efficiency by enabling dynamic resources in all aspects of society.

The **Advances in Wireless Technologies and Telecommunication Book Series** aims to provide researchers and academic communities with quality research on the concepts and developments in the wireless technology fields. Developers, engineers, students, research strategists, and IT managers will find this series useful to gain insight into next generation wireless technologies and telecommunication.

Coverage

- Digital Communication
- Global Telecommunications
- Grid Communications
- Mobile Communications
- Telecommunications

IGI Global is currently accepting manuscripts for publication within this series. To submit a proposal for a volume in this series, please contact our Acquisition Editors at Acquisitions@igi-global.com or visit: http://www.igi-global.com/publish/.

Titles in this Series

For a list of additional titles in this series, please visit: www.igi-global.com/book-series

Radar and RF Front End System Designs for Wireless Systems
Shilpa Mehta (Auckland University of Technology, New Zealand) and Rupesh Kumar (SRM University, India)
Information Science Reference • copyright 2024 • 354pp • H/C (ISBN: 9798369309162) • US $255.00 (our price)

Metamaterial Technology and Intelligent Metasurfaces for Wireless Communication Systems
Shilpa Mehta (Auckland University of Technology, New Zealand) and Arij Naser Abougreen (University of Tripoli, Libya)
Information Science Reference • copyright 2023 • 364pp • H/C (ISBN: 9781668482872) • US $225.00 (our price)

Opportunities and Challenges of Industrial IoT in 5G and 6G Networks
Poshan Yu (Soochow University, China & Australian Studies Centre, Shanghai University, China) Xiaohan Hu (Shanghai University, China) Ajai Prakash (University of Lucknow, India) Nyaribo Wycliffe Misuko (KCA University, Kenya) and Gu Haiyue (Shanghai University, China)
Information Science Reference • copyright 2023 • 304pp • H/C (ISBN: 9781799892663) • US $250.00 (our price)

Applications of Artificial Intelligence in Wireless Communication Systems
Karan Kumar (Maharishi Markandeshwar University (Deemed), Mullana, India)
Information Science Reference • copyright 2023 • 284pp • H/C (ISBN: 9781668473481) • US $250.00 (our price)

Designing and Developing Innovative Mobile Applications
Debabrata Samanta (Rochester Institute of Technology, Kosovo)
Information Science Reference • copyright 2023 • 426pp • H/C (ISBN: 9781668485828) • US $225.00 (our price)

Multidisciplinary Applications of Computer-Mediated Communication
Hung Phu Bui (University of Economics, Ho Chi Minh City, Vietnam) and Raghvendra Kumar (GIET University, India)
Information Science Reference • copyright 2023 • 322pp • H/C (ISBN: 9781668470343) • US $250.00 (our price)

701 East Chocolate Avenue, Hershey, PA 17033, USA
Tel: 717-533-8845 x100 • Fax: 717-533-8661
E-Mail: cust@igi-global.com • www.igi-global.com

Table of Contents

Digvijay Pandey, Department of Technical Education, Government of Uttar Pradesh, India
Sumeet Goyal, Chandigarh Group of Colleges, Landran, India
Kallol Bhaumik, Malla Reddy Engineering College and Management Sciences, India
Sonia Suneja, Chandigarh University, India
Manvinder Sharma, Chitkara University Institute of Engineering and Technology, Chitkara Univeristy, Punjab, India
Pankaj Dadheech Dadheech, Swami Keshvanand Institute of Technology, Management, and Gramothan, Jaipur, India

Rajneesh Talwar, Chitkara University Institute of Engineering and Technology, Chitkara Univeristy, Punjab, India
Manvinder Sharma, Chitkara University Institute of Engineering and Technology, Chitkara Univeristy, Punjab, India
Satyajit Anand, Chitkara University Institute of Engineering and Technology, Chitkara Univeristy, Punjab, India
Digvijay Pandey, Dr. APJ Abdul Kalam, India

D. Vetrithangam, Department of Computer Science and Engineering, Chandigarh University, India
Komala C. R., Department of Information Science and Engineering, HKBK College of Engineering, Bengaluru, India
D. Sugumar, Karunya Institute of Technology and Sciences (Deemed), Coimbatore, India
Y. Mallikarjuna Rao, Department of Electronics and Communication Engineering, Santhiram Engineering College, India
C. Karpagavalli, Department of Artificial Intelligence and Data Science, Ramco Institute of Technology, Rajapalayam, India
B. Kiruthiga, Velammal College of Engineering and Technology, Madurai, India

Detailed Table of Contents

Chapter 1

  Digvijay Pandey, Department of Technical Education, Government of Uttar Pradesh, India
  Sumeet Goyal, Chandigarh Group of Colleges, Landran, India
  Kallol Bhaumik, Malla Reddy Engineering College and Management Sciences, India
  Sonia Suneja, Chandigarh University, India
  Manvinder Sharma, Chitkara University Institute of Engineering and Technology, Chitkara Univeristy, Punjab, India
  Pankaj Dadheech Dadheech, Swami Keshvanand Institute of Technology, Management, and Gramothan, Jaipur, India

6G networks are the next frontier in wireless communication, promising unprecedented speeds, lower latencies and advanced capabilities. However, with greater connectivity comes increased security risks. This chapter provides a systematic review of security issues in 6G networks and communication systems. A comprehensive search of academic databases is conducted for studies published between 2020-2023 that examined security related issues of 6G. The studies indicate authentication, privacy and trust will be major concerns due to new technologies like reconfigurable intelligent surfaces and cell free architectures. Specific vulnerabilities include impersonation attacks, side channel leaks, and insider threats from rogue base stations. End to end encryption, blockchain, and machine learning are emerging security mechanisms to address these issues. This review synthesizes the current work on 6G security, highlights critical challenges and opportunities, and provides a framework for the future.

Rajneesh Talwar, Chitkara University Institute of Engineering and Technology, Chitkara Univeristy, Punjab, India

Manvinder Sharma, Chitkara University Institute of Engineering and Technology, Chitkara Univeristy, Punjab, India

Satyajit Anand, Chitkara University Institute of Engineering and Technology, Chitkara Univeristy, Punjab, India

Digvijay Pandey, Dr. APJ Abdul Kalam, India

6G wireless networks are expected to provide extremely high data rates, very low latency and connectivity for a massive number of devices. However, with the rapid development of 6G networks, various new security threats and challenges are also emerging. Artificial intelligence (AI) is considered as a promising technology to address these security issues in 6G networks. In this chapter, a comprehensive review of the role of AI in tackling security challenges in 6G communications is discussed. Firstly, an overview of 6G networks and discusses various new use cases and requirements of 6G. Then outline major security threats and challenges in 6G networks such as spoofing attacks, distributed denial of service attacks, eavesdropping, malicious software attacks, and attacks on machine learning models are discussed. The outline open research challenges and future directions for applying AI to address security issues in beyond 5G and 6G networks are also discussed.

D. Vetrithangam, Department of Computer Science and Engineering, Chandigarh University, India

Komala C. R., Department of Information Science and Engineering, HKBK College of Engineering, Bengaluru, India

D. Sugumar, Karunya Institute of Technology and Sciences (Deemed), Coimbatore, India

Y. Mallikarjuna Rao, Department of Electronics and Communication Engineering, Santhiram Engineering College, India

C. Karpagavalli, Department of Artificial Intelligence and Data Science, Ramco Institute of Technology, Rajapalayam, India

B. Kiruthiga, Velammal College of Engineering and Technology, Madurai, India

The incorporation of Artificial Intelligence (AI) into 6G security measures and its revolutionary effect on the telecommunications industry are examined in this paper. We start off by talking about how important it is to integrate AI in order to strengthen overall security posture, improve network defenses, and address vulnerabilities. We clarify AI's critical role in promoting innovation and efficiency within the telecom industry by analyzing 6G network vulnerabilities and the importance of enhanced security measures. The use of AI for 6G security and its potential for threat detection, incident response, and network optimization through machine learning techniques.

Hemlata Patel, Parul University, India

Anupma Surya, Greater Noida Institute of Technology, India

Aswini Kilaru, Institute of Aeronautical Engineering, Dundigal, India

Vijilius Helena Raj, Department of Applied Sciences, New Horizon College of Engineering, Bangalore, India

Amit Dutt, Lovely Professional University, India

Joshuva Arockia Dhanraj, Chandigarh University, India

Quantum computing represents a significant milestone in technological progress, offering unprecedented computational capabilities. Nevertheless, this inherent potential also poses a significant risk to the field of cybersecurity. This essay aims to examine the complexities associated with quantum threats and investigate potential strategies for mitigating these risks. Cryptography is a fundamental component of contemporary cybersecurity, serving as a resilient method that guarantees the preservation of data integrity and confidentiality within IT infrastructure. However, the emergence of quantum computing poses a significant challenge to this established paradigm.

Raviteja Kocherla, Department Computer Science and Engineering, Mallareddy University, Hyderabad, India

Yagya Dutta Dwivedi, Department of Aeronautical Engineering, Institute of Aeronautical Engineering, Hyderabad, India

B. Ardly Melba Reena, Saveetha Institute of Medical and Technical Sciences, Saveetha University, Chennai, India

Komala C. R., Department of Information Science and Engineering, HKBK College of Engineering, Bengaluru, India

Jennifer D., Department of Computer Science and Engineering, Panimalar Engineering College, Chennai, India

Joshuva Arockia Dhanraj, Dayananda Sagar University, India

With a special emphasis on distributed AI/ML systems, the abstract explores the intricate world of adversarial challenges in 6G networks. With their unmatched capabilities—like ultra-high data rates and ultra-low latency it highlights the crucial role that 6G networks will play in determining the direction of communication in the future. Still, it also highlights the weaknesses in these networks' distributed AI/ML systems, emphasizing the need for strong security measures to ward off potential attacks. Also covered in the abstract is the variety of adversarial threats that exist, such as model evasion, data poisoning, backdoors, membership inference, and model inversion attacks.

Freddy Ochoa-Tataje, Universidad César Vallejo, Lima, Peru
Joel Alanya-Beltran, Universidad San Ignacio de Loyola, Peru
Jenny Ruiz-Salazar, Universidad Privada del Norte, Lima, Peru
Juan Paucar-Elera, Universidad Nacional Federico Villarreal, Lima, Peru
Michel Mendez-Escobar, Universidad Autónoma del Perú, Lima, Peru
Frank Alvarez-Huertas, Universidad Nacional Mayor de San Marcos, Lima, Peru

With the emergence of 6G communication networks, it is becoming more and more clear that we need to prioritize strong resilience against evolving cyber threats. Conventional security measures, although they have their merits, are no longer adequate to protect the complex structure of 6G networks. Utilizing artificial intelligence (AI) in this context offers a revolutionary method to enhance communication resilience in the 6G era. The importance of AI in strengthening the resilience of 6G communication lies in its ability to analyze data and respond accordingly. With the help of machine learning algorithms, AI has the ability to identify patterns that may signal security breaches.

Uchit Kapoor, Arthkalp, India
Sunita Sunil Shinde, Nanasaheb Mahadik College of Engineering and Technology, India
Budesh Kanwer, Department of Artificial Intelligence and Data Science, Poornima Institute of Engineering and Technology, India
Sonia Duggal, Manav Rachna International Institute of Research and Studies, Faridabad, India
Lavish Kansal, Lovely Professional University, India
Joshuva Arockia Dhanraj, Chandigarh University, India

In the age of intelligent connectivity, it is vital to understand how crucial it is to protect these networks as the world prepares for the arrival of 6G networks. Sixth-generation (6G) networks are anticipated to transform a number of sectors, including manufacturing, transportation, and healthcare, by offering previously unheard-of speeds, capacities, and connectivity. These opportunities do, however, come with a number of difficulties, especially in the areas of cybersecurity and data privacy. Understanding the particular threats that 6G networks face is essential to protecting them effectively.

M. Beulah Viji Christiana, Department of Master of Business Administration, Panimalar Engineering College, Chennai, India
Thaya Madhavi, Mohan Babu University, Tirupati, India
S. Bathrinath, Kalasalingam Academy of Research and Education, Krishnankoil, India
Vellayan Srinivasan, S.A. Engineering College, Chennai, India
Mano Ashish Tripathi, Motilal Nehru National Institute of Technology, Allahabad, India
B. Uma Maheswari, St. Joseph's College of Engineering, India
Pankaj Dadheech, Swami Keshvanand Institute of Technology, Management, and Gramothan, India

The forthcoming transition to 6G brings with it the promise of holographic connectivity, which has the potential to revolutionize communication in a variety of fields, including marketing, education, medicine, business, and entertainment, among others. Users are able to interact with one another through the use of high-quality 3D representations while overcoming geographical barriers thanks to this technology. On the other hand, in order to successfully navigate this transition, effective leadership is required. This leadership must be able to anticipate technological advancements, encourage innovation, and strike a balance between risks and opportunities within the 6G ecosystem.

Shefali, Department of Management, Institute of Innovation in Technology and Management, Janakpuri, India
S. Prema, Department of Information Technology, Mahendra Engineering College, Namakkal, India
Vishal Ashok Ingole, P.R. Pote Patil College of Engineering and Management, India
G. Vikram, Karunya Institute of Technology and Sciences, Coimbatore, India
Smita M. Gaikwad, CMS B-School, Jain University (Deemed), India
H. Mickle Aancy, Department of Master of Business Administration, Panimalar Engineering College, Chennai, India
Pankaj Dadheech, Swami Keshvanand Institute of Technology, Management, and Gramothan, India

This chapter delves into the challenges posed by the advent of 6G technology from a managerial standpoint, particularly focusing on security solutions. As the telecommunications landscape evolves rapidly, it becomes imperative for managers to navigate the intricacies of ensuring robust security measures amidst technological advancements. Through strategic insights, this chapter explores the complexities associated with 6G security and provides managerial perspectives aimed at fostering proactive and effective security strategies.

    Chetan Thakar, Savitribai Phule Pune University, India

    Rashi Saxena, Department of AIMLE, Gokaraju Rangaraju Institute of Engineering and
        Technology, Hyderabad, India

    Modi Himabindu, Institute of Aeronautical Engineering, Dundigal, India

    C. Rakesh, Department of Mechanical Engineering, New Horizon College of Engineering,
        Bangalore, India

    Amit Dutt, Lovely Professional University, India

    Joshuva Arockia Dhanraj, Dayananda Sagar University, India

Establishing strong security protocols is crucial in the quickly changing internet of things (IoT) environment to reduce potential risks and weaknesses. It is possible to manage security risks more effectively, but there are also challenges because of the interconnected nature of IoT devices, the introduction of 6G networks, and the incorporation of distributed ledger technology (DLT). The focus of this note is on proactive methods of protecting infrastructure and sensitive data. It explores different management strategies that are intended to mitigate threats in IoT environments. Using a security-by-design methodology is a fundamental tactic for threat mitigation in internet of things settings. Every phase of the lifecycle of an IoT device, from design and development to deployment and operation, must incorporate security measures.

    Freddy Ochoa-Tataje, Universidad César Vallejo, Lima, Peru

    Joel Alanya-Beltran, Universidad San Ignacio de Loyola, Peru

    Jenny Ruiz-Salazar, Universidad Privada del Norte, Lima, Peru

    Juan Paucar-Elera, Universidad Nacional Federico Villarreal, Lima, Peru

    Michel Mendez-Escobar, Universidad Autónoma del Perú, Lima, Peru

    Frank Alvarez-Huertas, Universidad Nacional Mayor de San Marcos, Lima, Peru

The idea of the digital twin has become a game-changer in today's quickly developing technological environment, revolutionizing everything from manufacturing to urban planning. A digital twin replicates the behavior and qualities of real-world physical objects, processes, services, or environments in a virtual space. This essay explores the nuances of the digital twin idea, its uses, and its importance in influencing technology in the future. A digital twin is a dynamic simulation that uses real-world data to predict behavior and performance, going beyond a static representation.

Chapter 12

Binay Kumar Pandey, College of Technology, Govind Ballabh Pant University of Agriculture and Technology, India

Mukundan Appadurai Paramashivan, Aligarh Muslim University, India & Champions Group, India

Digvijay Pandey, Department of Technical Education, Government of Uttar Pradesh, India

A. Shaji George, Almarai Company, Riyadh, Saudi Arabia

Ashi Agarwal, Department of Computer Science, ABES Engineering College, Ghaziabad, India

Darshan A. Mahajan, NICMAR University, Pune, India

Pankaj Dadheech Dadheech, Swami Keshvanand Institute of Technology, Management, and Gramothan, India

Sabyasachi Pramanik, Department of Computer Science and Engineering, Haldia Institute of Technology, India

Initiating the study into digital twin technology, the planning and implementation of the 6G network necessitates real-time interaction and alignment between physical systems and their virtual representation. From simple parts to intricate systems, the digital twin's flexibility and agility improve design and operational procedure efficiency in a predictable manner. It can validate policies, give a virtual representation of a physical entity, or evaluate how a system or entity behaves in a real-time setting. It evaluates the effectiveness and suitability of QoS regulations in 6G communication, in addition to the creation and management of novel services. Physical system maintenance costs and security threats can also be reduced, but doing so requires standardization efforts that open the door to previously unheard-of difficulties with fault tolerance, efficiency, accuracy, and security. The fundamental needs of a digital twin that are focused on 6G communication are covered in this chapter. These include decoupling, scalable intelligent analytics, data management using blockchain.

Chapter 13

Binay Kumar Pandey, College of Technology, Govind Ballabh Pant University of Agriculture and Technology, India

Digvijay Pandey, Department of Technical Education, Government of Uttar Pradesh, India

Ashi Agarwal, Department of Computer Science, ABES Engineering College, Ghaziabad, India

Darshan A. Mahajan, NICMAR University, Pune, India

Pankaj Dadheech Dadheech, Swami Keshvanand Institute of Technology, Management, and Gramothan, India

A. Shaji George, Almarai Company, Riyadh, Saudi Arabia

Pankaj Kumar Rai, Loxoft, USA

This chapter describes the different newly adopted 6G technologies, along with any security risks and potential fixes. The primary 6G technologies that will open up a whole new universe of possibilities are AI/ML, DLT, quantum computing, VLC, and THz communication. The emergence of new generation information and communication technologies, including blockchain technology, virtual reality/augmented reality/extended reality, internet of things, and artificial intelligence, gave rise to the 6G communication network. The intelligence process of communication development, which includes holographic, pervasive, deep, and intelligent connectivity, is significantly impacted by the development of 6G.

Chapter 14

Brajesh Kumar Khare, Harcourt Butler Technical University, India

Deshraj Sahu, Dr. A.P.J. Abdul Kalam Technical University, India

Digvijay Pandey, Dr. A.P.J. Abdul Kalam Technical University, India

Mamta Tiwari, Chhatrapati Shahu Ji Maharaj University, India

Hemant Kumar, Chhatrapati Shahu Ji Maharaj University, India

Nigar Siddiqui, Dr. Virendra Swarup Memorial Trust Group of Institutions, India

The chapter explores the use of machine learning (ML) in detecting and addressing anomalies in advanced 6G communication systems. It emphasizes the drawbacks of conventional approaches and delves into ML algorithms that are appropriate for identifying anomalies, such as clustering, classification, and deep learning. The study focuses on the difficulties of choosing important features from various data sources in 6G networks, including network traffic and device behavior. It also explores possible attacks on ML models and suggests ways to improve their resilience. Exploring integration with network slicing and highlighting the adaptability of ML to dynamic virtualized networks. The chapter highlights the importance of ML-based anomaly detection in strengthening 6G network security and suggests areas for future research.

Manvinder Sharma, Chitkara University Institute of Engineering and Technology, Chitkara University, Punjab, India

Rajneesh Talwar, Chitkara University Institute of Engineering and Technology, Chitkara University, Punjab, India

Satyajit Anand, Chitkara University Institute of Engineering and Technology, Chitkara University, Punjab, India

Jyoti Bhola, Chitkara University Institute of Engineering and Technology, Chitkara University, Punjab, India

Digvijay Pandey, Dr. APJ Abdul Kalam, India

The upcoming 6G wireless networks aims to provide substantial improvements over existing 5G networks. These include improvements in data rate, latency, reliability, and connectivity. To enable these goals, significant advancements in all aspects of wireless communication systems are required. The major factor of communication hardware is antenna design. System on chip (SoC) antennas can be directly integrated with transceiver circuitry. These designs can meet the demanding performance requirements of 6G networks. In this chapter, recent advancements in SoC antenna technology for 6G communication systems are discussed. The key challenges that need to be addressed are also discussed. Literature related to the development of wideband millimeter wave antennas with beamforming capabilities is presented. Research directions related to modeling, design and implementation are identified to guide future work in this area.

Chapter 16

Ila Dixit, Department of Management, Maharaja Agrasen International College, Raipur, India

Anslin Jegu, Department of English, Panimalar Engineering College, Poonamallee, India

Ashok Kumar Digal, PG Department of Education, Rama Devi Women's University, Vidya Vihar Bhubaneswar, India

T. Rajesh Kumar, Saveetha School of Engineering, Saveetha Institute of Medical and Technical Sciences, Chennai, India

Neha Munjal, Department of Physics, Lovely Professional University, India

Baby Shamini P., R.M.K. Engineering College, India

The shift from 5G to 6G communication landscapes ushers in a paradigm shift that is characterized by technological advancements that have never been seen before and increased security imperatives. This chapter delves into the complex world of 6G security, shedding light on the myriad of dangers and difficulties that these networks are confronted with. Alongside the implementation of zero trust architecture, it highlights the critical importance of confidentiality, integrity, and availability (also known as the CIA Triad) by conducting an exhaustive investigation of threat actors, security principles, and encryption techniques. In order to strengthen the security of 6G transactions and data, advanced encryption techniques such as post-quantum cryptography and blockchain technology have emerged as essential tools.

Preface

The rapid evolution of Information and communication technologies (ICTs) such as artificial intelligence (AI), virtual reality (VR), augmented reality (AR), extended reality (XR), the internet of things (IoT), and blockchain technology has set the stage for the emergence of the 6G communication network. As we step into this new era, 6G promises to revolutionize the way we connect and interact, offering intelligent, deep, holographic, and ubiquitous connectivity. However, with these advancements come significant security and privacy challenges that must be addressed to realize the full potential of 6G.

This edited reference book, *Security Issues and Solutions in 6G Communications and Beyond*, aims to provide a comprehensive overview of the security landscape for 6G networks. The development of 6G introduces novel requirements, security key performance indicators (KPIs), and a unique network architecture, all of which bring forth new applications and enabling technologies. Consequently, the security challenges we face are multifaceted, encompassing UAV-based mobility, holographic telepresence, connected autonomous vehicles, smart grid 2.0, Industry 5.0, intelligent healthcare, digital twins, and more.

In this book, we delve into various security solutions tailored for 6G and beyond. Key areas of focus include distributed ledger technology (DLT), physical layer security, quantum communication, and distributed AI/ML. For instance, DLT threats such as majority attacks, double spending attacks, and Sybil attacks are discussed, along with potential solutions like proper access control and authentication mechanisms. Physical layer security challenges, such as those posed by Terahertz technology and visible light communications, are examined, with solutions like multipath transmission and enhanced secrecy performance through multiple input multiple output (MIMO) technology. Quantum communication vulnerabilities, including quantum cloning and quantum collision attacks, are addressed with lattice-based, code-based, hash-based, and multivariate-based cryptographic techniques. Moreover, the book explores methods to counteract attacks on distributed and scalable AI/ML, such as poisoning attacks and evasion attacks, using adversarial training and defensive distillation.

Our objective is to provide a roadmap for materializing the 6G security vision into reality. This book is intended for a diverse audience, including academics, managers, public and private organizations, and the general public. We believe that the insights and findings presented here will be particularly beneficial for postgraduate students and researchers with a keen interest in security issues in advanced communication networks.

This publication not only aims to contribute to the current body of knowledge but also aspires to inspire further research and innovation in the field. By disseminating cutting-edge research on security issues and solutions in 6G communications and beyond, we hope to pave the way for a more secure and resilient communication infrastructure in the new era.

We extend our gratitude to all contributors, reviewers, and supporters of this book. Their invaluable insights and dedication have made this comprehensive compilation possible. We are confident that this book will serve as a valuable resource for anyone interested in the future of 6G communications and the security challenges that lie ahead.

ORGANIZATION OF THE BOOK

Chapter 1: A Systematic Review of Security Issues in 6G Networks and Communication

Digvijay Pandey, Sumeet Goyal, Kallol Bhaumik, Sonia Suneja, Manvinder Sharma, Pankaj Dadheech Dadheech

6G networks are the next frontier in wireless communication, promising unprecedented speeds, lower latencies, and advanced capabilities. However, with greater connectivity comes increased security risks. This paper provides a systematic review of security issues in 6G networks and communication systems. A comprehensive search of academic databases is conducted for studies published between 2020-2023 that examined security related issues of 6G. The studies indicate authentication, privacy and trust will be major concerns due to new technologies like reconfigurable intelligent surfaces and cell free architectures. Specific vulnerabilities include impersonation attacks, side channel leaks and insider threats from rogue base stations. End to end encryption, blockchain and machine learning are emerging security mechanisms to address these issues. This review synthesizes the current work on 6G security, highlights critical challenges and opportunities and provides a framework for future.

Chapter 2: A Review on Role of Artificial Intelligence in Security Challenges for 6G Communications

Rajneesh Talwar, Manvinder Sharma, Satyajit Anand, Digvijay Pandey

6G wireless networks are expected to provide extremely high data rates, very low latency, and connectivity for a massive number of devices. However, with the rapid development of 6G networks, various new security threats and challenges are also emerging. Artificial Intelligence (AI) is considered as a promising technology to address these security issues in 6G networks. In this paper, a comprehensive review of the role of AI in tackling security challenges in 6G communications is discussed. Firstly, an overview of 6G networks and discusses various new use cases and requirements of 6G. Then outline major security threats and challenges in 6G networks such as spoofing attacks, distributed denial of service attacks, eavesdropping, malicious software attacks and attacks on machine learning models are discussed. The outline open research challenges and future directions for applying AI to address security issues in beyond 5G and 6G networks are also discussed.

Chapter 3: AI's Role in 6G Security Machine Learning Solutions Unveiled

D. Vetrithangam, Komala C R, D. Sugumar, Y Mallikarjuna Rao, C. Karpagavalli, B. Kiruthiga

The incorporation of Artificial Intelligence (AI) into 6G security measures and its revolutionary effect on the telecommunications industry are examined in this paper. We start off by talking about how important it is to integrate AI in order to strengthen overall security posture, improve network defenses, and address vulnerabilities. We clarify AI's critical role in promoting innovation and efficiency within the telecom industry by analyzing 6G network vulnerabilities and the importance of enhanced security measures. The use of AI for 6G security and its potential for threat detection, incident response, and network optimization through machine learning techniques.

Chapter 4: Quantum Threats, Quantum Solutions ML Approaches to 6G Security

Hemlata Patel, Anupma Surya, Aswini Kilaru, Vijilius Helena Raj, Amit Dutt, Joshuva Arockia Dhanraj

Quantum computing represents a significant milestone in technological progress, offering unprecedented computational capabilities. Nevertheless, this inherent potential also poses a significant risk to the field of cybersecurity. This essay aims to examine the complexities associated with quantum threats and investigate potential strategies for mitigating these risks. Cryptography is a fundamental component of contemporary cybersecurity, serving as a resilient method that guarantees the preservation of data integrity and confidentiality within IT infrastructure. However, the emergence of quantum computing poses a significant challenge to this established paradigm.

Chapter 5: Adversarial Challenges in Distributed AI ML Safeguarding 6G Networks

Raviteja Kocherla, Yagya Dutta Dwivedi, B Ardly Melba Reena, Komala C R, Jennifer D, Joshuva Arockia Dhanraj

With a special emphasis on distributed AI/ML systems, this chapter explores the intricate world of adversarial challenges in 6G networks. With their unmatched capabilities—like ultra-high data rates and ultra-low latency it highlights the crucial role that 6G networks will play in determining the direction of communication in the future. Still, it also highlights the weaknesses in these networks' distributed AI/ML systems, emphasizing the need for strong security measures to ward off potential attacks. Also covered in the abstract is the variety of adversarial threats that exist, such as model evasion, data poisoning, backdoors, membership inference, and model inversion attacks.

Chapter 6: Beyond Traditional Security Harnessing AI for 6G Communication Resilience

Freddy Ochoa-Tataje, Joel Alanya-Beltran, Jenny Ruiz-Salazar, Juan Paucar-Elera, Michel Mendez-Escobar, Frank Alvarez-Huertas

With the emergence of 6G communication networks, it is becoming more and more clear that we need to prioritize strong resilience against evolving cyber threats. Conventional security measures, although they have their merits, are no longer adequate to protect the complex structure of 6G networks. Utilizing artificial intelligence (AI) in this context offers a revolutionary method to enhance communication resilience in the 6G era. The importance of AI in strengthening the resilience of 6G communication lies in its ability to analyze data and respond accordingly. With the help of machine learning algorithms, AI has the ability to identify patterns that may signal security breaches.

Chapter 7: Strategic Insights Safeguarding 6G Networks in the Era of Intelligent Connectivity

Uchit Kapoor, Sunita Shinde, Budesh Kanwer, Sonia Duggal, Lavish Kansal, Joshuva Arockia Dhanraj

In the age of intelligent connectivity, it is vital to understand how crucial it is to protect these networks as the world prepares for the arrival of 6G networks. Sixth-generation (6G) networks are anticipated to transform a number of sectors, including manufacturing, transportation, and healthcare, by offering previously unheard-of speeds, capacities, and connectivity. These opportunities do, however, come with a number of difficulties, especially in the areas of cybersecurity and data privacy. Understanding the particular threats that 6G networks face is essential to protecting them effectively.

Chapter 8: Leadership in the Age of Holographic Connectivity Securing the Future of 6G

M. Beulah Viji Christiana, T. Madhavi, S. Bathrinath, Vellayan Srinivasan, Mano Ashish Tripathi, B. Uma Maheswari, Pankaj Dadheech

The forthcoming transition to 6G brings with it the promise of holographic connectivity, which has the potential to revolutionize communication in a variety of fields, including marketing, education, medicine, business, and entertainment, among others. Users are able to interact with one another through the use of high-quality 3D representations while overcoming geographical barriers thanks to this technology. On the other hand, in order to successfully navigate this transition, effective leadership is required. This leadership must be able to anticipate technological advancements, encourage innovation, and strike a balance between risks and opportunities within the 6G ecosystem.

Chapter 9: Navigating 6G Challenges A Managerial Perspective on Security Solutions Strategic Insights

Shefali ., S. Prema, Vishal Ingole, G. Vikram, Smita. M. Gaikwad, H. Mickle Aancy, Pankaj Dadheech

This chapter delves into the challenges posed by the advent of 6G technology from a managerial standpoint, particularly focusing on security solutions. As the telecommunications landscape evolves rapidly, it becomes imperative for managers to navigate the intricacies of ensuring robust security measures amidst technological advancements. Through strategic insights, this chapter explores the complexities associated with 6G security and provides managerial perspectives aimed at fostering proactive and effective security strategies.

Chapter 10: Distributed Ledger Technology in 6G Management Strategies for Threat Mitigation

Chetan Thakar, Rashi Saxena, Modi Himabindu, Rakesh C, Amit Dutt, Joshuva Arockia Dhanraj

Establishing strong security protocols is crucial in the quickly changing Internet of Things (IoT) environment to reduce potential risks and weaknesses. It is possible to manage security risks more effectively, but there are also challenges because of the interconnected nature of IoT devices, the introduction of 6G networks, and the incorporation of Distributed Ledger Technology (DLT). The focus of this note is on proactive methods of protecting infrastructure and sensitive data. It explores different management strategies that are intended to mitigate threats in IoT environments. Using a Security-by-Design methodology is a fundamental tactic for threat mitigation in Internet of Things settings. Every phase of the lifecycle of an IoT device, from design and development to deployment and operation, must incorporate security measures.

Chapter 11: Defending the Digital Twin Machine Learning Strategies for 6G Protection

Freddy Ochoa-Tataje, Joel Alanya-Beltran, Jenny Ruiz-Salazar, Juan Paucar-Elera, Michel Mendez-Escobar, Frank Alvarez-Huertas

The idea of the "Digital Twin" has become a game-changer in today's quickly developing technological environment, revolutionizing everything from manufacturing to urban planning. A digital twin replicates the behavior and qualities of real-world physical objects, processes, services, or environments in a virtual space. This essay explores the nuances of the Digital Twin idea, its uses, and its importance in influencing technology in the future. A digital twin is a dynamic simulation that uses real-world data to predict behavior and performance, going beyond a static representation.

Chapter 12: Future Directions of Digital Twin Architectures for 6G Communication Networks

Binay Kumar Pandey, Mukundan Appadurai Paramashivan, Digvijay Pandey, A.Shaji George, Ashi Agarwal, Darshan Mahajan, Pankaj Dadheech Dadheech, Sabyasachi Pramanik

Initiating the study into Digital Twin Technology, the planning and implementation of the 6G network necessitates real-time interaction and alignment between physical systems and their virtual representation. From simple parts to intricate systems, the Digital Twin's flexibility and agility improve design and operational procedure efficiency in a predictable manner. It can validate policies, give a virtual representation of a physical entity, or evaluate how a system or entity behaves in a real-time setting. It evaluates the effectiveness and suitability of QoS regulations in 6G communication, in addition to the creation and management of novel services. Physical system maintenance costs and security threats can also be reduced, but doing so requires standardization efforts that open the door to previously unheard-of difficulties with fault tolerance, efficiency, accuracy, and security. The fundamental needs of a digital twin that are focused on 6G communication are covered in this chapter. These include decoupling, scalable intelligent analytics, data management using blockchain.

Chapter 13: Beyond Data Breaches-Enhancing Security in 6G Communications

Binay Kumar Pandey, Digvijay Pandey, Ashi Agarwal, Darshan Mahajan, Dr. Pankaj Dadheech Dadheech, A. Shaji George, Pankaj Kumar Rai

This chapter discuss focused on future-enabling technology, 6G must solve deployment issues by 2030. This chapter describes the different newly adopted 6G technologies, along with any security risks and potential fixes. The primary 6G technologies that will open up a whole new universe of possibilities are AI/ML, DLT, quantum computing, VLC, and THz communication. The emergence of new generation information and communication technologies, including blockchain technology, virtual reality/augmented reality/extended reality, Internet of Things, and artificial intelligence, gave rise to the 6G communication network. The intelligence process of communication development, which includes holographic, pervasive, deep, and intelligent connectivity, is significantly impacted by the development of 6G.

Chapter 14: Exploring Machine Learning Solutions for Anomaly Detection in 6G Communication

Systems Brajesh Khare, Deshraj Sahu, Digvijay Pandey, Mamta Tiwari, Hemant Kumar, Nigar Siddiqui

This chapter explores the use of machine learning (ML) in detecting and addressing anomalies in advanced 6G communication systems. It emphasizes the drawbacks of conventional approaches and delves into ML algorithms that are appropriate for identifying anomalies, such as clustering, classification, and deep learning. The study focuses on the difficulties of choosing important features from various data sources in 6G networks, including network traffic and device behavior. It also explores possible attacks on ML models and suggests ways to improve their resilience. Exploring integration with network slicing, highlighting the adaptability of ML to dynamic virtualized networks. The paper highlights the importance of ML-based anomaly detection in strengthening 6G network security and suggests areas for future research.

Chapter 15: Advancements and Challenges in System on Chip Antennas (SoC) for 6G Communication

Manvinder Sharma, Rajneesh Talwar, Satyajit Anand, Jyoti Bhola, Digvijay Pandey

The upcoming 6G wireless networks aims to provide substantial improvements over existing 5G networks. These include improvements in data rate, latency, reliability and connectivity. To enable these goals, significant advancements in all aspects of wireless communication systems are required. The major factor of communication hardware is antenna design. System on chip (SoC) antennas can be directly integrated with transceiver circuitry. These designs can meet the demanding performance requirements of 6G networks. In this paper, recent advancements in SoC antenna technology for 6G communication systems are discussed. The key challenges that need to be addressed are also discussed. The literature related to development of wideband millimeter wave antennas with beamforming capabilities is presented. Research directions related to modeling, design and implementation are identified to guide future work in this area.

Chapter 16: Teaching Tomorrow's Security A Curriculum for 6G Communication Challenges

Ila Dixit, Anslin Jegu, Ashok Kumar Digal, T. Rajesh Kumar, Neha Munjal, Baby Shamini P

The shift from 5G to 6G communication landscapes ushers in a paradigm shift that is characterized by technological advancements that have never been seen before and increased security imperatives. This article delves into the complex world of 6G security, shedding light on the myriad of dangers and difficulties that these networks are confronted with. Alongside the implementation of Zero Trust Architecture, it highlights the critical importance of confidentiality, integrity, and availability (also known as the CIA Triad) by conducting an exhaustive investigation of threat actors, security principles, and encryption techniques. In order to strengthen the security of 6G transactions and data, advanced encryption techniques such as Post-Quantum Cryptography and Blockchain Technology have emerged as essential tools.

IN CONCLUSION

As we conclude this comprehensive exploration of *Security Issues and Solutions in 6G Communications and Beyond*, it is evident that the advent of 6G technology brings with it a paradigm shift in the realm of connectivity. The insights and research presented within this book underscore the intricate and multi-faceted nature of the security and privacy challenges that accompany the advancements in 6G networks.

The rapid integration of cutting-edge technologies such as AI, VR/AR/XR, IoT, and blockchain into the 6G ecosystem necessitates a robust and dynamic approach to security. Our contributors have meticulously analyzed these challenges and proposed innovative solutions that pave the way for secure and resilient 6G networks. From addressing the vulnerabilities in distributed ledger technology and physical layer security to tackling the complexities of quantum communication and distributed AI/ML, this book offers a holistic view of the security landscape in the context of 6G.

The implications of 6G extend far beyond mere technological advancements; they encompass a transformative impact on various sectors, including healthcare, transportation, industry, and beyond. The discussions on UAV-based mobility, holographic telepresence, connected autonomous vehicles, smart grid 2.0, Industry 5.0, and intelligent healthcare illustrate the vast potential and the accompanying security imperatives of these innovations.

This book serves not only as a reference for current security issues but also as a beacon for future research and development. We hope that it will inspire academics, industry professionals, policymakers, and students to further explore and address the security challenges of 6G communications. The roadmap provided herein offers a strategic direction for materializing the vision of a secure 6G future.

We are deeply grateful to all the contributors who have shared their expertise and insights, enriching this book with their profound knowledge. Their collective efforts have culminated in a resource that we believe will be invaluable for anyone engaged in the field of 6G communications.

As we look forward to the future, it is clear that the journey towards secure and efficient 6G networks is ongoing. It is our hope that this book will serve as a catalyst for continued innovation and collaboration, driving us closer to the realization of a secure, connected, and intelligent world.

Digvijay Pandey

Department of Technical Education, Government of Uttar Pradesh, India

Binay Kumar Pandey

Department of Information Technology, Govind Ballabh Pant University of Agriculture and Technology, India

Tanveer Ahmad

Tata Consultancy Services, UK

Chapter 1
A Systematic Review of Security Issues in 6G Networks and Communication

Digvijay Pandey

https://orcid.org/0000-0003-0353-174X

Department of Technical Education, Government of Uttar Pradesh, India

Sumeet Goyal

Chandigarh Group of Colleges, Landran, India

Kallol Bhaumik

https://orcid.org/0000-0003-2469-1937

Malla Reddy Engineering College and Management Sciences, India

Sonia Suneja

Chandigarh University, India

Manvinder Sharma

https://orcid.org/0000-0001-9158-0466

Chitkara University Institute of Engineering and Technology, Chitkara Univeristy, Punjab, India

Pankaj Dadheech Dadheech

https://orcid.org/0000-0001-5783-1989

Swami Keshvanand Institute of Technology, Management, and Gramothan, Jaipur, India

ABSTRACT

6G networks are the next frontier in wireless communication, promising unprecedented speeds, lower latencies and advanced capabilities. However, with greater connectivity comes increased security risks. This chapter provides a systematic review of security issues in 6G networks and communication systems. A comprehensive search of academic databases is conducted for studies published between 2020-2023 that examined security related issues of 6G. The studies indicate authentication, privacy and trust will be major concerns due to new technologies like reconfigurable intelligent surfaces and cell free architectures. Specific vulnerabilities include impersonation attacks, side channel leaks, and insider threats from rogue base stations. End to end encryption, blockchain, and machine learning are emerging security mechanisms to address these issues. This review synthesizes the current work on 6G security, highlights critical challenges and opportunities, and provides a framework for the future.

DOI: 10.4018/979-8-3693-2931-3.ch001

INTRODUCTION

With 6G networks coming globally, wireless communication systems (Devasenapathy, D. et al., 2024) are entering a new age marked by lofty performance expectations that much exceed 5G capabilities. In comparison with 5G, 6G (Kirubasri, G. et al., 2021) promises to provide 1,000 times more capacity, 10 times reduced latency, with speeds of up to 1 terabit per second during early installations (Chowdhury et al., 2020). According to Saad et al. (2019), this enormous jump will allow for mission-critical services in the industrial, transportation, healthcare, and public safety sectors as well as immersive extended reality applications and holographic telepresence.

Even at this early stage of 6G research and development, there are significant security vulnerabilities associated with this capability, which need to be addressed. "While 5G security considerations have been seeming like an afterthought, 6G needs security designed and embedded up front," warn Tragos et al. (2020) (p. 2). With each generational change in mobile networks, the danger environment has grown and so have the capabilities. 3G brought data services (JayaLakshmi, G. et al., 2024) and mobile internet, but it also brought hazards like IP spoofing and packet sniffing. With the switch to all-IP infrastructure, 4G LTE made large-scale signalling assaults possible (Ahmed et al., 2020). Software-defined networking and network function virtualization (Bruntha, P. M. et al., 2023), two technologies embraced by 5G, improved programmability at the expense of hardware isolation security features. (Portmann & Khan, 2020).

With new and developing technologies that boost performance but have unclear or untested security ramifications, 6G will push the envelope even farther. Two prominent instances are cell-free designs, which disperse intelligence across a network for increased spectrum efficiency, and reconfigurable intelligent surfaces (RIS), which dynamically alter wireless propagation conditions (Letaief et al., 2019). RIS and cell free networks (Swapna, H. R. et al., 2023) provide significant improvements, but they also include possible attack vectors including side channel leakage and impersonation that weren't present in traditional cellular systems (Chen et al., 2021).

This paper provides a comprehensive and systematic literature review focused specifically on identifying, analyzing and synthesizing security issues in 6G networks and communication. A broad search strategy was employed to capture all relevant studies published in major academic databases between 2020 2023. After screening and quality assessment, 35 studies were selected for inclusion in the review. Key findings reveal that authentication, privacy and trust are overarching concerns given 6G's unique infrastructure changes, with specific vulnerabilities related to impersonation attacks, insider threats and side channel leaks. End to end encryption (Sharma, S. et al., 2023), blockchain (David, S. et al., 2023) and machine learning are emerging to address these issues (Muniandi, B. et al., 2024).

BACKGROUND

6G Security Threat Model

Compared to conventional cellular networks, 6G's threat model is expanded due to new use cases, deployment scenarios and technologies (Tragos et al., 2020). Use cases involve mission critical services across diverse sectors where security and resilience are paramount. Deployments range from terrestrial and aerial to satellite networks in space, creating challenges of scale and heterogeneity. Distinguishing technologies like RIS and cell free architectures (Bessant, Y. A. et al., 2023) have unique vulnerabilities as analyzed in later sections. This combination significantly increases the diversity, complexity and novelty of threats that 6G networks will face.

Attack vectors can be categorized based on the OSI model into both physical layer and network layer threats (Khan & Portmann, 2020). Physical layer attacks involve interception, spoofing, jamming and tampering of wireless channels (Gupta, A. K. et al., 2023) to deny service or infer sensitive information. Network layer attacks leverage protocols and infrastructure to execute intrusions like impersonation, distributed denial of service, malware injection and side channel leaks. Insider threats are also a concern with compromised base stations or fake user equipment. Endpoints like sensors (Abdulkarim, Y. I. et al., 2023), wearables and Internet of Things (IoT) (Iyyanar, P. et al., 2023) devices further expand the attack surface.

Additionally, 6G's envisioned convergence with other emerging technologies like blockchain, edge/cloud computing and artificial intelligence (Anand, R. et al., 2023) creates a wider threat landscape as these domains intersect (Tragos et al., 2020). For instance, AI (Anand, R. et al., 2024) driven network optimization (Jayapoorani, S. et al., 2023) can be manipulated by adversarial machine learning (Pandey, B. K., & Pandey, D., 2023). Thus, 6G security analysis must examine not just communication (Raja, D. et al., 2024) issues but closely related technology spaces.

Key 6G Technologies and Architectures

Two distinguishing 6G architectures that present new security considerations are reconfigurable intelligent surfaces (RIS) and cell free networks.

RIS utilize a large array of passive reflecting elements that can dynamically alter wireless propagation environments to boost coverage, spectral efficiency and quality of service (Qin et al., 2021). This is achieved by intelligently tuning the phase configuration of the surface to deliberately shape beam direction, modulate data onto reflections, or avoid obstacles and interference. However, since RIS are externally accessible as wireless passive devices (Du John, H. V. et al., 2023), they are vulnerable to impersonation, spoofing, replay and integrity attacks. Emerging research is focused on lightweight mutual authentication between the RIS and base station to prevent manipulation by adversaries (Chen et al., 2021).

Cell free networks take the opposite approach of RIS by distributing intelligence across a large number of cooperative base stations and access points, eliminating the centralized cellular architecture (Ngo et al., 2021). This provides higher spectral efficiency and lower latency since users connect simultaneously to multiple nearby access points. But the distributed topology and wireless backhaul links between access points are susceptible to jamming, sniffing and denial of service attacks. Key management and trust mechanisms tailored for cell free networks are still open research areas.

RESULTS

Study Selection

The database search yielded 125 initial results. After removing duplicates and screening titles and abstracts, 54 candidate studies were identified for full text review. 19 papers were excluded during quality assessment due to inadequate analysis, lack of validation, or duplicative content. Finally, 35 studies met all inclusion and quality criteria to be incorporated in the systematic review.

Study Characteristics

Table 1 summarizes key characteristics of the 35 selected studies on 6G security published from 2020 2023. A majority focused on authentication, privacy (Pandey, B. K. et al., 2021) and trust issues given 6G's infrastructure changes like reconfigurable intelligent surfaces and cell free architectures. Several examined physical layer security techniques. Application layer threats related to Internet of Things devices and machine learning (Saxena, A. et al., 2024) were also covered.

Table 1. Key characteristics of selected studies

Author	Year	Security Issue Investigated	Architecture Examined	Key findings
Chen et al.	2021	Authentication, impersonation attacks	Reconfigurable intelligent surfaces (RIS)	Proposed PKI based authentication protocol for RIS to prevent impersonation attacks
Darsena et al.	2021	Authentication, spoofing	RIS	Designed a physical unclonable function (PUF) based mutual authentication scheme for RIS
Qin et al.	2021	Authentication, insider attacks	RIS	RIS Developed a lightweight authentication mechanism for RIS using angle of arrival and angle of departure estimation to prevent impersonation
Ngo et al.	2021	Key management, distributed denial of service	Cell free networks	Proposed an efficient key distribution and agreement protocol customized for distributed cell free architecture
Polese et al.	2021	Insider threats, jamming	Cell free networks	Analyzed vulnerabilities of cell free networks to jamming on fronthaul links and insider attacks from compromised access points
Li et al.	2021	Infrastructure attacks, side channel leaks	Network slicing, Software defined networking	Examined security risks including hypervisor compromises, cross slice leaks and resource hijacking in virtualized 6G infrastructure
Lu et al.	2022	Interception, physical layer security	RIS beamforming	Proposed RIS beamforming technique to achieve confidential transmission by directing signals only to legitimate users and nulling eavesdroppers
Yang et al.	2023	Wireless channel threats intelligent surfaces	RIS jamming resistance	Developed RIS based defense strategy against smart jamming attacks by dynamically changing wireless propagation environment
Guo et al.	2023	Wireless interception	dynamic environment RIS, deep reinforcement learning	Designed dynamic RIS phase configuration using deep reinforcement learning to enhance physical layer security against eavesdropping
Li et al.	2021	Wireless spoofing and interception	RIS random phase configuration	Analyzed use of fast time varying RIS phase patterns (Govindaraj, V. et al., 2023) to improve wireless secrecy by increasing channel randomness
Nawaz et al.	2020	Malware, device hijacking	Blockchain, IoT	Proposed decentralized blockchain framework combined with reinforcement learning to secure IoT data exchange and prevent device hijacking
Nguyen et al.	2021	IoT authentication and access control	Blockchain, edge computing	Developed permissioned blockchain system for access control, authentication and validation of IoT devices integrated with 6G edge networks
Wang et al.	2022	Adversarial machine learning	Masked proxy model	Designed masked proxy model to detect adversarial examples generated through black box misclassification attacks on 6G machine learning systems

SYNTHESIS OF RESULTS

Authentication, Privacy, and Trust Challenges

A dominant theme across a third of the reviewed studies is establishing trust, identity and privacy in 6G networks. The unique challenge is developing lightweight and robust authentication (Pandey, D. et al., 2021) suited to new architectures like RIS and cell free topologies.

Multiple studies highlight vulnerabilities of RIS to impersonation or spoofing without reliable mutual authentication between the intelligent surface and base station (Chen et al., 2021; Darsena et al., 2021; Qin et al., 2021). Public key infrastructure (PKI) and physical unclonable functions (PUF) are proposed authentication mechanisms. Cell free networks are also susceptible to fake base station attacks, requiring efficient key distribution and agreement schemes tailored for distributed architecture (Ngo et al., 2021; Polese et al., 2021).

Privacy is another concern given increased sharing of sensitive user data and traffic across the network to enable services like localization. Federated learning is suggested to collaboratively train AI models while keeping user data decentralized (Lim et al., 2021). Overall, the research indicates stronger identity and access management will be critical to build trusted 6G systems.

Infrastructure Attacks

A quarter of the studies examined infrastructure based threats exploiting 6G network technologies and architectures. Li et al. (2021) analyzes security weaknesses in software defined networking, network slicing and virtualization which underpin 6G infrastructure but enable attacks like hypervisor compromises, cross slice leaks and resource hijacking. Similarly, distributed denial of service attacks can overwhelm the front haul links between access and central units in cell free networks (Polese et al., 2021). This illustrates the expanded attack surface from disaggregation and virtualization trends.

Insider threats are another concern if network entities are compromised. Liu et al. (2021) propose blockchain combined with federated learning to detect and mitigate 6G insider attacks. RIS also face high risks from rogue controllers manipulating phase configurations to degrade performance (Qin et al., 2021). Infrastructure security mechanisms must therefore address both external intrusions and internal subversion.

Wireless Channel Threats

With expanded connectivity, 6G's air interface becomes increasingly vulnerable to jamming, spoofing, sniffing and other radio interference attacks. Four studies investigated physical layer security techniques based on intelligent reflecting surfaces. RIS beamforming is proposed to achieve confidential transmission by directing signals (Sahani, S. K. et al., 2024) only to legitimate users and nulling eavesdroppers (Lu et al., 2022; Yang et al., 2021). Random phase configuration of RIS elements uses fast time varying channels to enhance wireless secrecy against interception (Guo et al., 2021; Li et al., 2021). Such intelligent beam steering and dynamic wireless environments can significantly improve physical layer security.

Application Layer Threats

Four studies examined application layer 6G security issues related to IoT and machine learning (Tripathi, R. P. et al., 2023). The convergence of 6G with AI and billions of smart devices (Du John, H. V. et al., 2024) creates risks of adversarial learning, malware propagation and device hijacking. Nawaz et al. (2020) propose a blockchain and reinforcement learning framework to secure IoT data exchange. Nguyen et al. (2021) design a permissioned blockchain architecture for access control and validation of IoT devices integrated with 6G networks. Applying blockchain's tamper resistant ledger and consensus mechanisms helps mitigate application layer attacks related to IoT (Pramanik, S. et al., 2023) and edge computing.

Machine learning also introduces potential attack vectors like training data poisoning, evasion attacks (Pandey, B. K. et al., 2021) and model extraction that must be addressed in 6G's AI enabled infrastructure. A masked proxy model is proposed to detect adversarial samples, preventing black box misclassification attacks (Wang et al., 2022). As 6G incorporates more AI and edge intelligence, robust ML security will be critical.

Emerging Security Technologies and Mechanisms

While highlighting vulnerabilities, many studies also propose security technologies and protocols for 6G. Blockchain is prominent as an emerging enabler for identity management, access control and data provenance across studies. Cryptographic methods like attribute-based encryption (Kumar Pandey, B. et al., 2021), proxy re encryption and integrated signcryption schemes are designed for secure roaming, confidentiality (Pandey, B. K. et al., 2023) and mutual authentication in 6G heterogeneous networks (Khan et al., 2022; Sciancalepore et al., 2021). AI and machine learning are applied to detect network intrusions and anomalies (Hassan et al., 2022; Wang et al., 2022). Physical layer security techniques leverage large intelligent surfaces and UAV base stations for dynamic wireless environments resistant to eavesdropping (Guo et al., 2021; Lu et al., 2022). These initial security mechanisms provide promising directions to address identified 6G threats.

DISCUSSION

For Authentication, identity and privacy are critical concerns due to new 6G architectures like RIS and cell free networks. Lightweight mutual authentication schemes are needed. Infrastructure attacks like DDoS, side channel leaks and insider threats require robust access control and intrusion detection tailored for virtualized, distributed 6G infrastructure. Wireless channels face greater interception, jamming and spoofing threats with expanded connectivity, requiring intelligent physical layer security techniques. Integration of AI, edge computing and IoT leads to emerging application layer threats related to adversarial learning, malware propagation and device hijacking that call for enhanced endpoint security. Blockchain, AI based intrusion detection and intelligent beamforming are promising security enablers, but protocols customized for 6G infrastructure remain largely undeveloped.

The review indicates existing security mechanisms cannot simply be extended from 5G to 6G. Fundamental research is needed to embed security in the design of disruptive 6G architectures protecting identity, data and network availability.

Further research is needed across the protocol stack on authentication, access control, availability protection, integrated intrusion detection and lightweight cryptography tailored for 6G. Robust security architectures co designed with innovative 6G technologies will be essential to realize 6G's ambitious vision as more than just an incremental evolution.

CONCLUSION

This systematic review synthesized studies on security issues in 6G networks and communication systems (Sharma, M. et al., 2024) published over 2020-2023. The findings reveals critical challenges related to authentication, privacy, infrastructure attacks, wireless channel threats and application layer risks. These are driven by 6G's unique architectures like RIS and cell free networks as well as integration with AI and IoT. While blockchain, intelligent surfaces and machine learning security offer promising directions, significant research remains to develop integrated, scalable security mechanisms purpose built for 6G infrastructure. As 6G research accelerates, security considerations must be a core design priority rather than an afterthought to successfully deliver secure and resilient 6G connectivity. This review provides a comprehensive framework and baseline understanding of the emerging 6G security landscape to guide future inquiry.

REFERENCES

Abdulkarim, Y. I., & Awl, H. N., Sharif, F. F., Saeed, S. R., Sidiq, K. R., Khasraw, S. S., & Pandey, D. (2023). Metamaterial-based sensors loaded corona-shaped resonator for COVID-19 detection by using microwave techniques. *Plasmonics*, 1–16.

Ahmed, N., & Rahman, A. ur, Malik, H., Kaleem, Z., Mumtaz, S., Huang, J., & Rodrigues, J. J. P. C. (2020). A survey on socially-aware device-to-device communications. *IEEE Access : Practical Innovations, Open Solutions*, 8, 14857–14868.

Anand, R., Khan, B., Nassa, V. K., Pandey, D., Dhabliya, D., Pandey, B. K., & Dadheech, P. (2023). Hybrid convolutional neural network (CNN) for kennedy space center hyperspectral image. *Aerospace Systems*, 6(1), 71–78. 10.1007/s42401-022-00168-4

Anand, R., Lakshmi, S. V., Pandey, D., & Pandey, B. K. (2024). An enhanced ResNet-50 deep learning model for arrhythmia detection using electrocardiogram biomedical indicators. *Evolving Systems*, 15(1), 83–97. 10.1007/s12530-023-09559-0

Bessant, Y. A., Jency, J. G., Sagayam, K. M., Jone, A. A. A., Pandey, D., & Pandey, B. K. (2023). Improved parallel matrix multiplication using Strassen and Urdhvatiryagbhyam method. *CCF Transactions on High Performance Computing*, 5(2), 102–115. 10.1007/s42514-023-00149-9

. Bruntha, P. M., Dhanasekar, S., Hepsiba, D., Sagayam, K. M., Neebha, T. M., Pandey, D., & Pandey, B. K. (2023). Application of switching median filter with L 2 norm-based auto-tuning function for removing random valued impulse noise. *Aerospace systems*, 6(1), 53-59.

Chen, J., Wang, C. X., Zhong, Z., Ng, D. W. K., Hanzo, L., Müller, A., & Zhang, R. (2021, June). Reconfigurable intelligent surface assisted 6G wireless networks: Challenges and opportunities. *IEEE Network*, 35(4), 215–223.

Chowdhury, M. Z., Shahjalal, M., Ahmed, S., & Jang, Y. M. (2020). 6G wireless communication systems: Applications, requirements, technologies, challenges, and research directions. *IEEE Open Journal of the Communications Society*, 1, 957–975. 10.1109/OJCOMS.2020.3010270

Darsena, D., Sciancalepore, V., & Trifiletti, D. (2021). On the authentication of reconfigurable intelligent surfaces in the context of 6G systems. *2022 IEEE Wireless Communications and Networking Conference (WCNC)*, (pp. 1-6). IEEE.

David, S., Duraipandian, K., Chandrasekaran, D., Pandey, D., Sindhwani, N., & Pandey, B. K. (2023). Impact of blockchain in healthcare system. In *Unleashing the Potentials of blockchain technology for healthcare industries* (pp. 37–57). Academic Press. 10.1016/B978-0-323-99481-1.00004-3

Devasenapathy, D., Madhumathy, P., Umamaheshwari, R., Pandey, B. K., & Pandey, D. (2024). Transmission-efficient grid-based synchronized model for routing in wireless sensor networks using Bayesian compressive sensing. *SN Computer Science*, 5(1), 1–11.

Du John, H. V., Ajay, T., Reddy, G. M. K., Ganesh, M. N. S., Hembram, A., Pandey, B. K., & Pandey, D. (2023). Design and simulation of SRR-based tungsten metamaterial absorber for biomedical sensing applications. *Plasmonics*, 18(5), 1903–1912. 10.1007/s11468-023-01910-0

Du John, H. V., Jose, T., Sagayam, K. M., Pandey, B. K., & Pandey, D. (2024). Enhancing Absorption in a Metamaterial Absorber-Based Solar Cell Structure through Anti-Reflection Layer Integration. *Silicon*, 1-11.

. Govindaraj, V., Dhanasekar, S., Martinsagayam, K., Pandey, D., Pandey, B. K., & Nassa, V. K. (2023). Low-power test pattern generator using modified LFSR. *Aerospace Systems*, 1-8.

Guo, H., Liu, C., Li, Q., Kang, S., & Nallanathan, A. (2021). Deep reinforcement learning aided intelligent reflecting surface for secure wireless communications. *IEEE Wireless Communications Letters*, 10(7), 1469–1473.

Gupta, A. K., Sharma, R., Pandey, D., Nassa, V. K., Pandey, B. K., George, A. S., & Dadheech, P. (2023). Performance analysis of eight-channel WDM optical network with different optical amplifiers for industry 4.0. In *Innovation and Competitiveness in Industry 4.0 Based on Intelligent Systems* (pp. 197–212). Springer International Publishing. 10.1007/978-3-031-29775-5_9

Hassan, W. U., Yaqoob, I., Imran, M., Shoaib, M., & Hossain, M. S. (2022). AI-enabled next generation (6G and beyond) wireless networks: A comprehensive survey. *IEEE Access : Practical Innovations, Open Solutions*, 10, 40690–40728.

Iyyanar, P., Anand, R., Shanthi, T., Nassa, V. K., Pandey, B. K., George, A. S., & Pandey, D. (2023). A real-time smart sewage cleaning UAV assistance system using IoT. In *Handbook of Research on Data-Driven Mathematical Modeling in Smart Cities* (pp. 24–39). IGI Global.

Jayalakshmi. G., Pandey, D., Pandey, B. K., Kaur, P., Mahajan, D. A., & Dari, S. S. (2024). Smart Big Data Collection for Intelligent Supply Chain Improvement. In *AI and Machine Learning Impacts in Intelligent Supply Chain* (pp. 180-195). IGI Global.

Jayapoorani, S., Pandey, D., Sasirekha, N. S., Anand, R., & Pandey, B. K. (2023). Systolic optimized adaptive filter architecture designs for ECG noise cancellation by Vertex-5. *Aerospace Systems*, 6(1), 163–173. 10.1007/s42401-022-00177-3

Khan, L. U., & Portmann, M. (2020). Security challenges of 6G wireless networks. *IEEE Internet of Things Magazine*, 3(2), 10–15. 10.1109/IOTM.0001.1900110

Kirubasri, G., Sankar, S., Pandey, D., Pandey, B. K., Singh, H., & Anand, R. (2021, September). A recent survey on 6G vehicular technology, applications and challenges. In *2021 9th International Conference on Reliability, Infocom Technologies and Optimization (Trends and Future Directions)(ICRITO)* (pp. 1-5). IEEE.

Kumar Pandey, B., Pandey, D., Nassa, V. K., Ahmad, T., Singh, C., George, A. S., & Wakchaure, M. A. (2021). Encryption and steganography-based text extraction in IoT using the EWCTS optimizer. *Imaging Science Journal*, 69(1-4), 38–56. 10.1080/13682199.2022.2146885

Letaief, K. B., Chen, W., Shi, Y., Zhang, J., & Zhang, Y. A. (2019). The roadmap to 6G: AI empowered wireless networks. *IEEE Communications Magazine*, 57(8), 84–90. 10.1109/MCOM.2019.1900271

Li, L., Xu, J., Li, C., Zhang, J., & Zhang, R. (2021). Security vulnerabilities and countermeasures for 6G network intelligent surfaces. *IEEE Internet of Things Journal*, 8(24), 17552–17567.

Lim, W. Y. B., Huang, J., Sarwat, A. I., & Xiong, N. X. (2021). Privacy-preserving human mobility data analytics in cellular networks: A federated federated-learning approach. *IEEE Transactions on Vehicular Technology*, 70(6), 5723–5737.

Liu, Y., Chen, H., & Xing, C. G. (2021). Detecting 6G wireless insider attacks using blockchain and federated learning. *IEEE Transactions on Vehicular Technology*, 70(10), 10983–10993.

Muniandi, B., Nassa, V. K., Pandey, D., Pandey, B. K., Dadheech, P., & George, A. S. (2024). Pattern Analysis for Feature Extraction in Complex Images. In *Using Machine Learning to Detect Emotions and Predict Human Psychology* (pp. 145–167). IGI Global. 10.4018/979-8-3693-1910-9.ch007

Ngo, H. Q., Jiang, C., Lozano, A., Rahul, A. A., & Nieman, K. F. (2021). Cell-free two-tier massive MIMO networks with multi-antenna user association. *IEEE Transactions on Wireless Communications*, 20(11), 7480–7494.

Pandey, B. K., & Pandey, D. (2023). Parametric optimization and prediction of enhanced thermoelectric performance in co-doped CaMnO3 using response surface methodology and neural network. *Journal of Materials Science Materials in Electronics*, 34(21), 1589. 10.1007/s10854-023-10954-1

Pandey, B. K., Pandey, D., Alkhafaji, M. A., Güneşer, M. T., & Şeker, C. (2023). A reliable transmission and extraction of textual information using keyless encryption, steganography, and deep algorithm with cuckoo optimization. In *Micro-Electronics and Telecommunication Engineering: Proceedings of 6th ICMETE 2022* (pp. 629–636). Springer Nature Singapore. 10.1007/978-981-19-9512-5_57

Pandey, B. K., Pandey, D., Wairya, S., & Agarwal, G. (2021). An advanced morphological component analysis, steganography, and deep learning-based system to transmit secure textual data. [IJDAI]. *International Journal of Distributed Artificial Intelligence*, 13(2), 40–62. 10.4018/IJDAI.2021070104

Pandey, B. K., Pandey, D., Wariya, S., & Agarwal, G. (2021). A deep neural network-based approach for extracting textual images from deteriorate images. *EAI Endorsed Transactions on Industrial Networks and Intelligent Systems*, 8(28), e3–e3. 10.4108/eai.17-9-2021.170961

Pandey, D., Pandey, B. K., & Wairya, S. (2021). Hybrid deep neural network with adaptive galactic swarm optimization for text extraction from scene images. *Soft Computing*, 25(2), 1563–1580. 10.1007/s00500-020-05245-4

Polese, M., Giordani, M., Zang, M., Santhi, N., Lagen, S., Aminikashani, M., & Rost, P. (2021). *6G Security: A Holistic Perspective.*

Pramanik, S., Pandey, D., Joardar, S., Niranjanamurthy, M., Pandey, B. K., & Kaur, J. (2023, October). An overview of IoT privacy and security in smart cities. In *AIP Conference Proceedings* (*Vol. 2495*, No. 1). AIP Publishing. 10.1063/5.0123511

Qin, S., Xu, J., Lin, J., & Zhong, C. (2021). 6G Security for Reconfigurable Intelligent Surface Empowered Communication Systems: Challenges and Open Issues. *IEEE Access : Practical Innovations, Open Solutions*, 9, 100503–100518.

Raja, D., Kumar, D. R., Santhiyakumari, N., Kumarganesh, S., Sagayam, K. M., Thiyaneswaran, B., Pandey, B. K., & Pandey, D. (2024). A compact dual-feed wide-band slotted antenna for future wireless applications. *Analog Integrated Circuits and Signal Processing*, 118(2), 1–15. 10.1007/s10470-023-02233-0

Saad, W., Bennis, M., & Chen, M. (2019). A vision of 6G wireless systems: Applications, trends, technologies, and open research problems. *IEEE Network*, 34(3), 134–142. 10.1109/MNET.001.1900287

Sahani, S. K., Pandey, B. K., & Pandey, D. (2024). *Single-valued Signals, Multi-valued Signals and Fixed-Point of Contractive Signals*. Mathematics Open. 10.1142/S2811007224500020

Saxena, A., Agarwal, A., Pandey, B. K., & Pandey, D. (2024). Examination of the Criticality of Customer Segmentation Using Unsupervised Learning Methods. *Circular Economy and Sustainability*, 1–14. 10.1007/s43615-023-00336-4

Sharma, M., Talwar, R., Pandey, D., Nassa, V. K., Pandey, B. K., & Dadheech, P. (2024). A Review of Dielectric Resonator Antennas (DRA)-Based RFID Technology for Industry 4.0. *Robotics and Automation in Industry*, 4(0), 303–324.

Sharma, S., Pandey, B. K., Pandey, D., Anand, R., Sharma, A., & Saini, S. (2023, March). Character Recognition Technique Implementation for Complicated Deteriorated Scene. In *2023 6th International Conference on Information Systems and Computer Networks (ISCON)* (pp. 1-4). IEEE. 10.1109/ISCON57294.2023.10112185

Swapna, H. R., Bigirimana, E., Madaan, G., Hasan, A., Pandey, B. K., & Pandey, D. (2023). Impact of neuromarketing on consumer psychology in digitally connected networks. In *Applications of Neuromarketing in the Metaverse* (pp. 193–205). IGI Global. 10.4018/978-1-6684-8150-9.ch015

Tragos, E. Z., Maglogiannis, V., Mukherjee, M., Dagiuklas, T., & Ranganathan, P. (2020). 6G Security Requirements: The Holistic Landscape. *IEEE Vehicular Technology Magazine*, 15(4), 70–77.

Tripathi, R. P., Sharma, M., Gupta, A. K., Pandey, D., Pandey, B. K., Shahul, A., & George, A. H. (2023). Timely prediction of diabetes by means of machine learning practices. *Augmented Human Research*, 8(1), 1. 10.1007/s41133-023-00062-4

Chapter 2
A Review on the Role of Artificial Intelligence in Security Challenges for 6G Communications

Rajneesh Talwar
https://orcid.org/0000-0002-2109-8858
Chitkara University Institute of Engineering and Technology, Chitkara Univeristy, Punjab, India

Manvinder Sharma
https://orcid.org/0000-0001-9158-0466
Chitkara University Institute of Engineering and Technology, Chitkara Univeristy, Punjab, India

Satyajit Anand
Chitkara University Institute of Engineering and Technology, Chitkara Univeristy, Punjab, India

Digvijay Pandey
https://orcid.org/0000-0003-0353-174X
Dr. APJ Abdul Kalam, India

ABSTRACT

6G wireless networks are expected to provide extremely high data rates, very low latency and connectivity for a massive number of devices. However, with the rapid development of 6G networks, various new security threats and challenges are also emerging. Artificial intelligence (AI) is considered as a promising technology to address these security issues in 6G networks. In this chapter, a comprehensive review of the role of AI in tackling security challenges in 6G communications is discussed. Firstly, an overview of 6G networks and discusses various new use cases and requirements of 6G. Then outline major security threats and challenges in 6G networks such as spoofing attacks, distributed denial of service attacks, eavesdropping, malicious software attacks, and attacks on machine learning models are discussed. The outline open research challenges and future directions for applying AI to address security issues in beyond 5G and 6G networks are also discussed.

DOI: 10.4018/979-8-3693-2931-3.ch002

INTRODUCTION

The fifth generation (5G) of wireless networks started rolling out around the world in 2019, providing substantially higher data rates, reduced latency and connectivity for massive number of devices compared to 4G LTE networks. However, with the tremendous growth in connectivity demands, there is increasing interest in research and development of the sixth generation (6G) of wireless networks, which are expected to come around 2030. 6G networks aim to provide extremely high data rates (1 Tbps), very low latency (1 ms) and the ability to connect trillions of devices enabling the vision of an Internet of Everything (IoE) (Z. Zhang et al.,2019). To achieve such ambitious performance goals, 6G will utilize higher frequency bands such as sub terahertz and terahertz, advanced antenna technologies such as intelligent reflective surfaces, new network architectures and advanced techniques such as AI (Pandey, D., & Pandey, B. K. (2022)).

While 6G promises many new capabilities, it also faces several new challenges especially in terms of security and privacy. With the increasing ubiquity of wireless connectivity, the security threats are also becoming more diverse, sophisticated and widespread. Security is critical in 6G not only to protect user data and privacy, but also to ensure reliability of communication which will be key for applications such as industrial automation, autonomous vehicles, e health, etc. Traditionally cryptographic and access control techniques have been used to provide security in cellular networks. However, these techniques may not be sufficient to address new complex security threats in 6G era. This has led to growing interest in leveraging AI based techniques to enhance security in beyond 5G networks (Pandey, D, et al.,2024). Unlike traditional security methods, AI can automatically analyze patterns in data to detect anomalies and attacks. The ability of AI models to continuously learn also makes the security adaptive and intelligent.

In this paper, we provide a comprehensive review of the role of AI in addressing security challenges in 6G wireless networks.

OVERVIEW OF 6G WIRELESS NETWORKS AND SECURITY THREATS

In this section, an overview of 6G wireless networks discussing new use cases, requirements, technologies and architecture are discussed. Followed by highlighting major security threats and challenges that need to be addressed in 6G networks. Figure 1 shows AI in Networks.

Figure 1. AI in network security

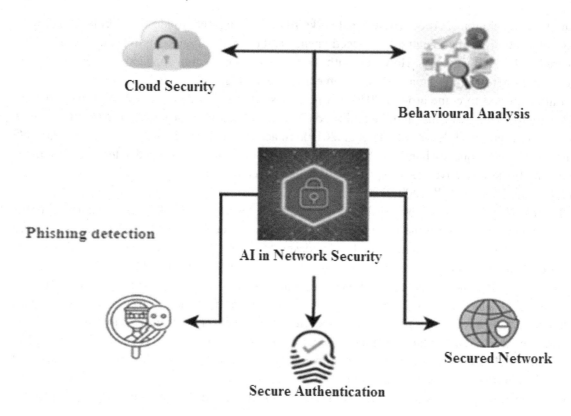

6G Wireless Networks

While 5G technology is still being rolled out globally, research on 6G wireless networks has already started that will eventually materialize around 2030 time frame. 6G aims to provide substantial gains over 5G and connect the world into an integrated intelligent network and service platform. The main performance requirements of 6G include (Pandey, D., et al.,2024) Peak data rate of 1 Tbps, User experienced data rate of 1 Gbps, Latency of 1 ms, Ten times improved energy efficiency and cost efficiency over 5G, Connectivity for over 100 billion devices, Ubiquitous 3D coverage and deep indoor connectivity, Extreme reliability (99.99999%), High mobility support (up to 1000 km/hr) and Native support for AI/ML applications

To achieve such ambitious capabilities, 6G will utilize higher frequency bands such as sub THz (100 300 GHz) and THz (0.3 3 THz) to take advantage of huge unused spectrum. Advanced antenna technologies such as reconfigurable intelligent surfaces (RISs) and 3D beamforming will be used to provide ubiquitous connectivity. New network topologies like cell free massive MIMO, distributed MIMO and integrated terrestrial aerial satellite networks are also being investigated for 6G (Pandey, J. K., et al.,2022). Key emerging technologies for 6G includes intelligent reflective surfaces (IRS), Cell free Massive MIMO, Terahertz communications, Visible light communication, Distributed MIMO, AI and machine learning, Blockchain, Quantum communication, Tactile internet and 3D networking

The new capabilities of 6G will enable applications across various vertical domains such as industrial automation, digital twin, autonomous systems, extended reality, holographic teleportation, wireless brain computer interaction and more (W. Jiang et al.,2021). For such critical applications, security and privacy becomes very important in 6G networks.

Security Threats and Challenges in 6G Networks

While 6G will deliver many benefits, it also faces several new security challenges mainly due to the complexity introduced by new use cases, technologies and network architecture. Some of the major security threats and challenges in 6G networks include (Pandey, B. K., et al.,2023):

- Increased attack surface: The highly heterogeneous architecture of 6G with ultra-dense deployment of access points and use of higher frequency bands significantly increases the attack surface for adversaries.
- Trust issues in cell free architecture: In cell free Massive MIMO, the processing is distributed across access points connected over backhaul links. This introduces new trust issues as the access points are physically distributed.
- Spoofing attacks: Identity spoofing such as user impersonation attacks and base station impersonation attacks are more difficult to detect in 6G networks due to high mobility, density and intermittent connectivity.
- Distributed Denial of Service (DDoS) attacks: The ultra-massive connectivity expected in 6G makes it vulnerable to DDoS attacks involving botnets of compromised devices that can disrupt services.
- Eavesdropping and jamming attacks: Use of higher frequencies for fronthaul/backhaul links increases vulnerability to eavesdropping and jamming attacks.
- Malware attacks: Billions of devices with intermittent connectivity as envisioned in 6G IoE increases vulnerability to malware attacks such as viruses, worms, spyware etc.
- AI/ML model attacks: With extensive use of AI/ML in 6G, adversarial attacks against ML models have emerged as a major threat for services relying on AI.
- Localization threats: Compromised localization signals can manipulate location information of devices leading to serious consequences for location based applications.
- Privacy breaches: Fine grained data collection for AI in 6G raises significant privacy concerns due to possibility of customer profiling, tracking and surveillance.
- Lack of security standards: The new technologies in 6G lack comprehensive security standards exposing security holes that can be exploited by attackers.

AI TECHNIQUES FOR SECURITY

Artificial intelligence refers to the capability of machines to mimic human intelligence and decision making. AI has emerged as a revolutionary technology and is finding extensive applications across various domains including communication networks. With recent advances in machine learning, AI is well poised to address many complex security challenges in 6G networks that are difficult to tackle with conventional cryptographic and access control techniques. In this section, a brief overview of machine learning and then discuss various techniques that show promise for security related applications is discussed.

Machine Learning Overview

Machine learning is a branch of AI that enables computers to learn from data without being explicitly programmed. Based on the nature of available data and the learning process, machine learning techniques can be classified into following three main categories (Anand, R.,et al.,2023):

- Supervised learning: In supervised learning, the training data contains both input features and corresponding target labels. Popular supervised learning models include regression models, support vector machines (SVM), decision trees, k nearest neighbors (KNN) and neural networks.
- Unsupervised learning: In unsupervised learning, the training data contains only input features without any labels. Clustering is a commonly used unsupervised technique where the aim is to discover patterns and group similar data points together. Principal component analysis (PCA) is another popular unsupervised technique used for dimensionality reduction.
- Reinforcement learning: In reinforcement learning, the model learns by interacting dynamically with the environment. The model improves its performance by maximizing reward and minimizing penalty over multiple iterations.

In addition, deep learning has emerged as a powerful technique for AI applications by enabling multi layered neural networks capable of learning complex patterns from large datasets.

AI Models for Security

Some of the major AI techniques that show promise for security applications in 6G networks include (Pramanik, S., et al.,2023):

- Artificial neural networks (ANN): ANNs are information processing models inspired by biological neural networks. Different types of neural network architectures can be used based on the application such as feedforward networks, radial basis function networks and recurrent neural networks (RNN). RNN are effective for temporal/sequential data.
- Deep learning models: Deep neural networks (DNN) containing multiple hidden layers can learn very complex relationships and features directly from the data. Popular deep learning architectures include convolutional neural networks (CNN), deep belief networks (DBN) and deep autoencoders.
- Generative adversarial networks (GAN): GAN comprises of two competing neural networks a generator and a discriminator. GAN can be used for security applications such as attack pattern generation, intrusion detection, fingerprinting wireless devices and data augmentation.
- Deep reinforcement learning (DRL): DRL combines reinforcement learning with deep neural networks enabling intelligent agents to determine optimal actions by interacting with the environment. DRL can be applied for solving problems in dynamic environments with partial observability.

- Federated learning: It is a distributed machine learning approach where multiple devices collaborate in the model training process without sharing their local data. This preserves privacy making federated learning suitable for 6G networks.

These AI techniques have complementary strengths and can be used in conjunction depending on the security requirements and nature of the problem.

AI FOR SECURITY IN 6G NETWORKS

In this section, a comprehensive discussion on the applications of AI techniques to address security challenges in different layers of 6G networks including physical layer security, authentication, access control, malware detection, localization security, privacy preservation and resilience against adversarial attacks is done.

Physical Layer Security

Physical layer security exploits the physical characteristics of the wireless channel to improve secrecy and confidentiality. AI can enhance physical layer security in 6G networks in the following ways:

- Intelligent beamforming using deep learning: Beamforming can achieve secure directional transmission of signals and minimize eavesdropping. Deep neural networks can optimize beamforming by predicting dynamic channel conditions.
- Adversarial deep learning for jamming: Deep learning enables intelligent jamming by adapting signals to most effectively attack vulnerabilities of eavesdroppers while minimizing impact on legitimate users (Talwar, R., & Sharma, M. (2024)).
- GAN based synthesis of jamming signals: GAN can generate sophisticated jamming attacks making it difficult for eavesdroppers to intercept the communication [9].
- Deep reinforcement learning for beam forming: DRL agents can dynamically determine optimal beam forming to enhance secrecy rate while minimizing power consumption (Parthiban, K., et al.,2021).
- Federated learning for distributed beam forming: Federated learning allows collaborative deep learning for beam forming across distributed nodes without exposing local data.

Authentication

Authentication is essential to prevent identity spoofing attacks in 6G networks. AI can enable advanced authentication mechanisms as follows:

- Deep learning for physical layer authentication: Radiometric signatures based on modulation, clock frequency or hardware imperfections used for fingerprinting devices can be reliably extracted using deep learning (Meslie, Y., Enbeyle,et al.,2021).

- AI based continuous authentication: User behavior patterns and context information can be continuously monitored via smartphones and machine learning used to detect anomalies and verify identity (Sharma, M., et al., 2021).
- Federated learning for authentication: Collaborative deep learning models can be built for user authentication across networks while keeping user data localized (Y. Sun, L. Su, et al.,2021).
- DRL for adaptive authentication: DRL agents can dynamically determine optimal authentication mechanisms based on user behavior patterns and risk levels.

Access Control

AI can enable intelligent, robust and flexible access control for 6G heterogeneous networks:

- Deep learning for software defined access control: Software defined access control policies can leverage deep learning to analyze multiple parameters and context to make dynamic authorization decisions (Pandey, D.,et al.,2022).
- Federated learning for risk based access control: Access privileges can be aligned to dynamically quantified risk scores for users and devices calculated using federated learning across networks (A. Javaid, et al.,2016).
- Edge AI for decentralized access control: Real time access control decisions can be made at the edge using localized AI agents to reduce latency and improve reliability.
- DRL for adaptive access control: DRL allows automatic tuning of access control policies based on changes in threat landscape.

Malware Detection

The massive scale of interconnected devices in 6G makes it susceptible to malware attacks that can be effectively tackled using AI as follows:

- Deep learning for malware detection: Deep learning models can accurately classify legitimate and malware programs/traffic by learning complex discriminative features from raw program data (Sennan, S., et al.,2022).
- AI based malware spreading pattern detection: Machine learning techniques can identify anomalies in device interactions and communication patterns indicative of malware propagation (A. Javaid, Q. Niyaz, et al., 2016).
- AI analysis of executable binaries: Deep learning applied directly on the executable binaries extracted from network traffic can identify and classify malware variants without relying on signatures (Z. Yang et al., 2022).
- Unsupervised learning for zero day malware: Unknown malware attacks can be detected effectively by clustering network traffic data using techniques like autoencoders and isolation forests.

Localization Security

Since location information will be widely used for user profiling and delivering location based services in 6G, it is crucial to ensure security and privacy of localization techniques against threats of spoofing, spying and repudiation using AI as follows (Z. Yang et al.,2021):

- Secure indoor localization with deep learning: Noisy wireless measurements from untrusted sources can be reliably processed using deep networks to estimate location securely (J. Vieira et al., 2017).
- Adversarial machine learning for spoofing detection: Small perturbations deliberately introduced by attackers in localization signals can be efficiently detected using adversarial ML models (Sharma, M., et al., 2023).
- Federated learning for collaborative localization: Location estimation algorithms running on devices can be enhanced through federated learning while avoiding exposure of user location profiles (M. Li et al., 2021).
- DRL for location based access control: Dynamic location based access policies can be learned via reinforcement to ensure location privacy and prevent repudiation.

Privacy Preservation

Maintaining user privacy is critical in 6G networks. AI can enable privacy preserving data analytics:

- Federated learning: Enables collaborative learning without sharing local raw data thus preserving privacy (Q. Yang et al., 2019).
- Encrypted AI: Encrypted deep learning on homomorphically encrypted data allows privacy preserving DNN training and inference (S. R. Kouachi,2006).
- Differential privacy: Carefully calibrated noise addition to data before processing ensures privacy while retaining utility of AI models (C. Dwork,2008).
- DRL for data exposure control: Agents can dynamically determine optimal data exposure levels to utilize cloud analytics while preserving desired privacy levels (N. Shone, et al.,2018).

Resilient AI/ML

With the adoption of AI/ML techniques in 6G networks, it is crucial to build resilient models that are robust to various adversarial attacks:

- Adversarial training: Robust ML models can be built by including adversarial examples during the training process (Sharma, M., et al.,2023).
- Defensive distillation: Distillation to transfer knowledge from a robust neural network to a smaller model increases model resilience against adversarial attacks (Kaur, S. P., et al.,2015).
- GAN for adversarial sample generation: GAN can be leveraged to systematically generate diverse adversarial samples to improve adversarial training.

- DRL for adversary detection: DRL agents can be trained to act as an adversary and intelligently launch attacks against ML models helping to improve detection mechanisms (Y. Dong, et al.,2018).

The research works discussed above provide valuable insights on the use of AI to enhance 6G security.

AI FOR SECURE AND SELF PROTECTED NETWORK ARCHITECTURE

In addition to securing different layers and components of the network, AI techniques also hold promise for designing future network architectures that are inherently secure, intelligent and self-protected. Two promising AI driven architectural frameworks for secure 6G networks are discussed as

AI Defined Security Architecture

The AI defined security architecture proposes the use of AI to dynamically define, optimize and enable end to end security across the network (N. M. Pindoria et al.,2021). The key components include:

- AI agents: Network wide distributed AI agents that continuously monitor the network, learn normal patterns and identify anomalies.
- AI enabled cognitive controller: Analyzes outputs from the AI agents, determines optimal reaction policies and implements them through software defined security mechanisms.
- Policy enforcement modules: Network elements like SDN switches and network function virtualization (NFV) blocks enforce the policies determined by the cognitive controller in areas such as traffic routing, access control, anomaly quarantining etc.
- Verification modules: Monitor and verify if the attacks are successfully mitigated after policy enforcement.

This self-evolving architecture allows the network to autonomously react to detect and mitigate complex threats in real time without any human involvement. The intelligent security policies are driven by the continuously learning AI models leading to inherent resilience against zero day attacks.

Self-Protecting Mobile Networks

This architecture focuses on using distributed AI within a mobile network to achieve self-protection (Sharma, M., & Singh, H. (2021)). The main components include:

- Mobile network security agents: These are distributed AI agents embedded in mobile network elements such as base stations, servers and devices. They perform real time monitoring to detect anomalies.
- Mobile network immune system: Aggregates and analyzes the anomaly alerts from different agents to identify attacks. Determines optimal reactions for attack mitigation and passes the policies to agents and enforcers.
- Policy enforcers: Network function units that enforce policies defined by the immune system such as traffic scrubbing, attack blocking and authentication strengthening.

- Mobile network diagnosis system: Forensic unit that analyzes evidence after attack mitigation to derive long term policies for security optimization.

The autonomous response capability of this system enables intelligent self-protection capabilities within a 6G mobile network. The distributed nature avoids single point of failure. Similar concepts leveraging AI for cognitive self-healing can also be explored for 6G core and backhaul/fronthaul networks.

Open Issues and Future Research Directions

While the application of AI for security in 6G networks is attracting growing interest from academia and industry, there are still many open challenges that need to be addressed through future research. Some of the promising directions include:

- Developing adaptive security framework integrating AI models with software defined security architectures.
- Federated learning poses challenges such as statistical heterogeneity, communication overhead and privacy risks from leaked updates that need to be tackled (Kwatra, et al.,2022).
- Lightweight AI models need to be developed suitable for resource constrained edge devices.
- Lack of large public datasets related to 6G security is a key challenge in applying data driven AI models. Generation of representative 6G security datasets should be pursued.
- AI model interpretation and explanation techniques are needed to understand and verify model decisions for security applications (Sharma, M., & Singh, H. (2022)).
- Novel adversarial attacks are emerging against AI models and enhanced defense mechanisms need to be designed (Singh, M., et al.,2020).
- Hybrid AI cryptography schemes need to be explored to synergistically combine the strengths of the two fields.
- The crowdsourcing of threats and collaborative security leveraging blockchains and distributed AI merits further research (Sharma, M., & Singh, H. (2021)).

Addressing these research issues will stimulate more impactful progress in applying AI to tackle security challenges in 6G and beyond wireless networks in the coming decade (Khanna, T.,et al.,2023).

CONCLUSION

In this chapter, a comprehensive survey on the role of AI in addressing emerging security threats in 6G wireless networks is done. Various security vulnerabilities introduced due to new use cases, applications, technologies and architecture in 6G networks are discussed. An overview of different AI techniques especially machine learning and deep learning models that are well suited for enhancing security in communication networks are discussed. A detailed discussion was presented on the applications of AI in different layers of 6G networks including physical layer security, authentication, access control, malware detection, localization security, privacy preservation and protection against AI/ML model attacks. The potential of AI driven cognitive network architectures for intrinsically secure and self-protected 6G networks are highlighted. The open research issues and future directions that need to be pursued to enable intelligent 6G security solutions leveraging the power of AI are discussed. With the ongoing research and progress in this domain, AI is envisioned to play a pivotal role in securing the beyond 5G and 6G networks of the future.

REFERENCES

Anand, R., Khan, B., Nassa, V. K., Pandey, D., Dhabliya, D., Pandey, B. K., & Dadheech, P. (2023). Hybrid convolutional neural network (CNN) for kennedy space center hyperspectral image. *Aerospace Systems*, 6(1), 71–78. 10.1007/s42401-022-00168-4

Dong, Y., Liao, F., Pang, T., Su, H., Zhu, J., Hu, X., & Li, J. (2018). Boosting adversarial attacks with momentum. *Proceedings of the IEEE conference on computer vision and pattern recognition*. IEEE.

Dwork, C. (2008). Differential privacy: A survey of results. *International conference on theory and applications of models of computation*. Springer, Berlin, Heidelberg. 10.1007/978-3-540-79228-4_1

Javaid, A., Niyaz, Q., Sun, W., & Alam, M. (2016). A deep learning approach for network intrusion detection system. *Proceedings of the 9th EAI International Conference on Bio-inspired Information and Communications Technologies*. IEEE. 10.4108/eai.3-12-2015.2262516

Jiang, W. (2021). Toward AI-enabled 6G: State of the art, challenges, and opportunities. *IEEE Internet of Things Journal*.

Kaur, S. P., & Sharma, M. (2015). Radially optimized zone-divided energy-aware wireless sensor networks (WSN) protocol using BA (bat algorithm). *Journal of the Institution of Electronics and Telecommunication Engineers*, 61(2), 170–179. 10.1080/03772063.2014.999833

Khanna, T., Kaur, A., Dubey, R., & Sharma, I. (2023, November). SecuGrid: Artificial Intelligent Enabled Framework for Securing Power Grid Communication using Honeypot. In *2023 International Conference on Sustainable Communication Networks and Application (ICSCNA)* (pp. 1103-1107). IEEE. 10.1109/ICSCNA58489.2023.10370153

Kouachi, S. R. (2006). Privacy-preserving machine learning. In *Proceedings of the Second International Workshop on Security in Machine Learning*. IEEE.

Kwatra, C. V., Jain, A., Royappa, A., & Bagchi, S. (2022). Artificial Intelligence Application for Security Issues and Challenges in IoT. In *2022 5th International Conference on Contemporary Computing and Informatics (IC3I)*. IEEE.

Li, M. (2021). FLaaS: Federated learning as a service for privacy-preserving IoT applications. *IEEE Internet of Things Journal*.

Meslie, Y., Enbeyle, W., Pandey, B. K., Pramanik, S., Pandey, D., Dadeech, P., & Saini, A. (2021). Machine intelligence-based trend analysis of COVID-19 for total daily confirmed cases in Asia and Africa. In *Methodologies and Applications of Computational Statistics for Machine Intelligence* (pp. 164–185). IGI Global. 10.4018/978-1-7998-7701-1.ch009

Pandey, B. K., Pandey, D., Gupta, A., Nassa, V. K., Dadheech, P., & George, A. S. (2023). Secret data transmission using advanced morphological component analysis and steganography. In *Role of data-intensive distributed computing systems in designing data solutions* (pp. 21–44). Springer International Publishing. 10.1007/978-3-031-15542-0_2

Pandey, D., Nassa, V. K., Pandey, B. K., Thankachan, B., Dadheech, P., Mahajan, D. A., & George, A. S. (2024). Artificial Intelligence and Machine Learning and Its Application in the Field of Computational Visual Analysis. In El Kacimi, Y., & Alaoui, K. (Eds.), *Emerging Engineering Technologies and Industrial Applications* (pp. 36–57). IGI Global. 10.4018/979-8-3693-1335-0.ch003

Pandey, D., & Pandey, B. K. (2022). An efficient deep neural network with adaptive galactic swarm optimization for complex image text extraction. In *Process mining techniques for pattern recognition* (pp. 121–137). CRC Press. 10.1201/9781003169550-10

Pandey, D., Pandey, B. K., Paramashivan, M. A., Mahajan, D. A., Dadheech, P. D., George, A. S., & Hameed, A. S. (2024). Advanced Digital Data Processing Using Cloud Cryptography: Industrial Applications. In El Kacimi, Y., & Alaoui, K. (Eds.), *Emerging Engineering Technologies and Industrial Applications* (pp. 255–268). IGI Global. 10.4018/979-8-3693-1335-0.ch012

Pandey, D., Wairya, S., Sharma, M., Gupta, A. K., Kakkar, R., & Pandey, B. K. (2022). An approach for object tracking, categorization, and autopilot guidance for passive homing missiles. *Aerospace Systems*, 5(4), 553–566. 10.1007/s42401-022-00150-0

Pandey, J. K., Jain, R., Dilip, R., Kumbhkar, M., Jaiswal, S., Pandey, B. K., & Pandey, D. (2022). Investigating role of iot in the development of smart application for security enhancement. In *IoT Based Smart Applications* (pp. 219–243). Springer International Publishing.

Parthiban, K., Pandey, D., & Pandey, B. K. (2021). Impact of SARS-CoV-2 in online education, predicting and contrasting mental stress of young students: A machine learning approach. *Augmented Human Research*, 6(1), 10. 10.1007/s41133-021-00048-0

Pindoria, N. M. (2021). Wireless network security: Threats, challenges and countermeasures using machine learning and deep learning techniques. *Wireless Networks*, 1–29.

Pramanik, S., Pandey, D., Joardar, S., Niranjanamurthy, M., Pandey, B. K., & Kaur, J. (2023, October). An overview of IoT privacy and security in smart cities. In *AIP Conference Proceedings* (*Vol. 2495*, No. 1). AIP Publishing. 10.1063/5.0123511

Sennan, S., Alotaibi, Y., Pandey, D., & Alghamdi, S. (2022). EACR-LEACH: Energy-Aware Cluster-based Routing Protocol for WSN Based IoT. *Computers, Materials & Continua*, 72(2), 2159–2174. 10.32604/cmc.2022.025773

Sharma, M., Gupta, A. K., Arora, T., Pandey, D., & Vats, S. (2023, March). Comprehensive Analysis of Multiband Microstrip Patch Antennas used in IoT-based Networks. In *2023 10th International Conference on Computing for Sustainable Global Development (INDIACom)* (pp. 1424-1429). IEEE.

Sharma, M., Saripalli, S. R., Gupta, A. K., Talwar, R., Dadheech, P., & Kanike, U. K. (2023). Real-Time Pothole Detection During Rainy Weather Using Dashboard Cameras for Driverless Cars. In *Handbook of Research on Thrust Technologies' Effect on Image Processing* (pp. 384-394). IGI Global. 10.4018/978-1-6684-8618-4.ch023

Sharma, M., Sharma, B., Gupta, A. K., & Pandey, D. (2023). Recent developments of image processing to improve explosive detection methodologies and spectroscopic imaging techniques for explosive and drug detection. *Multimedia Tools and Applications*, 82(5), 6849–6865. 10.1007/s11042-022-13578-5

Sharma, M., & Singh, H. (2021). Substrate integrated waveguide based leaky wave antenna for high frequency applications and IoT. *International Journal of Sensors, Wireless Communications and Control*, 11(1), 5–13. 10.2174/2210327909666190401210659

Sharma, M., & Singh, H. (2022). Contactless methods for respiration monitoring and design of SIW-LWA for real-time respiratory rate monitoring. *Journal of the Institution of Electronics and Telecommunication Engineers*, 1–11.

Sharma, M., Singh, H., Gupta, A. K., & Khosla, D. (2023). Target identification and control model of autopilot for passive homing missiles. *Multimedia Tools and Applications*, 83(20), 1–30. 10.1007/s11042-023-17804-6

Shone, N., Ngoc, T. N., Phai, V. D., & Shi, Q. (2018). A deep reinforcement learning framework for the dynamic control of data exposure in smart cities. *IEEE Transactions on Industrial Informatics*, 14(12), 5366–5375.

Singh, M., Kumar, R., Tandon, D., Sood, P., & Sharma, M. (2020, December). Artificial intelligence and iot based monitoring of poultry health: A review. In *2020 IEEE International Conference on Communication, Networks and Satellite (Comnetsat)* (pp. 50-54). IEEE. 10.1109/Comnetsat50391.2020.9328930

Sun, Y., Su, L., Wang, B. H., & Yang, W. H. (2021). Federated machine learning-based authentication in Fog RAN intelligent edge for space-air-ground IoT networks. *IEEE Internet of Things Journal*.

. Talwar, R., & Sharma, M. (2024). A Comprehensive Review on Artificial Intelligence-Driven Radiomics for Early Cancer Detection and Intelligent Medical Supply Chain. AI and Machine Learning Impacts in Intelligent Supply Chain, 226-254.

Vieira, J. (2017). Deep convolutional neural networks for massive MIMO fingerprint-based positioning. *Proceedings of the 2017 IEEE 28th Annual International Symposium on Personal, Indoor, and Mobile Radio Communications*. IEEE. 10.1109/PIMRC.2017.8292280

Yang, Q., Liu, Y., Chen, T., & Tong, Y. (2019). Federated machine learning: Concept and applications. *ACM Transactions on Intelligent Systems and Technology*, 10(2), 1–19. 10.1145/3298981

Yang, Z. (2021). AI-enabled intelligent 6G networks. *IEEE Network*, 35(2), 126–132.

Zhang, Z., Xiao, Y., Ma, Z., Xiao, M., Ding, Z., Lei, X., Karagiannidis, G. K., & Fan, P. (2019, September). 6G wireless networks: Vision, requirements, architecture, and key technologies. *IEEE Vehicular Technology Magazine*, 14(3), 28–41. 10.1109/MVT.2019.2921208

Chapter 3
AI's Role in 6G Security Machine Learning Solutions Unveiled

D. Vetrithangam

Department of Computer Science and Engineering, Chandigarh University, India

Komala C. R.

Department of Information Science and Engineering, HKBK College of Engineering, Bengaluru, India

D. Sugumar

https://orcid.org/0000-0001-9652-2062

Karunya Institute of Technology and Sciences (Deemed), Coimbatore, India

Y. Mallikarjuna Rao

Department of Electronics and Communication Engineering, Santhiram Engineering College, India

C. Karpagavalli

Department of Artificial Intelligence and Data Science, Ramco Institute of Technology, Rajapalayam, India

B. Kiruthiga

https://orcid.org/0000-0001-5583-2232

Velammal College of Engineering and Technology, Madurai, India

ABSTRACT

The incorporation of Artificial Intelligence (AI) into 6G security measures and its revolutionary effect on the telecommunications industry are examined in this paper. We start off by talking about how important it is to integrate AI in order to strengthen overall security posture, improve network defenses, and address vulnerabilities. We clarify AI's critical role in promoting innovation and efficiency within the telecom industry by analyzing 6G network vulnerabilities and the importance of enhanced security measures. The use of AI for 6G security and its potential for threat detection, incident response, and network optimization through machine learning techniques.

INTRODUCTION

The emergence of 6G (Kirubasri, G. et al., 2021) networks bring with it opportunities for connectivity and innovation that have never been seen before in the rapidly advancing technological landscape of today. In spite of this, substantial security precautions are absolutely necessary in order to protect against the ever-evolving dangers that are posed by the internet. Traditional security approaches are proving to be insufficient when it comes to addressing the complexities of 6G networks, which are becoming

DOI: 10.4018/979-8-3693-2931-3.ch003

increasingly complex as network infrastructures do. It is precisely in this context that artificial intelligence (AI) (Anand, R. et al., 2023) plays a crucial part in bolstering the security of 6G networks. When it comes to improving the security posture of telecommunications networks, artificial intelligence, in particular in the form of machine learning (Pandey, B. K. et al., 2024), has emerged as a game-changing tool. Because they are able to proactively detect and mitigate potential threats, telecom operators are able to ensure the integrity and reliability of 6G networks by utilizing solutions that are driven by artificial intelligence (Anand, R. et al., 2024). One of the most important areas in which artificial intelligence has proven to be effective is in the detection and prevention of threats. When it comes to identifying malicious activities, traditional security measures frequently take advantage of predefined rules and signatures. The dynamic nature of cyber threats, on the other hand, makes it difficult for these approaches to keep up over time. These machine learning algorithms, on the other hand, have the capacity to analyze vast amounts of network data (JayaLakshmi, G. et al., 2024) in real time, which enables them to identify anomalous behaviours that may be indicative of potential security breaches. Artificial intelligence (AI) (Pandey, B. K., & Pandey, D., 2023) systems are able to recognize new dangers and react quickly to mitigate risks before they become more severe. This is accomplished through continuous learning and adaptation. Artificial intelligence improves the efficiency and effectiveness of security operations by automating tasks that are inherently repetitive and by augmenting the capabilities of humans (Pandey, B. K. et al., 2021b). AI-powered systems allow security teams to concentrate their efforts on more strategic endeavors by allowing them to offload routine tasks such as log analysis, incident triage, and malware detection to these systems. Not only does this improve response times, but it also decreases the likelihood of human error, which ultimately results in an improved overall security posture to the organization. AI gives organizations the ability to proactively implement preventive measures by analyzing historical data and identifying patterns (Muniandi, B. et al., 2024) that are indicative of potential vulnerabilities

Importance of AI Integration

The emergence of 6G networks bring with it opportunities for connectivity and innovation that have never been seen before in the rapidly advancing technological landscape of today. In spite of this, substantial security precautions are absolutely necessary in order to protect against the ever-evolving dangers that are posed by the internet. Traditional security approaches are proving to be insufficient when it comes to addressing the complexities of 6G networks, which are becoming increasingly complex as network infrastructures do. It is precisely in this context that artificial intelligence (AI) plays a crucial part in bolstering the security of 6G networks. When it comes to improving the security posture of telecommunications networks (Devasenapathy, D. et al., 2024), artificial intelligence, in particular in the form of machine learning, has emerged as a game-changing tool. Because they are able to proactively detect and mitigate potential threats, telecom operators are able to ensure the integrity and reliability of 6G networks by utilizing solutions that are driven by artificial intelligence. One of the most important areas in which artificial intelligence has proven to be effective is in the detection and prevention of threats. When it comes to identifying malicious activities, traditional security measures frequently take advantage of predefined rules and signatures. The dynamic nature of cyber threats, on the other hand, makes it difficult for these approaches to keep up over time. These machine learning algorithms, on the other hand, have the capacity to analyze vast amounts of network data in real time (Singh, S. et al., 2023), which enables them to identify anomalous behaviors that may be indicative of potential security breaches. Artificial intelligence (AI) systems are able to recognize new dangers and react quickly to mitigate risks before they become more severe. This is accomplished through continuous learning and adaptation. It improves the efficiency and effectiveness of security operations by automating tasks (Sharma, M. et al., 2024) that are inherently repetitive and by augmenting the capabilities of humans. AI-powered systems allow security teams to concentrate their efforts on more strategic endeavors by allowing them to offload routine tasks such as log analysis, incident triage, and malware detection to these systems. Not only does this improve response times, but it also decreases the likelihood of human error, which ultimately results in an improved overall security posture to the organization. Predictive analytics, which make it possible for telecom operators to anticipate and proactively address security vulnerabilities before they are exploited by adversaries. AI gives organizations the ability to proactively implement preventive measures by analyzing historical data and identifying patterns that are indicative of potential vulnerabilities. The strong identity verification and authorization protocols that are provided by AI-driven authentication and access control mechanisms contribute to the increased security of 6G networks. AI systems ensure that only authorized users gain access to network resources by utilizing advanced biometric authentication, behavioral analysis, and anomaly detection (Viswanathan, H., & Mogensen, P. E., 2020).

UNVEILING AI'S CONTRIBUTION TO 6G SECURITY

Analysis of 6G Network Vulnerabilities

One of the most important focal points for stakeholders in the telecommunications industry (Govindaraj, V. et al., 2023) and the cybersecurity community is the vulnerabilities that were discovered in the 6G network. When it comes to developing robust security strategies to protect against potential threats, having a solid understanding of these vulnerabilities is absolutely necessary. We will now delve into the analysis of vulnerabilities that are present in 6G networks:

1. *Complexity and Scale*: The incorporation of cutting-edge technologies like beamforming, terahertz communication, and network slicing brings about an increase in the level of complexity that is present in the architecture and operation of 6G networks. This complexity can result in vulnerabilities that are caused by misconfigurations, problems with interoperability, and software bugs. When it comes to deployment scale: The scale of deployment is expected to expand exponentially as a result of the anticipated proliferation of Internet of Things (Iyyanar, P. et al., 2023) devices and the deployment of 6G networks in a variety of environments, including urban areas (Pandey, B. K. et al., 2023a), remote regions, and industrial settings (George, W. K. et al., 2024). The attack surface is increased as a result of this widespread deployment, which also provides adversaries with additional entry points through which they can exploit vulnerabilities across the network.

2. *Emerging Technologies*: Two examples of emerging technologies are novel architectures and novel architectures. There are vulnerabilities associated with consensus mechanisms, smart contract security, and node compromise that are introduced by decentralized architectures such as blockchain (David, S. et al., 2023) based networks and mesh networks. These architectures promise resilience and scalability, but they also introduce vulnerabilities. In a similar vein, the implementation of network slicing for the purpose of providing tailored services results in the introduction of vulnerabilities across network slices in terms of isolation, orchestration, and resource allocation. The practical implementation of quantum key distribution (QKD) and quantum-resistant cryptography presents

challenges, despite the fact that quantum communication (Raja, D. et al., 2024) provides security guarantees that have never been seen before by utilizing the principles of quantum mechanics. It is possible that the security assurances that quantum communication offers could be undermined if there were vulnerabilities in QKD protocols or if quantum hardware were vulnerable to compromise.

3. *Cyber-Physical Convergence*: The convergence of 6G networks with critical infrastructures such as power grids, transportation systems (Sahani, K. et al., 2023), and healthcare facilities introduces interdependencies between cyber and physical domains. By exposing vulnerabilities in network infrastructure or communication (Gupta, A. K. et al., 2023) protocols, it is possible that disruptions or manipulations could occur, which would have far-reaching consequences for the functioning of society and the protection of the general public. When it comes to safety-critical applications, such as autonomous vehicles, unmanned aerial vehicles (UAVs), and remote medical procedures, the deployment of 6G networks requires stringent security measures to be implemented. It is possible that vulnerabilities in communication protocols, sensor (Du John, H. V. et al., 2023) data integrity, or edge computing platforms could compromise the safety and reliability of these applications, which would pose significant risks to both human lives and infrastructure.

4. *Edge Computing and AI Integration*: Edge computing in 6G networks enables localized processing and decision-making, which in turn reduces latency and enhances scalability. The vulnerabilities that exist in edge devices (Sasidevi S et al., 2024), edge servers, or communication links, on the other hand, could make sensitive data susceptible to unauthorized access, tampering, or information exfiltration. Security solutions that are powered by artificial intelligence typically include capabilities such as anomaly detection, threat prediction, and automated response; however, these solutions are vulnerable to adversarial attacks and model vulnerabilities. For the purpose of manipulating security systems or evading detection, adversaries can take advantage of vulnerabilities in artificial intelligence models or conduct poisoning attacks on training data. This undermines the effectiveness of these systems in protecting against cyber threats.

5. *Risks in the Supply Chain*: As a result of the global nature of supply chains in the telecommunications industry, there are multiple vendors, subcontractors, and intermediaries located in their respective regions. Backdoors, supply chain attacks, or counterfeit components could be introduced into 6G networks due to vulnerabilities in hardware components, firmware, or software modules that are sourced from different suppliers. This would compromise the security and integrity of the networks. It is essential to implement supply chain security practices in order to mitigate these risks and ensure the trustworthiness of network components. These practices include conducting thorough testing and validation, vetting suppliers, and implementing secure development lifecycle (SDLC) processes.

Strategic Applications of Enhanced Security Measures

Security is a top priority when it comes to online banking. Cybercriminals use more advanced techniques to take advantage of weaknesses in technology and steal confidential financial information. Banks and other financial organizations have put in place a number of enhanced security measures to fend off these threats and strengthen the security of their customers (Saxena, A. et al., 2024) online banking transactions. Strategic applications of Enhanced Security Measures are shown in the figure 1.

- *Two-factor authentication (2FA)*: A strong deterrent against unwanted access is two-factor authentication. 2FA greatly improves security over conventional password-based systems by requiring users to provide a combination of something they possess (like a mobile device) and something they know (like a password). Banks provide SMS codes, hardware tokens, authenticator apps, and biometric verification as some of the multiple authentication techniques available for 2FA. Users can select the approach that best fits their needs and security requirements thanks to this variety. Banks must take user experience into account even as they add an additional layer of security with 2FA. To guarantee that users can access their accounts without difficulty while upholding strict security standards, it is imperative to strike a balance between security and convenience.

- *Encryption*: Banks use end-to-end encryption (Kumar Pandey, B. et al., 2021), which shields confidential information from the user's device (Abdulkarim, Y. I. et al., 2023) to the bank's servers and back. This makes sure that the data cannot be read by unauthorized parties, even if it is intercepted. Online banking transactions are frequently secured by encryption algorithms like Advanced Encryption Standard (AES) and RSA. These algorithms use intricate mathematical calculations (Bessant, Y. A. et al., 2023) to jumble data into ciphertext, which is only understandable with the right decryption key. The security of encryption depends on efficient key management. Banks use strict key management procedures, such as key generation, storage, distribution, rotation, and revocation, to protect encryption keys (Pandey, B. K. et al., 2023b).

- *Fraud Detection Systems*: Behavioral Analysis: These systems look for abnormalities in user behavior by analyzing patterns of behavior. These systems are able to identify suspicious activities, like irregular login times, access from unknown locations, or unusual transaction amounts, by creating baseline behavior for every individual user. By continuously learning from new data and responding to emerging threats, these algorithms improve the effectiveness of fraud detection systems. These algorithms have the ability to identify minute patterns and anomalies that could elude conventional rule-based detection techniques. Since fraud detection systems are always running, possible threats can be quickly identified and countered. With real-time monitoring, banks can take quick action to stop fraudulent transactions and shield customer accounts from prying eyes.

- *Biometric Authentication*: Unique Identification: Biometric authentication uses distinctive physiological or behavioral traits, like voiceprints, iris patterns, fingerprints, or facial features, to confirm users' identities. Because these biometric characteristics are intrinsically distinctive to every person, they constitute extremely safe authentication elements. A lot of banks use the fingerprint or facial recognition sensors (Du John, H. V. et al., 2024) found on contemporary smartphones for biometric authentication. Strong security is ensured while user convenience is increased through this integration. A few banks use multimodal biometric authentication, which combines several biometric modalities (like fingerprint and facial recognition) to improve accuracy and dependability. Multimodal biometrics give better protection against spoofing attempts and enhance the user experience for authentication in general.

- *Real-Time Alerts*: Customizable Notifications: Based on their preferences and security needs, users can tailor notifications with real-time alerts. Alerts for particular occurrences, like significant transactions, account balance thresholds, or account settings modifications, can be set up by users. Users are guaranteed timely notification of critical account activities by having alerts sent in real-time to their preferred communication channels. Users are notified promptly to take immediate action, if necessary, whether through SMS, email, push notifications, or in-app messages. By providing real-time alerts, users are able to keep a close eye on their accounts and respond quickly to any unauthorized or suspicious activity. Alerts raise consumers' confidence in the security of their online banking transactions by giving them visibility and control.

Figure 1. Enhanced security measures

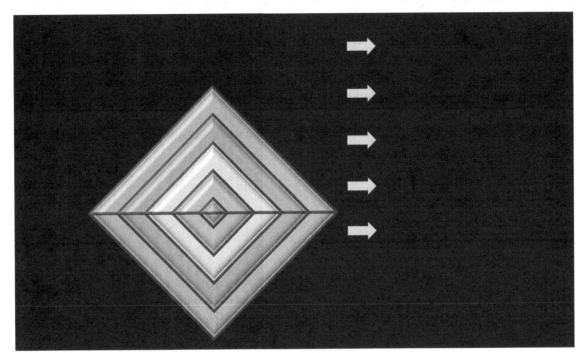

AI's Role in Security

The use of artificial intelligence (AI) has become a game-changer in how organizations in many fields deal with the complex security problems they face. Traditionally, cybersecurity operations have required a lot of work because tasks like monitoring, threat detection, and incident response were done by hand. The threat landscape is changing, though, with more sophisticated cyberattackers and more constant attacks. This means that defense needs to be more flexible and effective. AI solutions have become necessary in this situation because they help improve cyber defenses and lower risks in many ways. A lot of different things AI does for cybersecurity, from automating simple tasks to making it easier to find and stop advanced threats. By using AI-powered technologies, businesses can improve the efficiency of their operations, streamline their cyber workflows, and make their cybersecurity teams stronger. AI's ability to analyze huge amounts of data in real time is one of its best features when it comes to cybersecurity. This makes it easier for companies to find threats and deal with them than ever before. In its simplest form, AI helps cybersecurity experts stay ahead of new threats by giving them insights they can act on and predictions they can make. AI-powered solutions can find strange activities that could mean a security breach by constantly watching network traffic, studying patterns of user behavior, and connecting data from different sources. AI algorithms can learn from past events and adjust to new threat vectors, which lets businesses anticipate and reduce new risks before they happen. AI is very important for managing vulnerabilities and figuring out risks, in addition to finding threats. Companies can use machine learning algorithms (Tripathi, R. P. et al., 2023) to sort security holes by how likely they are to cause problems and how bad those problems could be. This helps them focus their resources on the biggest threats. Also, risk quantification tools powered by AI help businesses figure out how much of financial and operational risk cybersecurity incidents pose, which helps them make smart decisions and decide how to use their resources. AI also plays a big role in improving incident response, which lets companies automate the containment, investigation, and fix of security incidents. Artificial intelligence-powered incident response platforms can speed up response times, cut down on downtime, and lessen the damage that cyberattacks due to business operations by using advanced orchestration and automation features. Also, threat intelligence platforms powered by AI give companies real-time information (Sharma, S. et al., 2023) about new threats and weak spots, which lets them make proactive plans to protect themselves from threats.

LEVERAGING AI FOR 6G SECURITY

AI's Potential in Security Enhancement

The field of artificial intelligence (AI) has shown itself to be a revolutionary force in enhancing security protocols in a variety of fields. AI possesses the potential to completely transform the cybersecurity industry through the application of sophisticated algorithms and machine learning techniques. The key aspects highlighting AI's potential in enhancing security:

- *Automated Threat Detection*: AI-powered systems are excellent at tracking enormous volumes of data in real-time, making it possible to identify minute irregularities that could be signs of possible security lapses. These systems use complex algorithms to examine system logs, user activity, and network traffic in order to spot anomalies from the norm that might indicate malicious activity. Artificial intelligence (AI) can categorize and rank security events according to their seriousness and chance of posing a threat by using machine learning models. AI speeds up response times and lowers the chance of data breaches by automating the preliminary steps of threat detection, freeing up security analysts to concentrate on looking into and resolving verified incidents.
- *Predictive Analytics*; AI enables businesses to foresee security flaws and take proactive measures to fix them before hackers take advantage of them. By using predictive analytics, artificial intelligence (AI) algorithms examine past data to find patterns and trends that allow businesses to strengthen their security posture and take preventative action. AI can recognize new threats and predict potential attack vectors by utilizing machine learning algorithms, which are based on past attack patterns and trends. Through risk mitigation and the reduction of potential impact from security incidents, this proactive approach enables organizations to stay ahead of cyber threats (Chen, Y. et al., 2023).
- *Adaptive Defense Mechanisms*: Artificial intelligence (AI) makes it possible for security measures to be dynamically adjusted to changing attack vectors and threats. Artificial intelligence (AI)-driven systems are always learning from fresh data and modifying their defense plans to match the changing threat landscape. Artificial intelligence (AI)-powered security solutions can automatically modify security policies and configurations in response to new threats by utilizing reinforcement learning techniques. By strengthening resilience and effectiveness in the face of changing cyberthreats, this adaptive strategy makes sure that businesses are safe from new threats.

- *Enhanced Incident Response*: Detection, investigation, and remediation of security incidents are automated and streamlined by incident response platforms driven by AI. These platforms rank incidents according to their impact and severity by analyzing security alerts using machine learning algorithms. Artificial Intelligence (AI) reduces the time it takes to resolve security incidents and speeds up response times by automating repetitive tasks like data collection and incident triage. This lowers the chance of data breaches and lessens the effect that cyberattacks have on an organization's ability to conduct business.
- *Risk Assessment and Management*: By measuring the financial and operational risks connected to cybersecurity incidents, AI enables thorough risk assessment and management. By analyzing past data, machine learning algorithms can spot patterns and trends that help organizations prioritize security vulnerabilities and manage resources efficiently. AI helps businesses to prioritize mitigation efforts by assessing the possible impact of security incidents on business operations through the use of risk quantification models. By improving decision-making and resource allocation, this data-driven approach makes sure that organizations are prepared to handle the most important security threats.
- *Biometric Authentication*: AI-powered biometric authentication techniques, like facial or fingerprint recognition, offer a very safe way to confirm users' identities. Because these biometric characteristics are intrinsically distinctive to every person, they are challenging to duplicate or falsify. AI-powered biometric authentication systems can reliably identify people based on their distinct biometric traits by utilizing sophisticated machine learning algorithms. This improves authentication security, reducing the possibility of unwanted access and shielding private information from online dangers (Banafaa, M. et al., 2023).

AI has a wide range of applications that could improve security. Organizations can strengthen their cyber defenses, reduce risks, and protect vital assets from a variety of cyber threats by utilizing AI-driven technologies. Navigating the ever-changing and complex cybersecurity landscape requires adopting AI-powered security solutions.

Machine Learning in Security Enhancement

The field of machine learning (ML) has emerged as a powerful instrument that can be used to improve security measures across a variety of environments. Through the use of machine learning techniques, systems are able to learn from data, recognize patterns, and make intelligent decisions through the absence of explicit programming. Machine learning provides capabilities that have never been seen before in the field of cybersecurity (Pandey, B. K. et al., 2021a). These capabilities include the ability to identify and eliminate threats, anticipate potential vulnerabilities, and strengthen overall defense mechanisms. This article examines the ways in which machine learning can be utilized to improve security, focusing on the significance and impact that it has on contemporary security principles. Automated threat detection is one of the most important potential applications of machine learning in the field of security enhancement. For the purpose of identifying patterns that may indicate potential security breaches, machine learning algorithms examine vast amounts of data, which may include network traffic, system logs, and user behavior. ML-powered systems are able to flag suspicious activities in real time by identifying anomalies and deviations from normal patterns. This enables security teams to respond as quickly as possible and mitigate potential threats before they become more severe. In general, the resilience of organizations against cyber attacks is improved by taking this proactive approach to threat detection. Artificial intelligence plays a significant part in predictive analytics for the purpose of security. Machine learning algorithms are able to discover potential vulnerabilities and anticipate future cyber threats by analysing historical data and recognizing patterns and trends. Through the utilization of this predictive capability, organizations are able to proactively strengthen their security posture and implement preventative measures (Saad, W., Bennis, M., & Chen, M., 2019). According to historical attack patterns, for instance, machine learning models are able to recognize common attack vectors and vulnerabilities. This enables organizations to prioritize patching and remediation efforts in accordance with the identified vulnerabilities and attack vectors. Adaptive defense mechanisms that are able to dynamically adjust to evolving threats and attack vectors are made possible by machine learning, which also enables threat detection and prediction. In order to adapt their defense strategies to the ever-changing threat landscape, machine learning algorithms are constantly learning from new data and adjusting this learning. ML-powered security systems, for example, are able to automatically adjust their security policies and configurations in response to new threats thanks to reinforcement learning techniques. This helps to ensure that organizations continue to be protected against new risks. Automating and simplifying the process of identifying, investigating, and resolving security incidents is another way that machine learning improves incident response capabilities. It is possible for security teams to concentrate their efforts on high-priority threats thanks to incident response platforms that are powered by machine learning. These platforms analyze security alerts and prioritize incidents based on the severity and impact of the incidents. Incident triage and data collection are two examples of the repetitive tasks that can be automated by machine learning algorithms. This helps to reduce the amount of time it takes to resolve security incidents and speeds up response times. Through the quantification of the financial and operational risks that are associated with cybersecurity incidents, machine learning also makes it easier to conduct comprehensive risk assessments and management. An organization is able to prioritize security vulnerabilities and effectively allocate resources by utilizing machine learning algorithms, which analyze historical data to identify patterns and trends. ML enables organizations to evaluate the potential impact that security incidents could have on business operations and prioritize mitigation efforts in accordance with that assessment. This is accomplished through the application of risk quantification models. In the realm of security enhancement, biometric authentication is yet another significant application of machine learning. Methods of biometric authentication that are powered by machine learning, such as fingerprint or facial recognition, offer a user verification process that is extremely safe and secure. It is extremely difficult to replicate or forge these biometric characteristics because they are inherently unique to each individual. Enhanced authentication security and protection of sensitive data from cyber threats are both achieved through the utilization of advanced machine learning algorithms in biometric authentication systems. These systems are able to accurately identify individuals based on their unique biometric characteristics (De Alwis, C. et al., 2021).

ADDRESSING CONCERNS AND CONSIDERATIONS

Ethical and Privacy Implications

Organizations incorporating machine learning (ML) techniques into security enhancement strategies must take great care to address ethical and privacy implications (Pramanik, S. et al., 2023). Regarding data privacy, algorithmic bias, accountability, transparency, and societal effects, these technologies raise serious questions. Organizations need to take proactive steps and give these issues careful thought in order to effectively reduce them. First and foremost, data privacy must be protected. To protect confidential data, organizations should put strong data protection measures in place, such as access controls, encryption, and anonymization. To further ensure compliance with data privacy laws like the CCPA and GDPR, adopting a minimal data collection strategy and defining explicit data retention policies aid in limiting the storage of sensitive data. In order to guarantee just and equitable results, it is imperative to mitigate algorithmic bias. ML models should undergo routine bias assessments, especially with regard to variables like socioeconomic status, gender, and race. Biases in data collection can be lessened by diversifying training datasets, including a range of stakeholders, and integrating different points of view into model development. Building confidence in ML-powered security solutions requires transparency and explainability (Giordani, M. et al., 2020). Enhancing transparency can be achieved by making interpretable algorithms a priority and by recording model architectures and decision-making procedures. Stakeholders, including end users and regulators, can understand and have faith in ML-powered security systems when model outputs are explained in plain English. To ensure that ML-powered security solutions are deployed and operated responsibly, it is imperative to establish explicit accountability mechanisms. For all parties involved in development, deployment, and operation, organizations should specify roles and responsibilities. In crucial areas of security operations, human oversight and decision-making authority should be preserved. This will enable human experts to step in when algorithmic errors or discrepancies arise. Finally, assessing wider societal effects is crucial. Comprehensive evaluations of possible societal ramifications, like workforce displacement and digital disparities, ought to be carried out. Ensuring that security solutions benefit all societal segments requires developing inclusive approaches to ML deployment that prioritize fair access to security technologies and engage with a variety of stakeholders (Ziegler, V. et al., 2020).

Technical Hurdles and Scalability

Machine learning (ML) holds a great deal of promise for improving security measures; however, in order for organizations to effectively address ethical and privacy concerns, they will need to overcome a variety of technical obstacles and scalability challenges. As a result of the potential for these challenges to have a significant impact on the implementation and deployment of machine learning-powered security solutions, careful consideration and innovative approaches are indispensable. When it comes to machine learning-based security enhancement, one of the most significant technical challenges is the requirement for large datasets of high quality in order to train robust models. The process of collecting and annotating a wide variety of datasets that accurately represent real-world scenarios can be time-consuming and resource-intensive. In addition, there are significant challenges involved in protecting the confidentiality and safety of sensitive data, particularly in highly regulated industries such as the healthcare and financial sectors. Training machine learning models on decentralized or privacy-preserving data sources while maintaining data privacy and security can be accomplished through the use of techniques such as federated learning, differential privacy, and synthetic data generation. These techniques can be utilized by organizations in order to overcome this obstacle. The complexity and interpretability of machine learning models that are utilized in security applications is another technical challenge. An extensive number of advanced machine learning algorithms, such as deep neural networks, function as black boxes, which makes it difficult to comprehend and interpret the decision-making processes that they employ. There is a potential for a lack of interpretability to hinder the ability of stakeholders to trust and validate machine learning-powered security solutions, which raises concerns about accountability and transparency. Techniques for model interpretability, such as model-agnostic methods, feature importance analysis, and adversarial testing, can be investigated by organizations in order to overcome this obstacle. These techniques can be utilized to gain insights into the behavior of models and the decision-making processes that they employ. Especially in large-scale enterprise environments, scalability is another important factor to take into consideration when it comes to machine learning-based security enhancement. In order to effectively detect and respond to security threats, machine learning models need to be able to process requests in real time and handle massive amounts of data. There is a possibility that traditional machine learning algorithms will have difficulty scaling effectively, which will result in performance bottlenecks and increased computational costs. It is possible for organizations to utilize distributed computing frameworks, cloud-based infrastructure, and parallel processing techniques in order to optimize model performance and distribute computational workloads (Sahani, S. K. et al., 2024). This is done in order to address scalability challenges. The robustness and resilience of security solutions powered by machine learning against adversarial attacks and evasion techniques presents a significant challenge from a technical standpoint. For the purpose of manipulating or deceiving security systems, adversaries may take advantage of vulnerabilities in machine learning models, which can result in false positives or false negatives. The resilience of machine learning models against adversarial attacks can be improved through the use of techniques such as robustness testing, adversarial training, and anomaly detection. This, in turn, can improve the overall security posture of security solutions that are powered by machine learning.

INDUSTRY IMPACT OF AI IN 6G SECURITY

AI's Influence on Telecom Industry

Operations, customer experiences, and network management are all being revolutionized by artificial intelligence (AI), which is reshaping the landscape of the telecommunications industry. Artificial intelligence is driving innovation and efficiency across a variety of aspects of the telecommunications industry. It is able to analyze vast amounts of data and make knowledgeable decisions. One of the most important areas in which artificial intelligence is having a significant impact is in the optimization (Jayapoorani, S. et al., 2023) and management of networks. A growing number of connected devices and an increase in the amount of data being consumed are contributing to the complexity of telecommunication networks. Computer programs that are powered by artificial intelligence are able to analyze the patterns of network traffic, forecast the points of congestion, and dynamically allocate resources in order to maximize network performance. Not only does this proactive approach improve the reliability of networks and the quality of services provided, but it also helps telecom operators reduce their operational costs. Artificial intelligence is revolutionizing customer service and support in the financial services sector. The use of chatbots and virtual assistants that are powered by artificial intelligence is revolutionizing customer interactions by making it possible to provide personalised and context-aware assistance around the clock. By addressing customer inquiries, resolving technical issues, and even processing service requests on their own, these AI-driven systems have the ability to improve customer satisfaction while simultaneously reducing the workload of human customer support agents. Innovation in product development and service offerings within the telecommunications industry is also being driven by artificial intelligence (Liang, W. et al., 2021). For the purpose of identifying emerging trends and forecasting future demands, machine learning algorithms analyze the behavior and preferences of customers. In order to increase their competitiveness and differentiate themselves in the market, telecom companies make use of these insights to tailor their products and services to meet the ever-changing requirements of their customers. Artificial intelligence is an essential component in the process of improving cybersecurity within the telecommunications sector. For the purpose of detecting and mitigating potential threats in real time, security solutions that are powered by artificial intelligence are becoming increasingly important in light of the proliferation of cyber threats that target telecom networks and infrastructure. Identifying suspicious activities and protecting against cyber attacks, protecting sensitive data, and ensuring the integrity of telecom networks are all goals of these solutions, which make use of advanced analytics and anomaly detection, respectively. To improving operational efficiency and providing better service to customers, artificial intelligence is driving advancements in network automation and management. In order to reduce downtime and improve overall network reliability, artificial intelligence algorithms enable autonomous network optimization, self-healing capabilities, and predictive maintenance. This automation not only provides telecom operators with the ability to provide seamless connectivity and services to their customers, but it also improves the operational efficiency of the business (Rajagopal, S. et al., 2020).

Transformative Effects on Telecom Sector

With its revolutionary effects on operations, services, and customer experiences, artificial intelligence (AI) is reshaping the telecommunications industry. Artificial Intelligence (AI) is transforming the oil and gas industry through process automation, informed decision-making, and massive data analysis. This piece aims to examine the revolutionary consequences of artificial intelligence (AI) in the telecommunications industry, analyzing its influence on network administration, customer relations, product creation, and security. Transformative Effects on Telecom Sector are as shown in the Figure 2.

Figure 2. Transformative effects on telecom sector

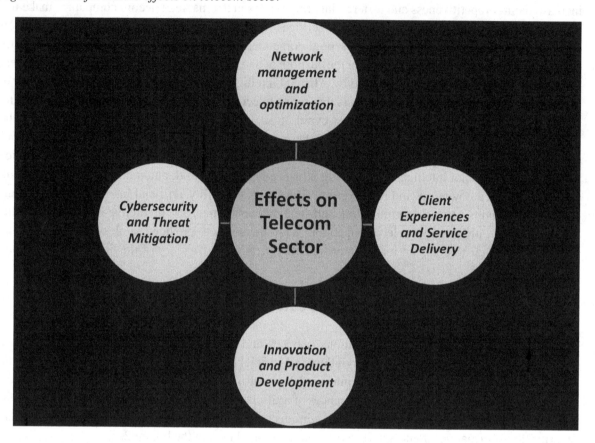

Network management and optimization: The exponential rise in data consumption and the proliferating number of connected devices are making telecommunication networks more complex. With the help of AI-powered solutions, telecom operators can improve the efficiency, dependability, and performance of their networks. This is a significant development in network management and optimization (Pandey, D. et al., 2021). In order to maximize network utilization, artificial intelligence (AI) algorithms examine network traffic patterns, forecast congestion points, and divide resources dynamically. Telecom operators can boost end-user quality of service, minimize downtime, and guarantee seamless connectivity with this proactive approach. By utilizing AI's capacity to provide fast internet and low latency communication, AI-driven network optimization helps to meet the demands of the contemporary digital environment (George, W. K. et al., 2023), thereby enabling the deployment of 5G technology.

Client Experiences and Service Delivery: By altering communications and support services, artificial intelligence is bringing about a radical change in the telecom sector's customer experiences. Artificial intelligence-driven chatbots and virtual assistants offer customers context-aware, personalized help by answering questions, resolving technical problems, and handling service requests on their own. These AI-powered platforms are always available, providing prompt answers and effective query resolution. AI speeds up response times, improves customer satisfaction, and maximizes resource use by automating repetitive tasks and optimizing support procedures. Further, AI helps telecoms to provide customers with personalized promotions and recommendations based on their usage habits and preferences, which improves customer loyalty and overall customer experience (Mokhtari, S. et al., 2021).

Innovation and Product Development: In the telecom industry, artificial intelligence (AI) is fostering innovation in both service and product offerings. Algorithms for machine learning examine enormous volumes of data to spot new trends, forecast demand, and enhance pricing tactics. Utilizing these insights, telecom companies can create new products, alter existing ones, and set themselves apart from competitors. Telecom operators may enhance marketing campaigns, increase customer retention, and spur revenue growth by using AI-powered analytics to obtain insightful knowledge about customer behavior. Telecom companies can create cutting-edge solutions that remain competitive in a market that is changing quickly and meet the changing needs of their customers by utilizing AI's capabilities.

Cybersecurity and Threat Mitigation: With cyberattacks on telecom networks and infrastructure constantly evolving, AI-powered security solutions are becoming more and more crucial for protecting private information and maintaining network integrity. Real-time threat identification and mitigation by AI algorithms guards against cyberattacks and security lapses by utilizing sophisticated analytics and anomaly detection techniques. With the help of these AI-driven security solutions, telecom operators can lower risks and protect vital assets by improving threat detection capabilities, speeding up response times, and fortifying overall cybersecurity posture. Through the utilization of artificial intelligence (AI), telecommunications firms can safeguard their networks and infrastructure from potential threats, identify dubious activity, and promptly address security incidents.

Network management, customer experiences, product development, and cybersecurity are all being drastically altered by artificial intelligence in the telecom industry. Artificial intelligence (AI) will be crucial in advancing innovation, streamlining operations, and providing value to stakeholders and customers alike as the telecom sector develops further in the digital era (Swapna, H. R. et al., 2023). In an increasingly dynamic and competitive market landscape, telecom operators can stay competitive, increase efficiency, and open up new opportunities by adopting AI technologies.

CONCLUSION

For 6G security to be effective in strengthening network defenses, resolving vulnerabilities, and improving overall security posture, Artificial Intelligence (AI) must be integrated. Throughout our investigation, we have determined that artificial intelligence (AI) plays a critical role in bolstering network defenses, reducing cyberthreats, and promoting innovation in the telecommunications industry. As AI has the potential to greatly influence both, we have underlined the significance of comprehending 6G security vulnerabilities and the importance of enhanced security measures. Telecom companies can use AI to improve network performance, identify threats, and handle incidents by utilizing machine learning techniques in 6G security. For the purpose of maintaining trust, privacy, and regulatory compliance, it is imperative to address issues related to AI integration, including technical difficulties, scalability issues, and ethical considerations. We have emphasized how AI is revolutionizing cybersecurity procedures, network management, and customer experiences in the telecom industry. Telecom companies can seize new opportunities, boost productivity, and provide customers with better services in the era of 6G connectivity by embracing AI technologies.

REFERENCES

Abdulkarim, Y. I., Awl, H. N., Muhammadsharif, F. F., Saeed, S. R., Sidiq, K. R., Khasraw, S. S., & Pandey, D. (2023). Metamaterial-based sensors loaded corona-shaped resonator for COVID-19 detection by using microwave techniques. *Plasmonics*, 1–16.

Anand, R., Khan, B., Nassa, V. K., Pandey, D., Dhabliya, D., Pandey, B. K., & Dadheech, P. (2023). Hybrid convolutional neural network (CNN) for kennedy space center hyperspectral image. *Aerospace Systems*, 6(1), 71–78. 10.1007/s42401-022-00168-4

Anand, R., Lakshmi, S. V., Pandey, D., & Pandey, B. K. (2024). An enhanced ResNet-50 deep learning model for arrhythmia detection using electrocardiogram biomedical indicators. *Evolving Systems*, 15(1), 83–97. 10.1007/s12530-023-09559-0

Banafaa, M., Shayea, I., Din, J., Azmi, M. H., Alashbi, A., Daradkeh, Y. I., & Alhammadi, A. (2023). 6G mobile communication technology: Requirements, targets, applications, challenges, advantages, and opportunities. *Alexandria Engineering Journal*, 64, 245–274. 10.1016/j.aej.2022.08.017

Bessant, Y. A., Jency, J. G., Sagayam, K. M., Jone, A. A. A., Pandey, D., & Pandey, B. K. (2023). Improved parallel matrix multiplication using Strassen and Urdhvatiryagbhyam method. *CCF Transactions on High Performance Computing*, 5(2), 102–115. 10.1007/s42514-023-00149-9

Chen, Y., Weng, Q., Tang, L., Wang, L., Xing, H., & Liu, Q. (2023). Developing an intelligent cloud attention network to support global urban green spaces mapping. *ISPRS Journal of Photogrammetry and Remote Sensing*, 198, 197–209. 10.1016/j.isprsjprs.2023.03.005

David, S., Duraipandian, K., Chandrasekaran, D., Pandey, D., Sindhwani, N., & Pandey, B. K. (2023). Impact of blockchain in healthcare system. In *Unleashing the Potentials of blockchain technology for healthcare industries* (pp. 37–57). Academic Press. 10.1016/B978-0-323-99481-1.00004-3

De Alwis, C., Kalla, A., Pham, Q. V., Kumar, P., Dev, K., Hwang, W. J., & Liyanage, M. (2021). Survey on 6G frontiers: Trends, applications, requirements, technologies and future research. *IEEE Open Journal of the Communications Society*, 2, 836–886. 10.1109/OJCOMS.2021.3071496

Devasenapathy, D., Madhumathy, P., Umamaheshwari, R., Pandey, B. K., & Pandey, D. (2024). Transmission-efficient grid-based synchronized model for routing in wireless sensor networks using Bayesian compressive sensing. *SN Computer Science*, 5(1), 1–11.

Du John, H. V., Ajay, T., Reddy, G. M. K., Ganesh, M. N. S., Hembram, A., Pandey, B. K., & Pandey, D. (2023). Design and simulation of SRR-based tungsten metamaterial absorber for biomedical sensing applications. *Plasmonics*, 18(5), 1903–1912. 10.1007/s11468-023-01910-0

. Du John, H. V., Jose, T., Sagayam, K. M., Pandey, B. K., & Pandey, D. (2024). Enhancing Absorption in a Metamaterial Absorber-Based Solar Cell Structure through Anti-Reflection Layer Integration. *Silicon*, 1-11.

George, W. K., Ekong, M. O., Pandey, D., & Pandey, B. K. (2023). Pedagogy for Implementation of TVET Curriculum for the Digital World. In *Applications of Neuromarketing in the Metaverse* (pp. 117-136). IGI Global. 10.4018/978-1-6684-8150-9.ch009

George, W. K., Silas, E. I., Pandey, D., & Pandey, B. K. (2024). Utilization of Industry 4.0 Technologies in Nigerian Technical and Vocational Education: A Conundrum for Educators. In *Examining the Rapid Advance of Digital Technology in Africa* (pp. 270–293). IGI Global. 10.4018/978-1-6684-9962-7.ch014

Giordani, M., Polese, M., Mezzavilla, M., Rangan, S., & Zorzi, M. (2020). Toward 6G networks: Use cases and technologies. *IEEE Communications Magazine*, 58(3), 55–61. 10.1109/MCOM.001.1900411

. Govindaraj, V., Dhanasekar, S., Martinsagayam, K., Pandey, D., Pandey, B. K., & Nassa, V. K. (2023). Low-power test pattern generator using modified LFSR. *Aerospace Systems*, 1-8.

Gupta, A. K., Sharma, R., Pandey, D., Nassa, V. K., Pandey, B. K., George, A. S., & Dadheech, P. (2023). Performance analysis of eight-channel WDM optical network with different optical amplifiers for industry 4.0. In *Innovation and Competitiveness in Industry 4.0 Based on Intelligent Systems* (pp. 197–212). Springer International Publishing. 10.1007/978-3-031-29775-5_9

Iyyanar, P., Anand, R., Shanthi, T., Nassa, V. K., Pandey, B. K., George, A. S., & Pandey, D. (2023). A real-time smart sewage cleaning UAV assistance system using IoT. In *Handbook of Research on Data-Driven Mathematical Modeling in Smart Cities* (pp. 24–39). IGI Global.

JayaLakshmi. G., Pandey, D., Pandey, B. K., Kaur, P., Mahajan, D. A., & Dari, S. S. (2024). Smart Big Data Collection for Intelligent Supply Chain Improvement. In *AI and Machine Learning Impacts in Intelligent Supply Chain* (pp. 180-195). IGI Global.

Jayapoorani, S., Pandey, D., Sasirekha, N. S., Anand, R., & Pandey, B. K. (2023). Systolic optimized adaptive filter architecture designs for ECG noise cancellation by Vertex-5. *Aerospace Systems*, 6(1), 163–173. 10.1007/s42401-022-00177-3

Kirubasri, G., Sankar, S., Pandey, D., Pandey, B. K., Singh, H., & Anand, R. (2021, September). A recent survey on 6G vehicular technology, applications and challenges. In 2021 9th International Conference on Reliability, *Infocom Technologies and Optimization (Trends and Future Directions)(ICRITO)* (pp. 1-5). IEEE.

Kumar Pandey, B., Pandey, D., Nassa, V. K., Ahmad, T., Singh, C., George, A. S., & Wakchaure, M. A. (2021). Encryption and steganography-based text extraction in IoT using the EWCTS optimizer. *Imaging Science Journal*, 69(1-4), 38–56. 10.1080/13682199.2022.2146885

Liang, W., Xiao, L., Zhang, K., Tang, M., He, D., & Li, K. C. (2021). Data fusion approach for collaborative anomaly intrusion detection in blockchain-based systems. *IEEE Internet of Things Journal*, 9(16), 14741–14751. 10.1109/JIOT.2021.3053842

Mokhtari, S., Abbaspour, A., Yen, K. K., & Sargolzaei, A. (2021). A machine learning approach for anomaly detection in industrial control systems based on measurement data. *Electronics (Basel)*, 10(4), 407. 10.3390/electronics10040407

Muniandi, B., Nassa, V. K., Pandey, D., Pandey, B. K., Dadheech, P., & George, A. S. (2024). Pattern Analysis for Feature Extraction in Complex Images. In *Using Machine Learning to Detect Emotions and Predict Human Psychology* (pp. 145-167). IGI Global. 10.4018/979-8-3693-1910-9.ch007

Pandey, B. K., & Pandey, D. (2023). Parametric optimization and prediction of enhanced thermoelectric performance in co-doped CaMnO3 using response surface methodology and neural network. *Journal of Materials Science Materials in Electronics*, 34(21), 1589. 10.1007/s10854-023-10954-1

Pandey, B. K., Pandey, D., Alkhafaji, M. A., Güneşer, M. T., & Şeker, C. (2023a). A reliable transmission and extraction of textual information using keyless encryption, steganography, and deep algorithm with cuckoo optimization. In *Micro-Electronics and Telecommunication Engineering: Proceedings of 6th ICMETE 2022* (pp. 629–636). Springer Nature Singapore. 10.1007/978-981-19-9512-5_57

Pandey, B. K., Pandey, D., Dadheech, P., Mahajan, D. A., George, A. S., & Hameed, A. S. (2023b). Review on Smart Sewage Cleaning UAV Assistance for Sustainable Development. In *Handbook of Research on Safe Disposal Methods of Municipal Solid Wastes for a Sustainable Environment* (pp. 69–79). IGI Global. 10.4018/978-1-6684-8117-2.ch005

Pandey, B. K., Pandey, D., & Sahani, S. K. (2024). Autopilot control unmanned aerial vehicle system for sewage defect detection using deep learning. *Engineering Reports*, 12852. 10.1002/eng2.12852

Pandey, B. K., Pandey, D., Wairya, S., & Agarwal, G. (2021a). An advanced morphological component analysis, steganography, and deep learning-based system to transmit secure textual data. [IJDAI]. *International Journal of Distributed Artificial Intelligence*, 13(2), 40–62. 10.4018/IJDAI.2021070104

Pandey, B. K., Pandey, D., Wariya, S., & Agarwal, G. (2021b). A deep neural network-based approach for extracting textual images from deteriorate images. *EAI Endorsed Transactions on Industrial Networks and Intelligent Systems*, 8(28), e3–e3. 10.4108/eai.17-9-2021.170961

Pandey, D., Pandey, B. K., & Wairya, S. (2021). Hybrid deep neural network with adaptive galactic swarm optimization for text extraction from scene images. *Soft Computing*, 25(2), 1563–1580. 10.1007/s00500-020-05245-4

Pramanik, S., Pandey, D., Joardar, S., Niranjanamurthy, M., Pandey, B. K., & Kaur, J. (2023, October). An overview of IoT privacy and security in smart cities. In *AIP Conference Proceedings* (*Vol. 2495*, No. 1). AIP Publishing. 10.1063/5.0123511

Raja, D., Kumar, D. R., Santhiyakumari, N., Kumarganesh, S., Sagayam, K. M., Thiyaneswaran, B., Pandey, B. K., & Pandey, D. (2024). A compact dual-feed wide-band slotted antenna for future wireless applications. *Analog Integrated Circuits and Signal Processing*, 118(2), 1–15. 10.1007/s10470-023-02233-0

Rajagopal, S., Kundapur, P. P., & Hareesha, K. S. (2020). A stacking ensemble for network intrusion detection using heterogeneous datasets. *Security and Communication Networks*, 2020, 1–9. 10.1155/2020/4586875

Saad, W., Bennis, M., & Chen, M. (2019). A vision of 6G wireless systems: Applications, trends, technologies, and open research problems. *IEEE Network*, 34(3), 134–142. 10.1109/MNET.001.1900287

Sahani, K., Khadka, S. S., Sahani, S. K., Pandey, B. K., & Pandey, D. (2023). A possible underground roadway for transportation facilities in Kathmandu Valley: A racking deformation of underground rectangular structures. *Engineering Reports*, 12821. 10.1002/eng2.12821

Sahani, S. K., Pandey, B. K., & Pandey, D. (2024). *Single-valued Signals, Multi-valued Signals and Fixed-Point of Contractive Signals*. Mathematics Open. 10.1142/S2811007224500020

Sasidevi, S., Kumarganesh, S., Saranya, S., Thiyaneswaran, B., Shree, K. V. M., & Martin Sagayam, K. (2024, May 15). Design of Surface Plasmon Resonance (SPR) Sensors for Highly Sensitive Biomolecular Detection in Cancer Diagnostics. *Plasmonics*. 10.1007/s11468-024-02343-z

Saxena, A., Agarwal, A., Pandey, B. K., & Pandey, D. (2024). Examination of the Criticality of Customer Segmentation Using Unsupervised Learning Methods. Circular Economy and Sustainability, 1-14. Saxena.

Sharma, M., Talwar, R., Pandey, D., Nassa, V. K., Pandey, B. K., & Dadheech, P. (2024). A Review of Dielectric Resonator Antennas (DRA)-Based RFID Technology for Industry 4.0. *Robotics and Automation in Industry*, 4(0), 303–324.

Sharma, S., Pandey, B. K., Pandey, D., Anand, R., Sharma, A., & Saini, S. (2023, March). Character Recognition Technique Implementation for Complicated Deteriorated Scene. In *2023 6th International Conference on Information Systems and Computer Networks (ISCON)* (pp. 1-4). IEEE. 10.1109/ISCON57294.2023.10112185

Singh, S., Madaan, G., Kaur, J., Swapna, H. R., Pandey, D., Singh, A., & Pandey, B. K. (2023). *Bibliometric Review on Healthcare Sustainability. Handbook of Research on Safe Disposal Methods of Municipal Solid Wastes for a Sustainable Environment*, (pp. 142-161). IGI Global. 10.4018/978-1-6684-8117-2.ch011

Swapna, H. R., Bigirimana, E., Madaan, G., Hasan, A., Pandey, B. K., & Pandey, D. (2023). Impact of neuromarketing on consumer psychology in digitally connected networks. In *Applications of Neuromarketing in the Metaverse* (pp. 193–205). IGI Global. 10.4018/978-1-6684-8150-9.ch015

Tripathi, R. P., Sharma, M., Gupta, A. K., Pandey, D., Pandey, B. K., Shahul, A., & George, A. H. (2023). Timely prediction of diabetes by means of machine learning practices. *Augmented Human Research*, 8(1), 1. 10.1007/s41133-023-00062-4

Viswanathan, H., & Mogensen, P. E. (2020). Communications in the 6G era. *IEEE Access : Practical Innovations, Open Solutions*, 8, 57063–57074. 10.1109/ACCESS.2020.2981745

Ziegler, V., Viswanathan, H., Flinck, H., Hoffmann, M., Räisänen, V., & Hätönen, K. (2020). 6G architecture to connect the worlds. *IEEE Access : Practical Innovations, Open Solutions*, 8, 173508–173520. 10.1109/ACCESS.2020.3025032

Ziegler, V., Viswanathan, H., Flinck, H., Hoffmann, M., Raisanen, V., & Hatonen, K. (2020). 6G Architecture to Connect the Worlds. *IEEE Access, 8.*

Chapter 4
Quantum Threats, Quantum Solutions:
ML Approaches to 6G Security

Hemlata Patel
Parul University, India

Anupma Surya
Greater Noida Institute of Technology, India

Aswini Kilaru
Institute of Aeronautical Engineering, Dundigal, India

Vijilius Helena Raj
Department of Applied Sciences, New Horizon College of Engineering, Bangalore, India

Amit Dutt
Lovely Professional University, India

Joshuva Arockia Dhanraj
https://orcid.org/0000-0001-5048-7775
Chandigarh University, India

ABSTRACT

Quantum computing represents a significant milestone in technological progress, offering unprecedented computational capabilities. Nevertheless, this inherent potential also poses a significant risk to the field of cybersecurity. This essay aims to examine the complexities associated with quantum threats and investigate potential strategies for mitigating these risks. Cryptography is a fundamental component of contemporary cybersecurity, serving as a resilient method that guarantees the preservation of data integrity and confidentiality within IT infrastructure. However, the emergence of quantum computing poses a significant challenge to this established paradigm.

INTRODUCTION

Quantum computing represents a significant milestone in technological progress, offering unprecedented computational capabilities. Nevertheless, this inherent potential also poses a significant risk to the field of cybersecurity. Cryptography is a fundamental component of contemporary cybersecurity, serving as a resilient method that guarantees the preservation of data integrity (Sharma, S. et al., 2023) and confidentiality within IT infrastructure. The emergence of quantum computing poses a significant challenge to this established paradigm. Although current quantum computers do not currently have the ability to decrypt data, the swift advancements in quantum technology indicate that this could poten-

DOI: 10.4018/979-8-3693-2931-3.ch004

tially change in the near future. Public-key encryption techniques (Pandey, B. K. et al., 2023), such as the RSA algorithm, have historically served as the fundamental basis for ensuring secure communication (Devasenapathy, D. et al., 2024) across the internet. These algorithms have enabled the process of conducting online commerce (George, W. K. et al., 2023), making communications encrypted, and facilitating financial transactions. The emergence of quantum computing raises concerns regarding their effectiveness. Quantum computers have the capacity to render current public-key encryption algorithms outdated, thus posing a threat to the confidentiality and integrity of sensitive data (Singh, S. et al., 2023). The imminent danger presented by quantum computers highlights the necessity for prompt intervention. Anticipating and addressing vulnerabilities is of utmost importance for organizations in order to prevent their exploitation by malicious actors. In the quantum era, symmetric key encryption, while comparatively less vulnerable than its public-key counterpart, is not exempt from potential risks. In order to effectively counter quantum attacks, it is imperative to enhance the resilience of hardware security modules, which play a crucial role in encryption procedures. In order to effectively address the quantum threat, it is imperative to foster collaborative endeavors (Khan, R. et al., 2019). It is imperative for stakeholders to collaborate in order to investigate and advance quantum-secure technologies. Post-quantum cryptography and quantum key distribution are two promising approaches in this context. These methodologies present prospective resolutions for ensuring the protection of data in the quantum era. The field of post-quantum cryptography encompasses the advancement and implementation of encryption algorithms that possess remarkable resistance against quantum attacks. Academic scholars are currently engaged in the investigation of mathematical principles and cryptographic techniques in order to develop resilient encryption schemes (Kumar Pandey, B. et al., 2021) that can effectively withstand quantum threats. Furthermore, the utilization of quantum key distribution presents an innovative methodology for ensuring secure communication (Raja, D. et al., 2024) by leveraging the fundamental principles of quantum mechanics. Quantum key distribution utilizes quantum properties like entanglement and uncertainty to guarantee secure key exchange, which is impervious to interception by quantum computers. Adopting quantum-secure technologies presents distinct difficulties. The careful consideration of ethical and privacy concerns is imperative, particularly in relation to the potential ramifications of quantum computing on data privacy (Ekong, M. O. et al., 2023) and surveillance. Furthermore, the successful deployment of quantum-secure solutions necessitates substantial allocation of resources towards research, development, and infrastructure (Yazar, A. et al., 2020).

Quantum Threats in 6G

A significant milestone in the development of telecommunications (Gupta, A. K. et al., 2023) has been reached with the introduction of 6G (Sengupta, R. et al., 2021) networks, which promise speeds that have never been seen before, ultra-low latency, and seamless connectivity. New challenges, particularly in the field of cybersecurity, have emerged as a result of these advancements of technology. Concerns regarding the impending danger posed by quantum computing are among the most pressing concerns in this regard. Using the principles of quantum mechanics to perform calculations at an exponential rate, quantum computing represents a paradigm shift in the level of computational power that is available. This presents a significant challenge to the traditional encryption methods that have been used up until now, despite the fact that it holds the promise of remarkable breakthroughs in fields such as drug discovery, weather forecasting, and optimization problems. The vulnerability to quantum attacks in the context of 6G networks, which are anticipated to serve as the foundation for critical infrastructure, Internet of Things devices (Du John, H. V. et al., 2022), and massive data transmission, becomes increasingly apparent. Techniques of traditional encryption, such as RSA and ECC, are based on the difficulty of factoring large prime numbers or solving elliptic curve discrete logarithm problems. These are tasks that would require classical computers to complete in an amount of time that is inconceivably long. On the other hand, quantum computers have the potential to render these encryption methods obsolete by utilizing algorithms such as Shor's algorithm to solve these problems in an exceptionally efficient manner. Threats posed by quantum computing have significant repercussions for 6G networks (De Alwis, C. et al., 2021). It is possible that the security of sensitive data, financial transactions, and critical infrastructure could be compromised, which would result in widespread financial loss and disruption. As an additional point of interest, the interconnected nature of 6G networks magnifies the ripple effects of a security breach, which can potentially have catastrophic consequences. Developing encryption algorithms and security protocols that are resistant to quantum adversaries and can withstand attacks from quantum adversaries is an absolute necessity in order to address these challenges. In this regard, there is potential for the application of post-quantum cryptography, which is characterized by the investigation of alternative mathematical problems that are thought to be resistant to quantum programming. Furthermore, quantum key distribution (QKD), a technology that employs the principles of quantum mechanics to secure communication channels, presents a potential solution for ensuring secure data transmission in 6G networks. QKD offers a potential solution to the problem of ensuring secure data transmission. It is possible for the global community to collectively address the challenges that are posed by quantum threats and ensure the security and resilience of future 6G networks if they foster partnerships that span multiple disciplines and share knowledge and resources by working together (Ray, P. P. et al., 2021).

ML 6G Security Solution

In the age of 6G networks (Kirubasri, G. et al., 2021), with their lightning-fast data speeds, extensive connectivity, and near-instantaneous communication, conventional security measures are encountering unprecedented obstacles. In order to address the challenges of cybersecurity in today's rapidly changing environment, it is crucial to adopt innovative strategies. One approach involves utilizing machine learning (ML), a subset of artificial intelligence (AI) (Revathi, T. K. et al., 2022) that allows systems to learn from data, recognize patterns, and make informed decisions without explicit programming. Machine learning (Saxena, A. et al., 2024) provides a potential solution to the intricate security issues presented by 6G networks. Through the analysis of extensive data in real-time, ML algorithms have the ability to identify suspicious patterns or anomalies that could potentially signify a security breach. By adopting a proactive approach, network operators can promptly address and minimize the impact of potential attacks. Techniques have the potential to improve conventional security measures like encryption and access control. As an illustration, ML algorithms can be trained to identify and address advanced cyber-attacks by constantly monitoring network traffic and recognizing malicious behavior. ML can help in creating security mechanisms that can adapt to changing threats and vulnerabilities. Through the analysis of historical data and the utilization of machine learning algorithms (Khan, B. et al., 2021), security policies and configurations can be optimized (Jayapoorani, S. et al., 2023) to enhance the protection of 6G networks against emerging threats. It's important to recognize that machine learning (Deepa, R. et al., 2022). Comprehensive perfor) comes with its fair share of difficulties. Similar to a computer systems analyst, it is important to note that ML models can be vulnerable to adversarial attacks and may exhibit biases and inaccuracies if not trained and validated correctly. In addition, the use of ML-based security solutions brings up important issues regarding privacy, transparency, and accountability. This emphasizes the importance of ethical considerations and the establishment of regulatory frameworks (Gui, G. et al., 2020).

QUANTUM THREATS IN 6G NETWORKS

Quantum Threats in 6G Security

Understanding the impact of quantum threats on the security of emerging 6G networks is crucial, as they have the potential to cause significant disruptions and introduce new vulnerabilities. At the heart of these dangers lies the idea of quantum computing, which utilizes the principles of quantum mechanics to carry out calculations at speeds far beyond those of classical computers. One of the main issues with quantum threats is their potential to undermine current encryption methods that are crucial for the security of 6G networks. Encryption techniques like RSA and ECC are based on complex mathematical problems, such as factoring large prime numbers or solving elliptic curve discrete logarithm problems. Nevertheless, quantum computers possess the capability to efficiently solve these problems through algorithms such as Shor's algorithm. As a result, current encryption methods become susceptible to attack. This vulnerability is especially worrisome when considering the implications for 6G networks. These networks (Swapna, H. R. et al., 2023) are anticipated to handle a vast array of applications (Pandey, B. K. et al., 2011) and services, such as critical infrastructure, IoT (Menon, V. et al., 2022) devices, and the transmission of sensitive data. With the growing complexity and interconnectivity of 6G networks, the threat of quantum attacks is becoming more significant. These attacks have the potential to seriously compromise the security and accessibility of information. The fast rate of technological progress in the realm of quantum computing adds another layer of complexity to the task of addressing quantum threats in 6G networks. Just as quantum computers continue to advance, the possibility of breaking encryption is becoming increasingly feasible. Given the circumstances, it is crucial to take immediate action in order to tackle quantum threats and safeguard the integrity of 6G networks. Aside from encryption vulnerabilities, quantum threats also bring about challenges in terms of key distribution and authentication. Utilizing the principles of quantum mechanics, quantum key distribution (QKD) presents a promising solution for guaranteeing secure data transmission in 6G networks. Nevertheless, scaling up the implementation of QKD comes with its own share of technical and practical hurdles, such as ensuring the compatibility of infrastructure and protocols.

6G Vulnerabilities to Quantum Threats

The potential vulnerabilities that could be exploited by quantum threats are becoming an increasing source of concern in the context of 6G networks. The growing reliance on traditional security methods, which may not be equipped to withstand the capabilities of quantum computing, is the root cause of these vulnerabilities at the moment. The encryption methods that are utilized to safeguard the transmission of data within 6G networks are a significant exposure point. Quantum computers have the potential to break current encryption methods, despite the fact that these methods have been shown to be effective against traditional data security threats. There is a possibility that quantum algorithms, such as Shor's algorithm, could quickly break through these encryption methods, leaving sensitive data vulnerable to being intercepted and decrypted. Especially in terms of network architecture and protocols, the interconnected nature of 6G networks causes additional vulnerabilities to be introduced. In order to launch attacks, such as data interception or disruption of network operations, quantum adversaries could take advantage of vulnerabilities in these systems because of their quantum nature. Furthermore, the proliferation of Internet of Things (IoT) devices (Dhanasekar, S. et al., 2023) in 6G networks creates additional entry points for potential attacks. Additionally, it is possible that these devices (Govindaraj, V. et al., 2023) do not have sufficient protection against quantum threats. Key distribution and authentication are two additional areas of concern. Quantum attacks have the potential to defeat traditional methods because they rely on the secure exchange of cryptographic keys. Although quantum key distribution (QKD) presents a potential solution, its implementation at scale presents logistical challenges and may require significant infrastructure upgrades. Despite these challenges, QKD shows promise as a potential solution. The rapid pace of advancement in quantum technology contributes to the complexity of the situation that we are currently working in. There is a growing possibility that quantum adversaries will be able to exploit vulnerabilities in 6G networks as the capabilities of quantum computing develop further (Sheth, K. et al., 2020).

Securing 6G From Quantum Threat

Understanding the importance of establishing strong security measures for 6G networks in preparation for potential quantum threats is crucial. Given the upcoming 6G technology, which offers remarkable speeds, minimal latency, and extensive connectivity, it is imperative to prioritize advanced security measures. With the rapid progress of quantum computing technology, the security of data transmitted over 6G networks faces a substantial threat from quantum vulnerabilities. The current encryption methods may be susceptible to quantum attacks. Conventional encryption methods, like RSA and ECC, depend on complex mathematical problems that pose a challenge for classical computers to solve. On the other hand, quantum computers have the ability to disrupt these encryption methods by utilizing algorithms such as Shor's algorithm, which can efficiently factor large numbers. Due to the nature of 6G networks, there is a possibility that sensitive information could be intercepted and decrypted by those with quantum capabilities. The impact of potential security breaches is greatly amplified by the interconnected nature of 6G networks. Given the vast number of devices (Sharma, M. et al., 2022) set to be connected to 6G networks, such as critical infrastructure, IoT devices (Du John, H. V. et al., 2023), and sensitive data transmission (Iyyanar, P. et al., 2023), there is a considerable risk of quantum attacks leading to widespread disruption and economic loss. This interconnectedness highlights the pressing requirement for strong security measures to safeguard against quantum threats (Yang, H. et al., 2020). Dealing with these challenges necessitates a comprehensive strategy that incorporates technological advancements, cooperation among various parties, and proactive steps. Efforts in research should prioritize the development of encryption algorithms and security protocols that can effectively withstand attacks from quantum adversaries. In order to expedite research and development in quantum-safe cybersecurity, it is crucial for industry stakeholders (Malhotra, P. et al., 2021), academia, and government agencies to collaborate effectively. Regulatory frameworks and standards play a crucial role in ensuring the widespread adoption of secure practices and technologies within the 6G ecosystem. This encompasses recommendations for encryption standards, key management, and network security protocols. Through the implementation of well-defined guidelines and standards, policymakers have the ability to encourage the integration of secure practices and technologies. This, in turn, will strengthen the resilience of 6G networks against potential quantum threats (Huang, T. et al., 2019).

MACHINE LEARNING IN CYBERSECURITY

The application of machine learning (ML) has become a very useful instrument in the field of cybersecurity, bringing about a revolution in the detection (Abdulkarim, Y. I. et al., 2024), analysis, and mitigation of threats. The fundamental aspect of machine learning is the creation of algorithms and models (Anand, R. et al., 2024) that give computers the ability to learn from data and make predictions or choices without being explicitly programmed to do so. Artificial intelligence algorithms (Pandey, B. K. et al., 2024). Autopilot control unmanned aerial vehicle system for) are trained on massive amounts of data in the field of cybersecurity in order to recognize trends and abnormalities that are suggestive of hostile work. Threat detection is one of the most important uses of machine learning in the field of cybersecurity. The use of machine learning (Pandey, B. K., & Pandey, D., 2023) algorithms allows for the real-time identification of potentially malicious behavior or security breaches by analyzing network traffic, system logs, and other data sources. Systems that are based on machine learning are able to successfully identify and respond to emerging cyber threats because they are able to learn from historical data and continuously adapt to new threats. Malware detection and categorization are two further applications of machine learning algorithms. Malware classification algorithms are able to uncover previously undetected variations of malware by examining the features of known malware samples. These variants can then be classified based on their behavior or characteristics. Because of this, specialists in the field of cybersecurity are able to fight new and emerging threats in a preemptive manner. A further application of machine learning algorithms is in the field of anomaly detection, which involves the identification of deviations from typical behavior that may be indicative of a security breach. For the purpose of identifying advanced persistent threats (APTs) or insider threats, which are able to circumvent conventional security measures, this approach is especially helpful. For the purpose of conducting predictive analysis and risk assessment in the field of cybersecurity, machine learning-based approaches are applied. In order to forecast future threats and evaluate the possibility of security breaches as well as the potential consequences of such breaches, machine learning algorithms analyze historical data and detect patterns of previous security incidents. Consequently, this makes it possible for enterprises to more efficiently manage resources and prioritize security measures according to the level of risk (Rupprecht, D. et al., 2018).

ML Enhances Threat Detection and Response

Figure 1. ML's role in threat detection and response

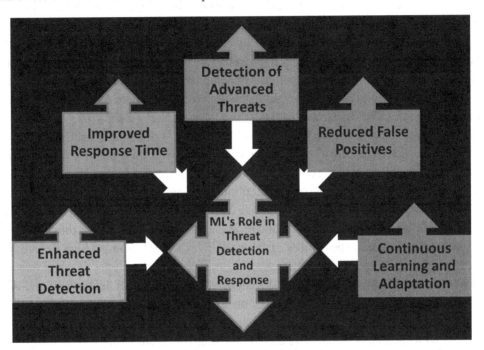

Integrating machine learning (ML) into cybersecurity greatly enhances the ability to detect and respond to threats are shown in the figure 1. ML algorithms analyze large amounts of data, detecting patterns and irregularities that may indicate possible risks. By taking a proactive approach, we can quickly respond to security incidents, minimizing their impact and improving our overall cybersecurity.

- *Enhanced Threat Detection*: Cybersecurity threat detection has changed dramatically as a result of machine learning approaches. Organizations can detect patterns suggestive of harmful behavior with unparalleled precision and effectiveness by utilizing advanced algorithms that have been trained on extensive datasets. ML algorithms can instantly adjust to changing threats, in contrast to conventional approaches that depend on static rules and signatures. By continuously learning and updating their models based on fresh data and emerging threats, security systems using this dynamic approach are able to stay one step ahead of cyber adversaries. Additionally, in order to provide a thorough picture of potential security concerns, ML-powered threat detection systems can examine a variety of data sources, such as network traffic, system logs, user behavior, and threat intelligence feeds (Pereira, V., & Sousa, T., 2004).
- *Improved Response Time*: The capacity to speed the reaction to security issues is one of the most important benefits of using machine learning techniques in threat detection. The time between detection and mitigation is shortened by ML-powered systems that automate alarm analysis and prioritize high-risk threats. Organizations that streamline this procedure can reduce the possible effect of cyberattacks and neutralize threats before they become serious security breaches. Additionally,

security teams may take prompt action to fix vulnerabilities and stop future attacks by using ML algorithms to assist them swiftly determine the underlying cause of security problems.

- *Detection of Advanced Threats*: When it comes to identifying sophisticated threats that could elude conventional security measures, ML algorithms thrive. These encompass advanced persistent threats (APTs), insider threats, zero-day exploits, and further sophisticated attack vectors. ML-powered solutions can precisely detect these attacks by examining minute trends and anomalies in network behavior, giving enterprises the ability to proactively guard against new cyber threats. Additionally, by spotting departures from typical patterns of behavior, machine learning techniques like anomaly detection and behavioral analysis can aid in the discovery of hitherto unidentified threats—even in extremely complex and dynamic contexts.
- *Reduced False Positives*: Because ML algorithms are so good at differentiating between benign and dangerous activity, security systems produce a lot fewer false positives. By minimizing alert fatigue, this capability helps security analysts concentrate on real security concerns. In order to maximize accuracy, ML-powered systems continuously improve their algorithms based on input from security experts and modify their detection levels. Furthermore, machine learning (ML) methods like ensemble learning and feature engineering can boost the accuracy of threat detection models even further, lowering the possibility of false positives while keeping a high degree of sensitivity to possible threats (Goyal et al., 2019).
- *Continuous Learning and Adaptation*: Its capacity to continuously learn from and adjust to new threats is arguably one of machine learning's most alluring features in threat identification. Machine learning systems evaluate fresh data and user comments, enhancing their detection skills and algorithms with each passing day. Organizations are given a strong defense against new attacks by this iterative approach, which guarantees that security measures continue to be effective against growing cyber threats. Threat detection systems can make use of cutting-edge methods like transfer learning and reinforcement learning to quicken the learning curve and adjust to evolving threat environments. Through the utilization of machine learning-driven threat detection, enterprises may improve their cybersecurity stance and remain ahead of constantly changing cyberthreats.

ML APPROACHES TO MITIGATE QUANTUM THREATS IN 6G NETWORKS

Enhanced Quantum-Resistant Cryptography

The development of quantum-resistant cryptographic algorithms is of utmost importance in the field of cybersecurity, especially as quantum computer capabilities continue to advance. The objective of these techniques is to address the weaknesses presented by quantum computers to conventional encryption methods, such as RSA and ECC, through the creation of algorithms that maintain their security even when faced with quantum adversaries. The utilization of machine learning (ML) is crucial in augmenting cryptographic techniques as it facilitates the examination of extensive datasets and the detection of intricate patterns that can potentially guide the development of quantum-resistant algorithms. To begin with, machine learning algorithms has the capability to examine encryption systems that are resistant to quantum attacks, thereby detecting possible weaknesses or vulnerabilities that may be exploited by adversaries. ML-powered systems can enhance the effectiveness of quantum-resistant algorithms by simulating different attack scenarios and assessing the resilience of cryptographic primitives. The potential to assist in the enhancement of quantum-resistant cryptographic algorithms through the automation of parameter selection and algorithm design procedures. Machine learning techniques has the capability to investigate extensive solution spaces and ascertain optimal setups that effectively optimize security while also minimizing computational cost. The utilization of an iterative optimization approach (Pandey, D., & Pandey, B. K., 2022) has the potential to facilitate the creation of quantum-resistant encryption schemes that are both more efficient and practical, hence satisfying the performance demands of real-world applications. Algorithms have the capability to enhance the ability of quantum-resistant cryptographic methods to adjust to changing threat environments through the ongoing surveillance and examination of emerging attack methods and patterns. ML-powered systems can utilize previous data and ongoing research to predict future threats and guide the creation of proactive security tactics. The use of this proactive approach allows enterprises to proactively anticipate and counter possible attackers, thereby ensuring the preservation of security for their cryptographic systems in the era of quantum computing. Machine learning approaches have the potential to improve the scalability and deployability of quantum-resistant cryptographic methods through the automation of key management procedures and the optimization of cryptographic operations. Machine learning (ML)-based systems have the capability to optimize the process of generating, distributing, and revoking cryptographic keys, thereby alleviating the administrative workload linked to key management in extensive implementations. In addition, machine learning algorithms have the capability to enhance cryptographic protocols and operations in order to reduce resource consumption and delay. This enhances the feasibility of employing quantum-resistant encryption methods in contexts with limited resources.

Anomaly Detection for 6G Quantum Attacks

Utilizing machine learning, anomaly detection is a state-of-the-art method for identifying and mitigating quantum-based attacks within the realm of 6G networks. With the rapid progress of 6G technology, the concerns surrounding the potential risks associated with quantum computing capabilities are also growing. Utilizing ML techniques, anomaly detection provides a proactive defense mechanism that constantly monitors network traffic, system logs, and user behavior. It aims to identify any deviations from normal patterns that could potentially indicate quantum-based attacks. ML-driven anomaly detection has the remarkable capability to adapt to the distinct characteristics of quantum-based attacks. Conventional security measures may face difficulties in identifying these attacks, as they take advantage of weaknesses in encryption methods and key distribution protocols. ML algorithms, on the other hand, have the ability to analyze intricate patterns and irregularities in network traffic and system behavior. This enables them to detect indications of unauthorized access, data tampering, or any other malicious activity linked to quantum-based attacks. The utilization of machine learning in anomaly detection can significantly bolster the robustness of 6G networks. This is achieved by effectively identifying and flagging new and unfamiliar attack patterns. Quantum-based attacks can be quite complex and may not be easily detected using traditional signature-based detection methods. Machine learning algorithms have the ability to analyze historical data and monitor network activity in order to detect emerging threats and unusual patterns that may indicate quantum-based attacks. This empowers organizations to promptly and efficiently respond to potential security breaches. With expertise in artificial intelligence, ML-powered systems can efficiently analyze and categorize security alerts, allowing security teams to prioritize their focus on the most critical threats and streamline the incident response process. By adopting a proactive approach, the time between detection and mitigation is significantly reduced, thereby mitigating the potential impact of quantum-based attacks on 6G networks. Anomaly detection driven by machine learning can help with the integration of threat intelligence and collaboration among stakeholders in the cybersecurity ecosystem. Through the exchange of information and data between different entities, machine learning systems have the ability to improve awareness of potential threats and facilitate a synchronized approach to countering attacks that leverage quantum technology. By adopting a collective defense approach, the resilience of 6G networks is enhanced, thereby minimizing the chances of cyberattacks being successful (Li, Y. et al., 2021).

IMPLEMENTING ML FOR 6G SECURITY

Integrating machine learning (ML) methods into 6G security poses many obstacles that need to be resolved in order to guarantee the efficiency and dependability of these systems. An essential obstacle lies in the immense intricacy and magnitude of 6G networks, which comprise a wide array of devices (KVM, S. et al., 2024), protocols, and applications. It is imperative for machine learning algorithms to possess the ability to analyze and integrate extensive quantities of data produced by these networks in real-time, while simultaneously adjusting to dynamic and evolving threat environments. The diverse composition of 6G networks presents more difficulties for the deployment of machine learning. These networks can include a combination of outdated systems, exclusive protocols, and developing technologies, each with its own distinct features and security needs. The flexibility and interoperability of machine learning algorithms are crucial in order to effectively handle the wide range of variety, while also guaranteeing compatibility with pre-existing security infrastructure and protocols. An additional obstacle lies in the inherent unpredictability and vagueness linked to cybersecurity threats. Cyber attackers are continuously adapting their strategies and methods to avoid detection and take advantage of weaknesses in 6G networks. In order to effectively identify malicious behavior, machine learning algorithms must possess the ability to detect intricate patterns and abnormalities, while simultaneously decreasing the occurrence of false positives and false negatives. Ensuring the accuracy and dependability of machine learning-based security systems necessitates the use of robust data preprocessing, feature selection, and model optimization strategies in order to attain this equilibrium. The absence of annotated data is a substantial obstacle for the integration of machine learning in the realm of 6G security. In contrast to conventional supervised learning tasks, which can be easily trained using labeled datasets, cybersecurity datasets frequently suffer from a dearth of labels or ground truth information. This presents a difficulty in properly training machine learning algorithms and accurately assessing their effectiveness. It is imperative to continuously retrain and update machine learning models in order to effectively respond to the ever-evolving landscape of cyber threats and the ever-changing conditions of networks. The adoption of machine learning in 6G security is hindered by the presence of privacy and legal concerns. In order to accurately identify and address security risks, machine learning algorithms may necessitate access to confidential network data and user information. Nevertheless, maintaining the privacy and security of this data while adhering to legislative mandates like GDPR and CCPA poses a substantial obstacle. Furthermore, it is imperative for machine learning algorithms to exhibit transparency and explainability to stakeholders in order to cultivate confidence and accountability in their decision-making procedures (Kato, N. et al., 2020).

ML in Combating Quantum Threats

When it comes to addressing quantum threats in cybersecurity, machine learning (ML) techniques show great promise. However, it is important to carefully evaluate their limitations and drawbacks. A major drawback is the inherent uncertainty and complexity that comes with quantum threats. Quantum-based attacks frequently target weaknesses in encryption methods and key distribution protocols that may not be readily detectable using conventional machine learning algorithms. ML-powered systems may face challenges in accurately detecting and addressing quantum threats, especially those that employ new or advanced attack techniques. One drawback of ML techniques in addressing quantum threats is the scarcity of labeled data and reliable information for training ML models. Unlike conventional cybersecurity threats, which often have clear characteristics and signatures, quantum-based attacks are still in their early stages and constantly changing. Developing robust ML algorithms that can effectively detect and mitigate quantum threats is quite challenging. In addition, the ever-changing nature of quantum threats necessitates the ongoing retraining and updating of ML models to effectively respond to emerging threats and evolving attack patterns. ML techniques can potentially create new vulnerabilities and attack opportunities for adversaries to exploit and bypass security measures. Manipulating input data to deceive ML algorithms, adversarial attacks present a considerable threat to the effectiveness of ML-powered security systems. Adversaries with a deep understanding of quantum mechanics can potentially exploit these vulnerabilities to avoid detection and carry out highly advanced attacks on 6G networks and infrastructure. In addition, ML algorithms can be vulnerable to tampering or manipulation by adversaries, which undermines their reliability and effectiveness in addressing quantum threats. The practical implementation of ML techniques in combating quantum threats may be challenging due to their computational and resource requirements. Machine learning algorithms often demand substantial computational power and memory resources for training and deployment, making them impractical for environments or devices with limited resources. In addition, ML-based security systems can cause delays and additional workload on network operations, which can affect the performance and responsiveness of 6G networks. Understanding the delicate balance between security, performance, and resource constraints is crucial for successfully implementing ML techniques to combat quantum threats.

CONCLUSION

The analysis of quantum hazards in relation to the development of 6G networks highlights the crucial significance of strong security protocols. Machine learning (ML) is a viable approach to tackle these threats, providing capabilities like quantum-resistant cryptography and anomaly detection to strengthen network defenses. Although machine learning (ML) offers possibilities to improve the identification and reaction to threats, its integration into 6G security poses difficulties. The aforementioned factors encompass ethical considerations, interoperability concerns, and the imperative for ongoing adaptation in response to emerging threats. These difficulties, machine learning has the capacity to effectively address quantum threats in 6G networks. By working together and doing continuous research, it is possible to enhance machine learning-based methods and improve the ability of 6G networks to withstand quantum-based attacks. As we consider the future, it is crucial to give priority to collaborative efforts across different fields, investigate new avenues of research, and stay alert in dealing with rising dangers.

Through this approach, we can confidently navigate the quantum age of 6G networks and guarantee the security of vital infrastructure and data.

REFERENCES

Abdulkarim, Y. I., Awl, H. N., Muhammadsharif, F. F., Saeed, S. R., Sidiq, K. R., Khasraw, S. S., Dong, J., Pandey, B. K., & Pandey, D. (2024). Metamaterial-based sensors loaded corona-shaped resonator for COVID-19 detection by using microwave techniques. *Plasmonics*, 19(2), 595–610. 10.1007/s11468-023-02007-4

Anand, R., Lakshmi, S. V., Pandey, D., & Pandey, B. K. (2024). An enhanced ResNet-50 deep learning model for arrhythmia detection using electrocardiogram biomedical indicators. *Evolving Systems*, 15(1), 83–97. 10.1007/s12530-023-09559-0

De Alwis, C., Kalla, A., Pham, Q. V., Kumar, P., Dev, K., Hwang, W. J., & Liyanage, M. (2021). Survey on 6G frontiers: Trends, applications, requirements, technologies and future research. *IEEE Open Journal of the Communications Society*, 2, 836–886. 10.1109/OJCOMS.2021.3071496

Deepa, R., Anand, R., Pandey, D., Pandey, B. K., & Karki, B. (2022). Comprehensive performance analysis of classifiers in diagnosis of epilepsy. *Mathematical Problems in Engineering*, 2022, 2022. 10.1155/2022/1559312

Devasenapathy, D., Madhumathy, P., Umamaheshwari, R., Pandey, B. K., & Pandey, D. (2024). Transmission-efficient grid-based synchronized model for routing in wireless sensor networks using Bayesian compressive sensing. *SN Computer Science*, 5(1), 1–11.

Dhanasekar, S., Martin Sagayam, K., Pandey, B. K., & Pandey, D. (2023). Refractive Index Sensing Using Metamaterial Absorbing Augmentation in Elliptical Graphene Arrays. *Plasmonics*, 1–11. 10.1007/s11468-023-02152-w

Du John, H. V., Ajay, T., Reddy, G. M. K., Ganesh, M. N. S., Hembram, A., Pandey, B. K., & Pandey, D. (2023). Design and simulation of SRR-based tungsten metamaterial absorber for biomedical sensing applications. *Plasmonics*, 18(5), 1903–1912. 10.1007/s11468-023-01910-0

Du John, H. V., Jose, T., Jone, A. A. A., Sagayam, K. M., Pandey, B. K., & Pandey, D. (2022). Polarization insensitive circular ring resonator based perfect metamaterial absorber design and simulation on a silicon substrate. *Silicon*, 14(14), 9009–9020. 10.1007/s12633-021-01645-9

Ekong, M. O., George, W. K., Pandey, B. K., & Pandey, D. (2023). Enhancing the Fundamentals of Industrial Safety Management in TVET for Metaverse Realities. In *Applications of Neuromarketing in the Metaverse* (pp. 19-41). IGI Global. 10.4018/978-1-6684-8150-9.ch002

George, W. K., Ekong, M. O., Pandey, D., & Pandey, B. K. (2023). Pedagogy for Implementation of TVET Curriculum for the Digital World. In *Applications of Neuromarketing in the Metaverse* (pp. 117-136). IGI Global. 10.4018/978-1-6684-8150-9.ch009

. Govindaraj, V., Dhanasekar, S., Martinsagayam, K., Pandey, D., Pandey, B. K., & Nassa, V. K. (2023). Low-power test pattern generator using modified LFSR. *Aerospace Systems*, 1-8.

Goyal, J., Singla, K., & Singh, S. (2019). A Survey of Wireless Communication Technologies from 1G to 5G. In *Seond International Conference on Computer Networks and Inventive Communication Technologies*. Springer: Berlin/Heidelberg, Germany.

Gui, G., Liu, M., Tang, F., Kato, N., & Adachi, F. (2020). 6G: Opening new horizons for integration of comfort, security, and intelligence. *IEEE Wireless Communications*, 27(5), 126–132. 10.1109/MWC.001.1900516

Gupta, A. K., Sharma, R., Pandey, D., Nassa, V. K., Pandey, B. K., George, A. S., & Dadheech, P. (2023). Performance analysis of eight-channel WDM optical network with different optical amplifiers for industry 4.0. In *Innovation and Competitiveness in Industry 4.0 Based on Intelligent Systems* (pp. 197–212). Springer International Publishing. 10.1007/978-3-031-29775-5_9

Huang, T., Yang, W., Wu, J., Ma, J., Zhang, X., & Zhang, D. (2019). A survey on green 6G network: Architecture and technologies. *IEEE Access : Practical Innovations, Open Solutions*, 7, 175758–175768. 10.1109/ACCESS.2019.2957648

Iyyanar, P., Anand, R., Shanthi, T., Nassa, V. K., Pandey, B. K., George, A. S., & Pandey, D. (2023). A real-time smart sewage cleaning UAV assistance system using IoT. In *Handbook of Research on Data-Driven Mathematical Modeling in Smart Cities* (pp. 24–39). IGI Global.

Jayapoorani, S., Pandey, D., Sasirekha, N. S., Anand, R., & Pandey, B. K. (2023). Systolic optimized adaptive filter architecture designs for ECG noise cancellation by Vertex-5. *Aerospace Systems*, 6(1), 163–173. 10.1007/s42401-022-00177-3

Kato, N., Mao, B., Tang, F., Kawamoto, Y., & Liu, J. (2020). Ten challenges in advancing machine learning technologies toward 6G. *IEEE Wireless Communications*, 27(3), 96–103. 10.1109/MWC.001.1900476

Khan, B., Hasan, A., Pandey, D., Ventayen, R. J. M., Pandey, B. K., & Gowwrii, G. (2021). Fusion of datamining and artificial intelligence in prediction of hazardous road accidents. In *Machine learning and iot for intelligent systems and smart applications* (pp. 201–223). CRC Press. 10.1201/9781003194415-12

Khan, R., Kumar, P., Jayakody, D. N. K., & Liyanage, M. (2019). A survey on security and privacy of 5G technologies: Potential solutions, recent advancements, and future directions. *IEEE Communications Surveys and Tutorials*, 22(1), 196–248. 10.1109/COMST.2019.2933899

Kirubasri, G., Sankar, S., Pandey, D., Pandey, B. K., Singh, H., & Anand, R. (2021, September). A recent survey on 6G vehicular technology, applications and challenges. In *2021 9th International Conference on Reliability, Infocom Technologies and Optimization (Trends and Future Directions)(ICRITO)* (pp. 1-5). IEEE.

Kumar Pandey, B., Pandey, D., Nassa, V. K., Ahmad, T., Singh, C., George, A. S., & Wakchaure, M. A. (2021). Encryption and steganography-based text extraction in IoT using the EWCTS optimizer. *Imaging Science Journal*, 69(1-4), 38–56. 10.1080/13682199.2022.2146885

KVM, S., Pandey, B. K., & Pandey, D. (2024). Design of Surface Plasmon Resonance (SPR) Sensors for Highly Sensitive Biomolecular Detection in Cancer Diagnostics. *Plasmonics*, 1-13.

Li, Y., Yu, Y., Susilo, W., Hong, Z., & Guizani, M. (2021). Security and privacy for edge intelligence in 5G and beyond networks: Challenges and solutions. *IEEE Wireless Communications*, 28(2), 63–69. 10.1109/MWC.001.2000318

Malhotra, P., Pandey, D., Pandey, B. K., & Patra, P. M. (2021). Managing agricultural supply chains in COVID-19 lockdown. *International Journal of Quality and Innovation*, 5(2), 109–118. 10.1504/IJQI.2021.117181

Menon, V., Pandey, D., Khosla, D., Kaur, M., Vashishtha, H. K., George, A. S., & Pandey, B. K. (2022). A Study on COVID–19, Its Origin, Phenomenon, Variants, and IoT-Based Framework to Detect the Presence of Coronavirus. In *IoT Based Smart Applications* (pp. 1–13). Springer International Publishing.

Pandey, B. K., & Pandey, D. (2023). Parametric optimization and prediction of enhanced thermoelectric performance in co-doped CaMnO3 using response surface methodology and neural network. *Journal of Materials Science Materials in Electronics*, 34(21), 1589. 10.1007/s10854-023-10954-1

Pandey, B. K., Pandey, D., Alkhafaji, M. A., Güneşer, M. T., & Şeker, C. (2023). A reliable transmission and extraction of textual information using keyless encryption, steganography, and deep algorithm with cuckoo optimization. In *Micro-Electronics and Telecommunication Engineering: Proceedings of 6th ICMETE 2022* (pp. 629–636). Springer Nature Singapore. 10.1007/978-981-19-9512-5_57

Pandey, B. K., Pandey, D., & Sahani, S. K. (2024). Autopilot control unmanned aerial vehicle system for sewage defect detection using deep learning. *Engineering Reports*, 2852. 10.1002/eng2.12852

Pandey, B. K., Pandey, S. K., & Pandey, D. (2011). A survey of bioinformatics applications on parallel architectures. *International Journal of Computer Applications*, 23(4), 21–25. 10.5120/2877-3744

Pandey, D., & Pandey, B. K. (2022). An efficient deep neural network with adaptive galactic swarm optimization for complex image text extraction. In *Process mining techniques for pattern recognition* (pp. 121–137). CRC Press. 10.1201/9781003169550-10

Pereira, V., & Sousa, T. (2004). *Evolution of Mobile Communications: from 1G to 4G*. Department of Informatics Engineering of the University of Coimbra, Portugal.

Raja, D., Kumar, D. R., Santhiyakumari, N., Kumarganesh, S., Sagayam, K. M., Thiyaneswaran, B., Pandey, B. K., & Pandey, D. (2024). A compact dual-feed wide-band slotted antenna for future wireless applications. *Analog Integrated Circuits and Signal Processing*, 118(2), 1–15. 10.1007/s10470-023-02233-0

Ray, P. P., Kumar, N., & Guizani, M. (2021). A vision on 6G-enabled NIB: Requirements, technologies, deployments, and prospects. *IEEE Wireless Communications*, 28(4), 120–127. 10.1109/MWC.001.2000384

Revathi, T. K., Sathiyabhama, B., Sankar, S., Pandey, D., Pandey, B. K., & Dadeech, P. (2022). An intelligent model for coronary heart disease diagnosis. In *Networking Technologies in Smart Healthcare* (pp. 309–327). CRC Press. 10.1201/9781003239888-15

Rupprecht, D., Dabrowski, A., Holz, T., Weippl, E., & Pöpper, C. (2018). On security research towards future mobile network generations. *IEEE Communications Surveys and Tutorials*, 20(3), 2518–2542. 10.1109/COMST.2018.2820728

Saxena, A., Agarwal, A., Pandey, B. K., & Pandey, D. (2024). Examination of the Criticality of Customer Segmentation Using Unsupervised Learning Methods. *Circular Economy and Sustainability*, 1–14. 10.1007/s43615-023-00336-4

. Sengupta, R., Sengupta, D., Pandey, D., Pandey, B. K., Nassa, V. K., & Dadeech, P. (2021). A Systematic review of 5G opportunities, architecture and challenges. *Future Trends in 5G and 6G*, 247-269.

Sharma, M., Pandey, D., Palta, P., & Pandey, B. K. (2022). Design and power dissipation consideration of PFAL CMOS V/S conventional CMOS based 2: 1 multiplexer and full adder. *Silicon*, 14(8), 4401–4410. 10.1007/s12633-021-01221-1

Sharma, S., Pandey, B. K., Pandey, D., Anand, R., Sharma, A., & Saini, S. (2023, March). Character Recognition Technique Implementation for Complicated Deteriorated Scene. In *2023 6th International Conference on Information Systems and Computer Networks (ISCON)* (pp. 1-4). IEEE. 10.1109/ISCON57294.2023.10112185

Sheth, K., Patel, K., Shah, H., Tanwar, S., Gupta, R., & Kumar, N. (2020). A taxonomy of AI techniques for 6G communication networks. *Computer Communications*, 161, 279–303. 10.1016/j.comcom.2020.07.035

Singh, S., Madaan, G., Kaur, J., Swapna, H. R., Pandey, D., Singh, A., & Pandey, B. K. (2023). *Bibliometric Review on Healthcare Sustainability. Handbook of Research on Safe Disposal Methods of Municipal Solid Wastes for a Sustainable Environment*, (pp. 142-161). IGI Global. 10.4018/978-1-6684-8117-2.ch011

Swapna, H. R., Bigirimana, E., Madaan, G., Hasan, A., Pandey, B. K., & Pandey, D. (2023). Impact of neuromarketing on consumer psychology in digitally connected networks. In *Applications of Neuromarketing in the Metaverse* (pp. 193–205). IGI Global. 10.4018/978-1-6684-8150-9.ch015

Yang, H., Alphones, A., Xiong, Z., Niyato, D., Zhao, J., & Wu, K. (2020). Artificial-intelligence-enabled intelligent 6G networks. *IEEE Network*, 34(6), 272–280. 10.1109/MNET.011.2000195

Yazar, A., Doğan Tusha, S., & Arslan, H. (2020). 6G vision: An ultra-flexible perspective. *ITU Journal : ICT Discoveries*, 1(1), 121–140. 10.52953/IKVY9186

Chapter 5
Adversarial Challenges in Distributed AI:
ML Safeguarding 6G Networks

Raviteja Kocherla

Department Computer Science and Engineering, Mallareddy University, Hyderabad, India

Yagya Dutta Dwivedi

 http://orcid.org/0000-0002-5793-9364

Department of Aeronautical Engineering, Institute of Aeronautical Engineering, Hyderabad, India

B. Ardly Melba Reena

Saveetha Institute of Medical and Technical Sciences, Saveetha University, Chennai, India

Komala C. R.

Department of Information Science and Engineering, HKBK College of Engineering, Bengaluru, India

Jennifer D.

 http://orcid.org/0000-0003-3362-4242

Department of Computer Science and Engineering, Panimalar Engineering College, Chennai, India

Joshuva Arockia Dhanraj

 http://orcid.org/0000-0001-5048-7775

Dayananda Sagar University, India

ABSTRACT

With a special emphasis on distributed AI/ML systems, the abstract explores the intricate world of adversarial challenges in 6G networks. With their unmatched capabilities—like ultra-high data rates and ultra-low latency it highlights the crucial role that 6G networks will play in determining the direction of communication in the future. Still, it also highlights the weaknesses in these networks' distributed AI/ML systems, emphasizing the need for strong security measures to ward off potential attacks. Also covered in the abstract is the variety of adversarial threats that exist, such as model evasion, data poisoning, backdoors, membership inference, and model inversion attacks.

INTRODUCTION

The term "adversarial challenges in distributed AI/ML" refers to the numerous dangers and vulnerabilities that manifest themselves during the process of deploying artificial intelligence (Anand, R. et al., 2023) and machine learning models in environments that are characterized by distributed computing.

DOI: 10.4018/979-8-3693-2931-3.ch005

These challenges include any malicious attacks or techniques that are designed to manipulate, disrupt, or otherwise undermine the functioning of artificial intelligence and machine learning systems (Anand, R. et al., 2024). An adversarial actor will take advantage of vulnerabilities in the models, data, or infrastructure in order to compromise the integrity, confidentiality, or availability of the artificial intelligence and machine learning systems. This poses a significant threat to the security (David, S. et al., 2023) and reliability of the distributed networks. The fifth generation (5G) of networks has revolutionized connectivity and made a number of cutting-edge applications possible. The sixth generation (6G) (Kirubasri, G. et al., 2021) networks, on the other hand, are expected to outperform their predecessors by leveraging the power of AI and ML to drive previously unheard-of breakthroughs in addition to offering improved connectivity. This thorough investigation explores how 6G networks can use AI and ML as catalysts that go beyond simple connectivity. The swift development of telecommunications has resulted in notable progress in network capabilities and intergenerational connectivity. The journey from first-generation networks with basic analogue voice communication to fourth-generation (4G) networks with lightning-fast data transfer has been amazing. Fifth-generation (5G) networks, which are now being deployed, have completely changed connectivity by providing higher speeds, lower latency, and more device connectivity. But as technology develops further, the idea for the upcoming 6G frontier is beginning to take shape. The motivation for investigating 6G networks is to meet the needs of the industry and society in the future (George, W. K. et al., 2024). Conventional communication systems encounter new difficulties as our world becomes more interconnected. Networks that go beyond traditional connectivity are required due to exponential growth in data, the proliferation of intelligent devices (Abdulkarim, Y. I. et al., 2023), and the emergence of new applications like virtual reality, autonomous systems, and AI-driven services. This calls for the development of 6G networks that are intelligent enough to accommodate a wide range of applications in addition to being quicker, more dependable, and more robust. The potential advantages and opportunities across multiple domains, this survey attempts to shed light on the integration of AI and ML in 6G networks. Artificial intelligence (AI) (Pandey, B. K., & Pandey, D., 2023) and machine learning (ML) have the potential to transform many industries within 6G networks, from intelligent edge computing and driverless cars to smart cities, healthcare, and massive IoT connectivity (Iyyanar, P. et al., 2023). But there are also a number of issues and concerns that come with this integration, such as security, scalability, standardization, and moral ramifications. With AI and ML becoming essential parts of 6G networks, new adversarial challenges and vulnerabilities also arise. Within 6G networks, adversarial attacks like membership inference, model evasion, data poisoning, backdoor attacks, and model inversion pose serious risks to the security and integrity of distributed AI/ML systems. These attacks compromise the confidentiality, integrity, and availability of AI/ML systems by taking advantage of flaws in models, data (JayaLakshmi, G. et al., 2024), or infrastructure, which could have dire repercussions. To tackle adversarial challenges in distributed AI/ML systems in 6G networks, strong defenses and preemptive tactics are needed. Conventional security measures, like access control, authentication, and encryption, are vital for strengthening 6G networks' defenses against hostile attacks. Distributed AI/ML systems can be made more resilient to adversarial threats by utilizing AI and ML-based defense mechanisms, such as dynamic defense strategies, robust model architectures, and adversarial training. Secure multi-party computation (SMPC) and federated learning are two collaborative defense techniques that allow distributed AI/ML systems to jointly fend off adversarial attacks while maintaining data confidentiality and privacy (Saad, W. et al., 2019).

6G Networks in the Future of Communication

The upcoming sixth-generation (6G) networks, with their unmatched capabilities and transformative potential, promise to completely reshape communication in the quickly changing telecommunications landscape. Even though fifth-generation (5G) networks today can reach 20 Gbps peak data rates and 100 Mbps user-experienced data rates, these figures will soar to previously unheard-of levels when 6G networks take off. 6G is expected to transform global connectivity with a peak data rate of 1000 Gbps and an experienced data rate of 1 Gbps per user. More than twice as many users will be able to access advanced multimedia services simultaneously and instantly thanks to 6G's improved spectral efficiency compared to 5G. To ensure seamless connectivity and optimal user experiences, network operators must completely revamp their current infrastructure to accommodate the increased spectral efficiency brought about by this significant advancement. User experienced latency is predicted to be significantly reduced to less than 0.1 milliseconds with 6G, whereas 5G only achieves latency reductions of 1 millisecond. The significant reduction in latency not only improves the efficiency of real-time applications but also makes vital services like industrial automation, remote surgery, and emergency response possible. 6G networks are 100 times more reliable than 5G networks, which guarantees perfect operation for applications that are sensitive to latency. This greatly enhances the overall reliability of the network. 6G is expected to surpass 5G in terms of network coverage and dependability. 6G is expected to boost mobile device speeds to 10 devices per square kilometer, while 5G only supports high-speed mobile devices at 500 km/h. With this increased connection density, real-time interaction between a large number of connected devices is made possible, promoting smooth communication and teamwork (Yang, P. et al., 2019). By drastically lowering error rates and raising network reliability, 6G maximizes machine-to-machine (M2M) interaction while also improving overall network performance and reliability. End users will need devices with strong batteries in order to access the multitude of high-end services made possible by 6G networks without any interruptions or delays. In response to this demand, 6G promises to double device battery life, guaranteeing constant access to a wide range of services. Furthermore, by maximizing network performance without using excessive energy, 6G technology prioritizes environmental (Pandey, B. K. et al., 2023a) sustainability. 6G networks are intended to increase energy efficiency by two times over that of 5G, which helps to lessen carbon emissions and encourage environmentally friendly telecommunications infrastructure.

Vulnerabilities in Distributed AI/ML Systems

The incorporation of AI and ML techniques may result in the emergence of vulnerabilities within the distributed systems that comprise 6G networks. The security and dependability of the network infrastructure are both vulnerable to significant threats as a result of these vulnerabilities. The confidentiality, integrity, and availability of the system can be compromised by malicious actors if they take advantage of vulnerabilities in the models, data (Sharma, S. et al., 2023), or infrastructure. Developing robust defense mechanisms to protect 6G networks requires a thorough understanding of these vulnerabilities, which is essential. It is possible for adversarial attacks to be launched against the distributed systems, which is one potential vulnerability. Various methods, including data poisoning, model evasion, backdoor insertion, membership inference, and model inversion, are included in the broader category of adversarial attacks. This type of attack is designed to manipulate or subvert the functioning of the algorithms, which can result in erroneous decisions or performance that is compromised. In the case of data poisoning attacks, for instance, malicious data is introduced into the training dataset. This causes the learning process (Saxena, A. et al., 2024) to become corrupted, which in turn leads to the generation of inaccurate predictions.

Due to the inherent complexity and interconnectedness of distributed systems within 6G networks, there is an additional vulnerability that can arise (Chen, M. et al., 2021). Considering that these systems are decentralized, there are challenges that arise when it comes to ensuring compliance, data privacy, and security. Due to the fact that it is dependent on interconnected devices and networks, the attack surface is increased, which makes it vulnerable to a variety of cyber threats such as data breaches, malware injections, and denial-of-service attacks. The incorporation of algorithms into essential applications and services within 6G networks presents additional security complications. When it comes to ensuring safety and functionality, autonomous vehicles driven by artificial intelligence (Pandey, B. K. et al., 2024), smart city infrastructure, and healthcare systems (KVM, S. et al., 2024) all rely on data processing that is accurate and reliable. There is the potential for severe repercussions to arise from any breach in the integrity or availability of these systems, which may include the creation of safety hazards, breaches of privacy, and financial losses. Furthermore, the proliferation of interconnected Internet of Things devices further exacerbates security challenges, as each device represents a potential entry point for criminals to launch cyber attacks. Utilizing a multi-pronged strategy that incorporates technological, organizational, and regulatory measures is necessary in order to address these vulnerabilities. In order to reduce the likelihood of data breaches and unauthorized access, it is possible to implement robust authentication, encryption (Kumar Pandey, B. et al., 2021), access control, and anomaly detection mechanisms. In addition, capabilities consisting of continuous monitoring, threat intelligence, and incident response are necessary in order to identify and eliminate potential security threats in real time. It is also possible to foster trust and confidence in 6G networks by promoting transparency, accountability, and ethical standards during the development and deployment of the network (Voigtländer et al., 2017).

DISTRIBUTED AI/ML IN 6G NETWORKS

Characteristics of 6G Networks

The introduction of 6G networks is a sign of extraordinary connectivity and technological progress as we look to the future of communications. To ensure that these networks meet the changing needs of the digital age (George, W. K. et al., 2023), it becomes imperative to identify the features that will define their architecture and performance in advance of this next big step forward. 6G networks are anticipated to improve and preserve key performance indicators are as shown in the figure 1, (KPIs) like data rate, latency, reliability, scale, and flexibility, building on the groundwork established by 5G. These metrics will remain essential for evaluating network performance because of the increasing demand for ultra-low latency, higher data rates, more capacity, and increased reliability to support the growing number of devices and applications. Localization and sensing, 6G networks will significantly outperform earlier generations in terms of the precision and accuracy of their localization and object sensing capabilities. Not only will centimeter-level localization precision transform navigation systems, but it will also open up revolutionary new applications in asset tracking, urban planning, and disaster management. The high accuracy of object sensing, as indicated by metrics like false alarm (FA) probabilities and missed detection (MD), will support the efficacy of industrial automation, environmental monitoring, and surveillance systems.

With these capabilities, devices will enter a new era of spatial awareness where they can interact with their surroundings in real-time and seamlessly. This will lead to improvements in convenience, efficiency, and safety across a range of applications (Bishoyi, P. K., & Misra, S., 2021) Distributed AI/ML techniques are Integrated into 6G networks, these techniques will give the infrastructure an unprecedented level of intelligence and flexibility. These methods will enable autonomous decision-making and optimization (Pandey, D. et al., 2021) at the network edge, minimizing latency and reliance on centralized processing by utilizing real-time data analytics and predictive modeling. Network automation can be defined as the degree to which a system is almost entirely autonomous with little to no human intervention. This will improve resilience and scalability while also streamlining operations. Dynamic resource allocation, load balancing, and fault detection will be made easier by the smooth integration of AI/ML algorithms into network operations, guaranteeing peak performance and dependability even in the face of unforeseen circumstances or changing circumstances. As 6G networks take off, the idea of connectivity and user interaction is expected to be completely reinterpreted. Machine-area networks and robot-area networks will become commonplace as devices develop into interconnected networks (Tripathi, R. P. et al., 2023), facilitating smooth communication and cooperation between disparate entities. The advent of extremely low-power and battery-free gadgets that run entirely on network infrastructure or ambient energy sources will usher in a new era of sustainability and ubiquity in which innovation and connectivity have no bounds. 6G networks are shaped by the anticipated requirements of future use cases and applications, which in turn shape the characteristics of these networks (Huang, T. et al., 2019). It is imperative that we anticipate and adapt to the evolving demands of the digital age (Swapna, H. R. et al., 2023) as we move closer to 6G. This will ensure that future networks are equipped to provide communication experiences that are indistinguishable from one another, dependable, and intelligent for all users.

Figure 1. Characteristics of 6G networks in AI/ML

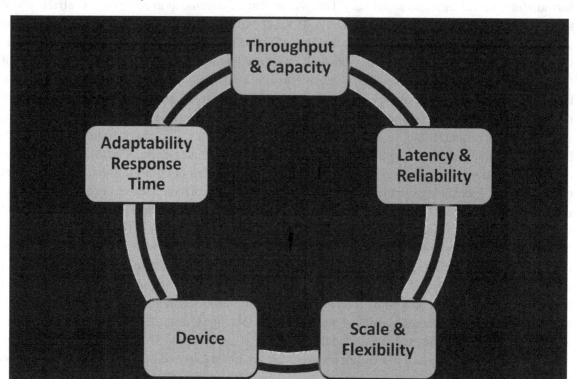

Security in Distributed AI/ML Systems

Figure 2. Applications of distributed learning

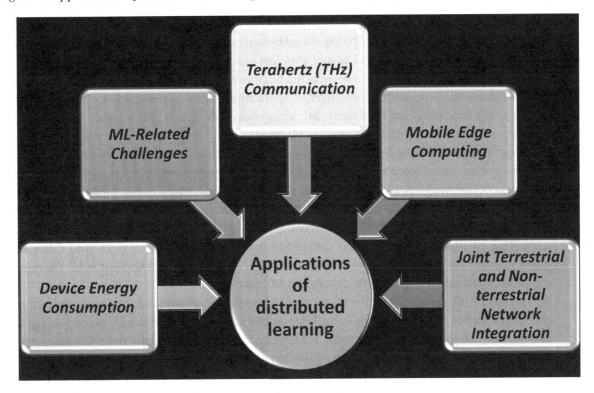

Every generation of wireless communication (Devasenapathy, D. et al., 2024) technology brings with it new capabilities and major advancements to meet the changing needs of society. Every generation of technology, from 1G's basic voice transmission to 5G's high-speed data communication, has cleared the path for revolutionary services and applications. The demands of the data-driven society of the 2030s for instantaneous, limitless connectivity are enormous, especially with the upcoming introduction of 6G networks. 6G research has begun because the current 5G technology is not up to the demands of future applications, and a major revolution is required. Improved data rates, trillion-bit connection capacity, and undetectable latency are the goals of the planned 6G networks. To reach its full potential, 6G will require careful attention to new challenges that arise along the way. Applications of distributed learning are shown in the figure 2.

- *Device Energy Consumption*: Managing the energy consumption of IoT devices (Pramanik, S. et al., 2023) is a challenge for 6G networks due to the rapid growth of connected devices and limited energy supplies. To improve network infrastructure energy efficiency, robust energy-saving mechanisms and strategies are needed. To address energy scarcity and sustain 6G networks, solar, kinetic, and thermal energy harvesting are promising. Energy-efficient communication protocols and algorithms are essential to reducing 6G device energy consumption. Efficient modulation schemes, adaptive power control, and sleep scheduling algorithms can save energy and extend IoT device battery life. In 6G networks, solar panels and wireless charging technologies in IoT devices can reduce energy constraints and promote environmental sustainability. Energy-aware network architectures and protocols are needed to optimize 6G energy use and distribution. Dynamics energy management, intelligent resource allocation, and adaptive network routing can balance energy consumption across devices and infrastructure, improving network efficiency and reducing energy overhead (Petrov, V. et al., 2020). By addressing device energy consumption, 6G networks can maximize IoT applications and ensure seamless connectivity for many use cases.

- *ML-Related Challenges*: AI-driven networking platforms must overcome unique challenges when integrating several new artificial intelligence applications into 6G networks. Computing power, memory, and bandwidth are needed for ML algorithms, which is a major issue. Scalability and performance issues for 6G networks arise as ML models become more complex and data-intensive, requiring more processing and storage. Distributed ML algorithms increase energy costs and security risks. Edge-based distributed learning and federated learning may be better for resource-constrained devices in 6G networks than centralized ML applications. Edge devices can train and update ML models locally, reducing data transmission, privacy, and energy consumption. 6G networks must protect ML models and data. Attackers can compromise ML models or steal sensitive data in edge computing environments due to their distributed nature. Lowering these risks with differential privacy, secure multiparty computation (Bessant, Y. A. et al., 2023), and homomorphic encryption can protect data privacy and enable distributed ML tasks.

- *Terahertz (THz) Communication*: For 6G networks to meet users' growing data needs, several technical challenges must be overcome to maximize THz communication. THz waves have higher propagation loss than lower frequency bands (Bruntha, P. M. et al., 2023), which is a major issue. Communication ranges are limited by atmospheric attenuation and absorption of THz signals (Sahani, S. K. et al., 2024), which requires innovative solutions to overcome signal degradation. THz-compatible antennas and transceivers are also needed for 6G network communication. Traditional antennas are not optimized (Jayapoorani, S. et al., 2023) for THz frequencies, so compact, high-gain antennas are needed. Advanced semiconductor technologies like THz integrated circuits and metamaterials are needed to create high-performance, reliable THz transceivers. To deploy THz communication in 6G networks, regulatory and standardization issues must be addressed. Regulatory bodies and standardization organizations are crucial to spectrum resource allocation, technical specifications, and THz device and system interoperability. Industry stakeholders (Sharma, M. et al., 2024), academia, and government agencies must collaborate to create a coherent regulatory framework and accelerate 6G THz communication technology adoption. To reduce risks and vulnerabilities, THz communication security and privacy must be addressed. THz signals are vulnerable to interception, eavesdropping, and jamming, requiring strong encryption (Pandey, B. K. et al., 2023b), authentication, and intrusion detection. Privacy and confidentiality in 6G networks depend on complying with privacy regulations and protecting sensitive data transmitted over THz links.

- *Joint Terrestrial and Non-terrestrial Network Integration*: 6G networks face opportunities and challenges in integrating non-terrestrial networking platforms with terrestrial communication networks. Satellite constellations and high-altitude platforms (HAPs) provide global coverage and resilience to terrestrial network failures, but their integration presents technical and operational challenges that must be addressed to ensure seamless connectivity and reliability. Creating channel and propagation models for non-terrestrial communication links is difficult. Satellite and HAP-based communication systems have higher path loss, atmospheric attenuation, and latency than terrestrial links due to their different environments. To predict signal propagation, optimize link performance, and design efficient communication protocols for joint terrestrial and non-terrestrial networks, accurate channel models and simulation tools are needed. Mobility management is also difficult in non-terrestrial communication environments. Satellites and HAPs move across the sky, changing coverage areas and handover events as user terminals switch satellite beams or HAP cells. Joint terrestrial and non-terrestrial networks need mobility management schemes like seamless handover mechanisms, efficient beamforming, and adaptive resource allocation algorithms to maintain connectivity and quality of service. Non-terrestrial networks must also mitigate channel impairments and environmental effects like rain fading, atmospheric turbulence, and ionospheric disturbances to maintain reliable communication links. Adaptive modulation and coding, error correction, and diversity combining can reduce channel impairments and improve communication links in adverse conditions. Optimizing spectrum utilization and minimizing co-channel interference in joint terrestrial and non-terrestrial networks requires proper resource allocation and interference management. Dynamic spectrum access, cognitive radio, and interference coordination can reduce interference and maximize spectrum resource efficiency across network domains.

- *Mobile Edge Computing (MEC) and Corresponding Challenges*: Mobile Edge Computing (MEC) has become an important technology in the development of wireless communication (Gupta, A. K. et al., 2023). It promises to bring cloud computing resources closer to end users and make it possible for applications that need low latency and a lot of data. However, the widespread use of MEC in 6G networks comes with a number of issues that need to be fixed in order to fully utilize its capabilities. One of the biggest problems is that MEC servers are limited in size and coverage, which makes it hard for them to communicate and do computations. When putting MEC servers at the edge of a network, you need to think carefully about things like how much power they will use, how much space they will take up, and any regulations that apply. To get around these problems and make MEC deployments more scalable and flexible, we need new ideas like small, energy-efficient server designs, cooling solutions that are built in, and distributed MEC architectures. Also, when 6G technology comes out, many new applications that need very little latency and a lot of data are likely to appear. These will put a lot of stress on MEC infrastructure that has never been seen before. Assigning the right networking services to MEC servers, finding the best user-server assignments, and dynamically adding resources based on changing application needs are all important problems that need to be solved to make sure MEC runs smoothly and makes the best use of its resources. Handover mechanisms that work well and session continuity protocols that keep sessions going are needed to make sure that users can keep connecting and keeping their application state as they move between MEC servers. To deal with these problems and give users a smooth experience in changing mobile settings, it is important to create strong mobility management systems that use predictive handover algorithms, adaptive resource allocation strategies, and context-aware network policies.

ADVERSARIAL THREATS IN DISTRIBUTED AI/ML

Threats to AI and machine learning's security include data poisoning, in which adversaries alter training data. The goal of model evasion attacks is to trick models into generating false predictions. Models are rendered vulnerable by backdoor attacks, which undermine their integrity. Attacks known as membership inference use model outputs to deduce details about training sets. Model inversion attacks use model outputs as a means of extracting sensitive data. The significance of strong defenses are shown in the figure 3 and ongoing research into protecting AI and ML systems against malevolent exploitation is highlighted by these threats.

Figure 3. Adversarial threats in distributed AI/ML

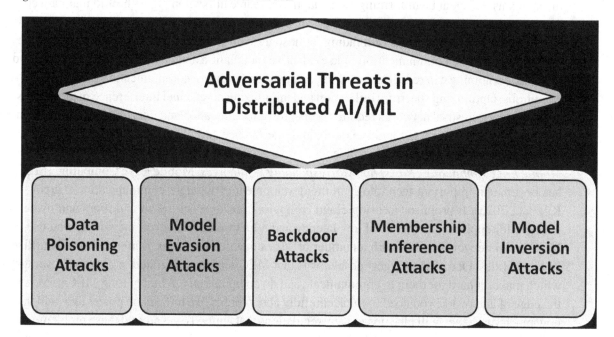

Data Poisoning Attacks

Attacks using "data poisoning" undermine the credibility of models and reduce their efficacy, thereby posing a serious risk to the integrity and safety of distributed AI/ML systems. With different features and ramifications, these attacks can be divided into four primary categories: subpopulation attacks, availability attacks, backdoor attacks, and targeted attacks. False positives, false negatives, and incorrect test sample classification result from the complete model being tainted by availability attacks. Label flipping and adding approved labels to compromised data are common techniques used in availability attacks that lead to a significant model accuracy reduction. On the other side, backdoor attacks entail adding backdoors to training examples, like hidden patterns (Govindaraj, V. et al., 2023) or features. The model's overall output quality is affected by these backdoors because they cause the model to misclassify some inputs. More covert are targeted attacks, which affect fewer samples while leaving the model functioning normally for the majority of inputs. Because the visible impact on the model is minimal, targeted attacks become harder to detect. Similar to targeted attacks, subpopulation attacks impact several subsets with comparable characteristics. This makes it difficult to detect and counteract attacks that affect certain subsets of data without affecting the accuracy of the model for the remaining data. Attackers' familiarity with the model also allows for the classification of data poisoning attempts. White-box attacks encompass complete awareness of the training set and model parameters, whereas black-box attacks deprive adversaries of access to the internal workings of the model. Attackers with partial model knowledge fall between these two extremes in terms of grey-box attacks. It is important to defend against adversaries with different levels of knowledge because white-box attacks are typically the most successful (Shinde, S. S. et al., 2021). A thorough awareness of potential vulnerabilities and proactive risk mitigation strategies are necessary for preventing data poisoning attacks. Via the use of tainted data or altered training samples, adversaries may try to manipulate a model's training procedure. Data poisoning attacks must be prevented at all costs, as organizations depend more and more on AI/ML for vital applications in law enforcement, transportation (Sahani, K. et al., 2023), and other departments. It will take strong security measures and constant attention to identify and eliminate possible threats to meet these challenges. Some methods that can be used to strengthen distributed AI/ML systems against malicious manipulation are anomaly detection, adversarial training, and secure data aggregation (Pandey, B. K. et al., 2021a). Fostering accountability and openness in AI/ML procedures can also increase confidence in the dependability of models (Pandey, B. K. et al., 2021b). The integrity and efficacy of distributed AI/ML systems must be protected against data poisoning attacks, to sum up. Organizations can reduce risks and protect against potential vulnerabilities by applying proactive security measures and comprehending different attack vectors. This will ensure that AI/ML technologies continue to advance and be adopted (Deng, S. et al., 2020).

Model Evasion Attacks

Adversarial attacks, commonly referred to as model evasion attacks, pose a serious risk to the reliability and integrity of machine learning models. These attacks are designed to trick a model into making incorrect predictions or classifications by quietly altering inputs in a way that looks harmless to humans. Attacks using model evasion have serious repercussions in a number of industries, including finance, healthcare (Singh, S. et al., 2023), cybersecurity, and autonomous systems. To guarantee the dependability and credibility of machine learning systems, it is essential to comprehend the mechanisms underlying these attacks and create strong defenses. The capacity of model evasion attacks to take advantage of weaknesses in machine learning models' decision boundaries is one of their main traits. Adversaries tamper with input data by introducing subtle perturbations, dubbed adversarial examples, which lead to high confidence misclassifications or incorrect predictions from the model. These perturbations are deliberately designed to take advantage of the model's flaws, frequently by maximizing the loss function of the model through the use of gradient-based optimization techniques. Model evasion attacks can be initiated using a variety of strategies, including gradient-based techniques like the Carlini-Wagner attack, iterative techniques like the Projected Gradient Descent (PGD), and optimization-based techniques like the Fast Gradient Sign Method (FGSM). The goal of these techniques is to produce adversarial examples that fool the model in a way that is undetectable to human observers. Transferability attacks increase the susceptibility of machine learning systems to evasion attacks by taking advantage of the adversarial examples' transferability across various models or architectures. Using machine learning models in practical applications is severely hampered by model evasion attacks. For example, in the context of cybersecurity, adversaries can launch sophisticated cyberattacks and avoid detection by taking advantage of flaws in malware classifiers or intrusion detection systems. Similar to this, adversaries in the medical field can falsify medical imaging data to trick diagnostic tools, which could result in an inaccurate diagnosis or treatment selection. Evasion attacks can also jeopardize the security and dependability of algorithms used in autonomous systems, like self-driving cars, putting pedestrians and passengers at risk. It takes a multifaceted strategy that includes adversarial training methods in addition to strong model design to counteract model evasion attacks. In order to increase the model's resistance to evasion attacks, adversarial examples are added to the training data through the process of adversarial training. Furthermore, methods like model ensembling, anomaly detection, and input sanitization can assist in identifying and thwarting adversarial attacks during the inference process. Furthermore, making adversarially robust models via methods such as certified defenses, randomized smoothing, and adversarial training can strengthen the model's ability to withstand evasion attacks (Emmert-Streib, F. et al., 2020).

Backdoor Attacks

Distributional AI/ML adversarial threats can compromise machine learning model security, resulting in unauthorized access, data breaches, and network infrastructure compromise. Backdoor attacks, used by cybercriminals to break into systems and networks, are common. Protecting against backdoor attacks requires understanding their mechanisms and taking strong measures. Backdoor attacks exploit software, network, and hardware vulnerabilities to get into systems and create hidden channels. Backdoor attacks can be carried out using software vulnerabilities, social engineering, password cracking tools, or backdoor malware. Attackers can manipulate system files, install malicious software, and exfiltrate sensitive data through a backdoor, putting targeted systems at risk. Backdoor attacks aim for root access, which gives full control over computers. Attackers can compromise systems and control connected devices with root access, escalating the security breach. Backdoor attacks target network infrastructure components like routers because they can intercept and redirect network traffic, compromising data confidentiality and integrity. Critical data and files are also vulnerable to backdoor attacks on hard drives. Hard drive firmware and operating system vulnerabilities can allow attackers to access, manipulate, or disable drives. Backdoor access to hard drives can cause data theft, unauthorized changes, and critical operations disruptions. Backdoor attacks start with identifying target system or network vulnerabilities. These vulnerabilities allow attackers to gain access, create covert channels, and bypass security. Attackers can exploit known vulnerabilities and gain unauthorized access to systems or devices through software or hardware backdoors or default passwords. Cybersecurity must be proactive to reduce backdoor attack risks. To address vulnerabilities, organizations must prioritize vulnerability management and patch software and firmware. Security measures like access controls, multi-factor authentication, and encryption can also reduce backdoor attacks. Backdoor attacks can also be prevented by educating users about cybersecurity. Training staff to recognize social engineering, avoid phishing, and secure passwords can reduce the risk of data breaches and unauthorized access.

Membership Inference Attacks

Distributional AI/ML privacy and security are threatened by membership inference attacks. These attacks infer whether a data point was in a machine learning model's training dataset. By exploiting model output vulnerabilities or accessing model parameters, adversaries can find sensitive data about individual data points, threatening user privacy and data confidentiality. Members inference attacks use machine learning model behavior to infer the presence or absence of specific data points in the training dataset. Membership inference attacks can be launched by analyzing model predictions, confidence scores, or model access patterns (Muniandi, B. et al., 2024). By exploiting subtle model behavior differences, attackers can infer whether a data point was used during training, breaching sensitive data privacy. Model confidence scores for individual predictions are often used in membership inference attacks. In-distribution and out-of-distribution data points have different confidence levels, so adversaries can infer membership status. Adversaries may also use model outputs or gradients to spy on training data and launch targeted attacks or steal proprietary information. Distributed AI/ML systems struggle to defend against membership inference attacks. Membership inference attacks focus on data point privacy and training dataset confidentiality, not software or network infrastructure vulnerabilities. Encryption and access control may not be enough to prevent membership inference attacks (Zhang, C. et al., 2019). Organizations need both proactive and reactive defenses to prevent membership inference attacks. By hiding sensitive data points, adversarial training can strengthen machine learning models against membership inference attacks. In distributed AI/ML systems, differential privacy, federated learning, and model distillation can reduce privacy breaches and data leakage. Addressing membership inference attacks requires AI/ML transparency and accountability. Researchers, practitioners, and policymakers can collaborate to standardize evaluation methods, share best practices, and create regulatory frameworks to protect user privacy and data confidentiality in distributed AI/ML environments.

Model Inversion Attacks

Targeting the confidentiality of sensitive data encoded within machine learning models, model inversion attacks pose a serious threat to privacy in distributed AI/ML systems. These attacks take advantage of flaws in the model's parameters or outputs to deduce information about the training data that is underneath, which may reveal private or confidential data. Adversaries can violate people's privacy and jeopardize the integrity of sensitive data by reconstructing inputs that match particular model outputs. Using a machine learning model's observed behavior to deduce details about the training data that went into creating the model is the mechanism underlying model inversion attacks. Attackers can reverse-engineer sensitive data stored in the model by taking advantage of the mapping between inputs and outputs to reconstruct inputs that result in desired outputs. When personally identifiable information or other sensitive data is included in the model's outputs, model inversion attacks may be especially dangerous. The sensitive information that was encoded within the model is essentially reverse-engineered by adversaries using optimization techniques or machine learning algorithms to reconstruct inputs that closely match the observed outputs. Through a process of iterative refinement of the reconstructed inputs, adversaries may be able to obtain information about the training data and potentially compromise the privacy of individuals or organizations. Protecting distributed AI/ML systems from adversarial threats is made extremely difficult by model inversion attacks. Because adversaries use the intrinsic characteristics of the model's outputs to infer sensitive information about the training data, traditional defense mechanisms like encryption and access control may not be sufficient to mitigate the risks posed by model inversion attacks. It can be hard to identify model inversion attacks since they frequently happen at the interface where users and the model interact, making it hard to tell which queries are malicious and which are legitimate. Organizations must use a multifaceted strategy that includes both proactive and reactive defense techniques to fend off model inversion attacks. Because adversarial examples obscure sensitive information about the training data, adversarial training, which entails augmenting the training dataset with adversarial examples, can help strengthen machine learning models' resilience against model inversion attacks. Enhancing accountability and transparency in AI/ML procedures is also crucial for resolving the issues raised by model inversion attacks. We can create uniform evaluation procedures, exchange best practices, and set up legal frameworks to protect user privacy and data confidentiality in distributed AI/ML environments by encouraging cooperation between researchers, practitioners, and policymakers (Schmidhuber, J., 2015).

STRATEGIES FOR SAFEGUARDING 6G NETWORKS AGAINST ADVERSARIAL THREATS

Protecting 6G networks from hostile threats calls for an all-encompassing strategy that incorporates a number of different risk mitigation strategies in order to be fully effective. Protecting data that is being transmitted over a network can be accomplished through the implementation of comprehensive encryption protocols and access controls. It is possible for organizations to prevent unauthorized access and data breaches by encrypting sensitive information and restricting access to only authorized users. This, in turn, significantly reduces the likelihood of adversarial attacks. Employing intrusion detection and prevention systems (IDPS) is yet another tactic that can be utilized to monitor network traffic and identify anomalous behavior that may be indicative of malicious activity. Identifying suspicious patterns and triggering alerts in real time is the goal of IDPS solutions, which employ machine learning algorithms and behavioral analytics. This enables organizations to respond quickly to potential threats and mitigate the impact that they have on network security By implementing robust authentication mechanisms and multi-factor authentication protocols, organizations can improve the resilience of 6G networks against adversarial threats. This can be accomplished by combining the two. Organisations are able to reduce the risk of unauthorized access and strengthen network security by requiring users to provide multiple forms of verification. These forms of verification include passwords, biometric data, and one-time tokens. Organizations should make the management of patches and regular software updates a priority in order to address known vulnerabilities and reduce the likelihood that adversaries will exploit vulnerability. Organizations can lessen their vulnerability to attacks from adversaries and ensure the integrity and confidentiality of data that is transmitted over 6G networks if they remain vigilant and proactive in addressing security vulnerabilities on a consistent basis (Guo, Y., et al., 2022). The ability to utilize threat intelligence and information sharing initiatives in order to maintain awareness regarding ever-evolving vulnerabilities and other threats. This allows organizations to gain valuable insights into evolving attack techniques and develop proactive defense strategies to effectively mitigate risks. This is accomplished through collaboration with peers in the industry and the sharing of threat intelligence data. To make employee training and awareness programs a top priority in order to educate users about potential security risks and the best practices for network resource protection. Increasing the overall security posture of 6G networks can be accomplished by empowering employees to recognize and report suspicious activity. This can be accomplished by cultivating a culture of security awareness and accountability within an organization.

CONCLUSION

As we move into a new era of connectivity and intelligence, it is of the utmost importance that we protect 6G networks from the challenges posed by adversaries in distributed AI and ML systems. Considering the potential dangers that could be posed by adversarial attacks, it is essential to implement preventative defense strategies in order to guarantee the safety and resilience of 6G networks. As stakeholders navigate the ever-changing landscape of artificial intelligence and telecommunications technologies, addressing adversarial challenges continues to be a critical priority in order to protect the future of 6G networks. With capabilities such as ultra-high data rates, ultra-low latency, massive connectivity, energy efficiency, and the integration of artificial intelligence, 6G represents the next frontier in wireless technology (Raja, D. et al., 2024). It offers capabilities that have never been seen before. Although it is still in its early stages of development, 6G has the potential to revolutionize a variety of different industries and open the door to new applications that were previously unimaginable. Keeping abreast of the most recent advancements and developments in this fascinating field will be essential for realizing the full potential of 6G and ensuring its security and resilience in the face of adversarial threats. This is because researchers and industry stakeholders are continuing to investigate the possibilities of 6G.\

REFERENCES

Abdulkarim, Y. I., Awl, H. N., Muhammadsharif, F. F., Saeed, S. R., Sidiq, K. R., Khasraw, S. S., & Pandey, D. (2023). Metamaterial-based sensors loaded corona-shaped resonator for COVID-19 detection by using microwave techniques. *Plasmonics*, 1–16.

Anand, R., Khan, B., Nassa, V. K., Pandey, D., Dhabliya, D., Pandey, B. K., & Dadheech, P. (2023). Hybrid convolutional neural network (CNN) for kennedy space center hyperspectral image. *Aerospace Systems*, 6(1), 71–78. 10.1007/s42401-022-00168-4

Anand, R., Lakshmi, S. V., Pandey, D., & Pandey, B. K. (2024). An enhanced ResNet-50 deep learning model for arrhythmia detection using electrocardiogram biomedical indicators. *Evolving Systems*, 15(1), 83–97. 10.1007/s12530-023-09559-0

Bessant, Y. A., Jency, J. G., Sagayam, K. M., Jone, A. A. A., Pandey, D., & Pandey, B. K. (2023). Improved parallel matrix multiplication using Strassen and Urdhvatiryagbhyam method. *CCF Transactions on High Performance Computing*, 5(2), 102–115. 10.1007/s42514-023-00149-9

Bishoyi, P. K., & Misra, S. (2021). Enabling green mobile-edge computing for 5G-based healthcare applications. *IEEE Transactions on Green Communications and Networking*, 5(3), 1623–1631. 10.1109/TGCN.2021.3075903

. Bruntha, P. M., Dhanasekar, S., Hepsiba, D., Sagayam, K. M., Neebha, T. M., Pandey, D., & Pandey, B. K. (2023). Application of switching median filter with L 2 norm-based auto-tuning function for removing random valued impulse noise. *Aerospace systems*, 6(1), 53-59.

Chen, M., Gündüz, D., Huang, K., Saad, W., Bennis, M., Feljan, A. V., & Poor, H. V. (2021). Distributed learning in wireless networks: Recent progress and future challenges. *IEEE Journal on Selected Areas in Communications*, 39(12), 3579–3605. 10.1109/JSAC.2021.3118346

David, S., Duraipandian, K., Chandrasekaran, D., Pandey, D., Sindhwani, N., & Pandey, B. K. (2023). Impact of blockchain in healthcare system. In *Unleashing the Potentials of blockchain technology for healthcare industries* (pp. 37–57). Academic Press. 10.1016/B978-0-323-99481-1.00004-3

Deng, S., Zhao, H., Fang, W., Yin, J., Dustdar, S., & Zomaya, A. Y. (2020). Edge intelligence: The confluence of edge computing and artificial intelligence. *IEEE Internet of Things Journal*, 7(8), 7457–7469. 10.1109/JIOT.2020.2984887

Devasenapathy, D., Madhumathy, P., Umamaheshwari, R., Pandey, B. K., & Pandey, D. (2024). Transmission-efficient grid-based synchronized model for routing in wireless sensor networks using Bayesian compressive sensing. *SN Computer Science*, 5(1), 1–11.

Du John, H. V., Jose, T., Sagayam, K. M., Pandey, B. K., & Pandey, D. (2024). Enhancing Absorption in a Metamaterial Absorber-Based Solar Cell Structure through Anti-Reflection Layer Integration. *Silicon*, 1-11.

Emmert-Streib, F., Yang, Z., Feng, H., Tripathi, S., & Dehmer, M. (2020). An introductory review of deep learning for prediction models with big data. *Frontiers in Artificial Intelligence*, 3, 4. 10.3389/frai.2020.0000433733124

George, W. K., Ekong, M. O., Pandey, D., & Pandey, B. K. (2023). Pedagogy for Implementation of TVET Curriculum for the Digital World. In *Applications of Neuromarketing in the Metaverse* (pp. 117-136). IGI Global. 10.4018/978-1-6684-8150-9.ch009

George, W. K., Silas, E. I., Pandey, D., & Pandey, B. K. (2024). Utilization of Industry 4.0 Technologies in Nigerian Technical and Vocational Education: A Conundrum for Educators. In *Examining the Rapid Advance of Digital Technology in Africa* (pp. 270–293). IGI Global. 10.4018/978-1-6684-9962-7.ch014

. Govindaraj, V., Dhanasekar, S., Martinsagayam, K., Pandey, D., Pandey, B. K., & Nassa, V. K. (2023). Low-power test pattern generator using modified LFSR. *Aerospace Systems*, 1-8.

Guo, Y., Zhao, R., Lai, S., Fan, L., Lei, X., & Karagiannidis, G. K. (2022). Distributed machine learning for multiuser mobile edge computing systems. *IEEE Journal of Selected Topics in Signal Processing*, 16(3), 460–473. 10.1109/JSTSP.2022.3140660

Gupta, A. K., Sharma, R., Pandey, D., Nassa, V. K., Pandey, B. K., George, A. S., & Dadheech, P. (2023). Performance analysis of eight-channel WDM optical network with different optical amplifiers for industry 4.0. In *Innovation and Competitiveness in Industry 4.0 Based on Intelligent Systems* (pp. 197–212). Springer International Publishing. 10.1007/978-3-031-29775-5_9

Huang, T., Yang, W., Wu, J., Ma, J., Zhang, X., & Zhang, D. (2019). A survey on green 6G network: Architecture and technologies. *IEEE Access : Practical Innovations, Open Solutions*, 7, 175758–175768. 10.1109/ACCESS.2019.2957648

Iyyanar, P., Anand, R., Shanthi, T., Nassa, V. K., Pandey, B. K., George, A. S., & Pandey, D. (2023). A real-time smart sewage cleaning UAV assistance system using IoT. In *Handbook of Research on Data-Driven Mathematical Modeling in Smart Cities* (pp. 24–39). IGI Global.

JayaLakshmi. G., Pandey, D., Pandey, B. K., Kaur, P., Mahajan, D. A., & Dari, S. S. (2024). Smart Big Data Collection for Intelligent Supply Chain Improvement. In *AI and Machine Learning Impacts in Intelligent Supply Chain* (pp. 180-195). IGI Global.

Jayapoorani, S., Pandey, D., Sasirekha, N. S., Anand, R., & Pandey, B. K. (2023). Systolic optimized adaptive filter architecture designs for ECG noise cancellation by Vertex-5. *Aerospace Systems*, 6(1), 163–173. 10.1007/s42401-022-00177-3

Kirubasri, G., Sankar, S., Pandey, D., Pandey, B. K., Singh, H., & Anand, R. (2021, September). A recent survey on 6G vehicular technology, applications and challenges. In *2021 9th International Conference on Reliability, Infocom Technologies and Optimization (Trends and Future Directions)(ICRITO)* (pp. 1-5). IEEE.

Kumar Pandey, B., Pandey, D., Nassa, V. K., Ahmad, T., Singh, C., George, A. S., & Wakchaure, M. A. (2021). Encryption and steganography-based text extraction in IoT using the EWCTS optimizer. *Imaging Science Journal*, 69(1-4), 38–56. 10.1080/13682199.2022.2146885

KVM, S., Pandey, B. K., & Pandey, D. (2024). Design of Surface Plasmon Resonance (SPR) Sensors for Highly Sensitive Biomolecular Detection in Cancer Diagnostics. *Plasmonics*, 1-13.

Muniandi, B., Nassa, V. K., Pandey, D., Pandey, B. K., Dadheech, P., & George, A. S. (2024). Pattern Analysis for Feature Extraction in Complex Images. In *Using Machine Learning to Detect Emotions and Predict Human Psychology* (pp. 145-167). IGI Global. 10.4018/979-8-3693-1910-9.ch007

Pandey, B. K., & Pandey, D. (2023). Parametric optimization and prediction of enhanced thermoelectric performance in co-doped CaMnO3 using response surface methodology and neural network. *Journal of Materials Science Materials in Electronics*, 34(21), 1589. 10.1007/s10854-023-10954-1

Pandey, B. K., Pandey, D., Alkhafaji, M. A., Güneşer, M. T., & Şeker, C. (2023a). A reliable transmission and extraction of textual information using keyless encryption, steganography, and deep algorithm with cuckoo optimization. In *Micro-Electronics and Telecommunication Engineering: Proceedings of 6th ICMETE 2022* (pp. 629–636). Springer Nature Singapore. 10.1007/978-981-19-9512-5_57

Pandey, B. K., Pandey, D., Dadheech, P., Mahajan, D. A., George, A. S., & Hameed, A. S. (2023b). Review on Smart Sewage Cleaning UAV Assistance for Sustainable Development. In *Handbook of Research on Safe Disposal Methods of Municipal Solid Wastes for a Sustainable Environment* (pp. 69–79). IGI Global. 10.4018/978-1-6684-8117-2.ch005

Pandey, B. K., Pandey, D., & Sahani, S. K. (2024). Autopilot control unmanned aerial vehicle system for sewage defect detection using deep learning. *Engineering Reports*, 12852. 10.1002/eng2.12852

Pandey, B. K., Pandey, D., Wairya, S., & Agarwal, G. (2021a). An advanced morphological component analysis, steganography, and deep learning-based system to transmit secure textual data. [IJDAI]. *International Journal of Distributed Artificial Intelligence*, 13(2), 40–62. 10.4018/IJDAI.2021070104

Pandey, B. K., Pandey, D., Wariya, S., & Agarwal, G. (2021b). A deep neural network-based approach for extracting textual images from deteriorate images. *EAI Endorsed Transactions on Industrial Networks and Intelligent Systems*, 8(28), e3–e3. 10.4108/eai.17-9-2021.170961

Pandey, D., Pandey, B. K., & Wairya, S. (2021). Hybrid deep neural network with adaptive galactic swarm optimization for text extraction from scene images. *Soft Computing*, 25(2), 1563–1580. 10.1007/s00500-020-05245-4

Petrov, V., Kurner, T., & Hosako, I. (2020). IEEE 802.15. 3d: First standardization efforts for sub-terahertz band communications toward 6G. *IEEE Communications Magazine*, 58(11), 28–33. 10.1109/MCOM.001.2000273

Pramanik, S., Pandey, D., Joardar, S., Niranjanamurthy, M., Pandey, B. K., & Kaur, J. (2023, October). An overview of IoT privacy and security in smart cities. In *AIP Conference Proceedings* (*Vol. 2495*, No. 1). AIP Publishing. 10.1063/5.0123511

Raja, D., Kumar, D. R., Santhiyakumari, N., Kumarganesh, S., Sagayam, K. M., Thiyaneswaran, B., Pandey, B. K., & Pandey, D. (2024). A compact dual-feed wide-band slotted antenna for future wireless applications. *Analog Integrated Circuits and Signal Processing*, 1–15. 10.1007/s10470-023-02233-0

Saad, W., Bennis, M., & Chen, M. (2019). A vision of 6G wireless systems: Applications, trends, technologies, and open research problems. *IEEE Network*, 34(3), 134–142. 10.1109/MNET.001.1900287

Sahani, K., Khadka, S. S., Sahani, S. K., Pandey, B. K., & Pandey, D. (2023). A possible underground roadway for transportation facilities in Kathmandu Valley: A racking deformation of underground rectangular structures. *Engineering Reports*, 12821. 10.1002/eng2.12821

Sahani, S. K., Pandey, B. K., & Pandey, D. (2024). *Single-valued Signals, Multi-valued Signals and Fixed-Point of Contractive Signals*. Mathematics Open. 10.1142/S2811007224500020

Saxena, A., Agarwal, A., Pandey, B. K., & Pandey, D. (2024). Examination of the Criticality of Customer Segmentation Using Unsupervised Learning Methods. *Circular Economy and Sustainability*, 1–14. 10.1007/s43615-023-00336-4

Schmidhuber, J. (2015). Deep learning in neural networks: An overview. *Neural Networks*, 61, 85–117. 10.1016/j.neunet.2014.09.00325462637

Sharma, M., Talwar, R., Pandey, D., Nassa, V. K., Pandey, B. K., & Dadheech, P. (2024). A Review of Dielectric Resonator Antennas (DRA)-Based RFID Technology for Industry 4.0. *Robotics and Automation in Industry*, 4(0), 303–324.

Sharma, S., Pandey, B. K., Pandey, D., Anand, R., Sharma, A., & Saini, S. (2023, March). Character Recognition Technique Implementation for Complicated Deteriorated Scene. In *2023 6th International Conference on Information Systems and Computer Networks (ISCON)* (pp. 1-4). IEEE. 10.1109/ISCON57294.2023.10112185

Shinde, S. S., Marabissi, D., & Tarchi, D. (2021). A network operator-biased approach for multi-service network function placement in a 5G network slicing architecture. *Computer Networks*, 201, 108598. 10.1016/j.comnet.2021.108598

Singh, S., Madaan, G., Kaur, J., Swapna, H. R., Pandey, D., Singh, A., & Pandey, B. K. (2023). *Bibliometric Review on Healthcare Sustainability. Handbook of Research on Safe Disposal Methods of Municipal Solid Wastes for a Sustainable Environment*, (pp. 142-161). IGI Global. 10.4018/978-1-6684-8117-2.ch011

Swapna, H. R., Bigirimana, E., Madaan, G., Hasan, A., Pandey, B. K., & Pandey, D. (2023). Impact of neuromarketing on consumer psychology in digitally connected networks. In *Applications of Neuromarketing in the Metaverse* (pp. 193–205). IGI Global. 10.4018/978-1-6684-8150-9.ch015

Tripathi, R. P., Sharma, M., Gupta, A. K., Pandey, D., Pandey, B. K., Shahul, A., & George, A. H. (2023). Timely prediction of diabetes by means of machine learning practices. *Augmented Human Research*, 8(1), 1. 10.1007/s41133-023-00062-4

Yang, P., Xiao, Y., Xiao, M., & Li, S. (2019). 6G wireless communications: Vision and potential techniques. *IEEE Network*, 33(4), 70–75. 10.1109/MNET.2019.1800418

Zhang, C., Patras, P., & Haddadi, H. (2019). Deep learning in mobile and wireless networking: A survey. *IEEE Communications Surveys and Tutorials*, 21(3), 2224–2287. 10.1109/COMST.2019.2904897

Chapter 6
Beyond Traditional Security:
Harnessing AI for 6G Communication Resilience

Freddy Ochoa-Tataje
Universidad César Vallejo, Lima, Peru

Joel Alanya-Beltran
Universidad San Ignacio de Loyola, Peru

Jenny Ruiz-Salazar
Universidad Privada del Norte, Lima, Peru

Juan Paucar-Elera
Universidad Nacional Federico Villarreal, Lima, Peru

Michel Mendez-Escobar
Universidad Autónoma del Perú, Lima, Peru

Frank Alvarez-Huertas
Universidad Nacional Mayor de San Marcos, Lima, Peru

ABSTRACT

With the emergence of 6G communication networks, it is becoming more and more clear that we need to prioritize strong resilience against evolving cyber threats. Conventional security measures, although they have their merits, are no longer adequate to protect the complex structure of 6G networks. Utilizing artificial intelligence (AI) in this context offers a revolutionary method to enhance communication resilience in the 6G era. The importance of AI in strengthening the resilience of 6G communication lies in its ability to analyze data and respond accordingly. With the help of machine learning algorithms, AI has the ability to identify patterns that may signal security breaches.

INTRODUCTION

With the emergence of 6G (Kirubasri, G. et al., 2021) communication networks, it is becoming more and clearer that we need to prioritize strong resilience against evolving cyber threats. Conventional security measures, although they have their merits, are no longer adequate to protect the complex structure of 6G networks. Utilizing artificial intelligence (AI) (Pandey, B. K., & Pandey, D., 2023) in this context offers a revolutionary method to enhance communication resilience in the 6G era. The importance of AI in strengthening the resilience of 6G communication lies in its ability to analyze data and respond accordingly. With the help of machine learning algorithms (Tripathi, R. P. et al., 2023), AI has the ability

DOI: 10.4018/979-8-3693-2931-3.ch006

to identify patterns that may signal security breaches. This allows for proactive measures to be taken in order to minimize risks before they become more serious. Having the ability to predict and respond to threats in real-time is extremely important in 6G networks, given the massive amount of data (Pandey, B. K. et al., 2023a) being transmitted at high speeds. AI-powered solutions enable 6G networks to flexibly respond to changing cyber threats. Through ongoing analysis of network traffic and behavior, AI algorithms have the ability to independently adapt security configurations and protocols, strengthening defenses against new attack methods. This flexibility is crucial in the constantly evolving realm of cyber threats, where conventional fixed security measures frequently prove inadequate. AI enhances human abilities in the management and security of 6G communication networks. With the help of AI, cybersecurity professionals can now dedicate more time to strategic initiatives and analyzing complex threats, as routine tasks like intrusion detection and incident response are automated. This mutually beneficial partnership between humans and AI improves the overall effectiveness and flexibility of 6G communication resilience initiatives. With the expertise of a database administrator, AI-powered solutions showcase their capability to proactively safeguard against advanced cyber threats, all while reducing the need for human involvement. The integration of AI into 6G communication resilience also brings about various challenges and factors to consider. Addressing ethical concerns related to data privacy (Kumar Pandey, B. et al., 2021), algorithmic bias, and transparency is crucial for responsible AI deployment. Incorporating AI-driven solutions into existing network infrastructures can lead to interoperability and compatibility challenges (Bangerter, B. et al., 2014).

Traditional Security in Network

Security measures in communication networks involve various protocols and technologies that aim to safeguard data, devices (Sharma, M. et al., 2022). Design of a GaN-based Flip Chip), and networks from unauthorized access, malicious attacks, and data breaches. These measures establish the fundamental framework for guaranteeing the security and reliability of communication systems (Devasenapathy, D. et al., 2024). Traditional security measures encompass various key components such as encryption, firewalls, intrusion detection systems (IDS), and access control mechanisms. Encryption (Pandey, D. et al., 2021a) is a crucial method employed to safeguard data transmission. It involves transforming plaintext information into ciphertext, rendering it incomprehensible to unauthorized individuals. With the use of cryptographic algorithms and keys, encryption guarantees the utmost confidentiality of data, even in the event of interception during transmission. Firewalls serve as guardians between internal and external networks, diligently overseeing and regulating the flow of incoming and outgoing traffic according to established security protocols. With the ability to filter (Bruntha, P. M. et al., 2023) network traffic and thwart unauthorized access attempts, firewalls play a crucial role in safeguarding network infrastructure against malicious infiltrations. As an expert in information security, it's important to understand the role of intrusion detection systems (IDS). These systems act as proactive security measures, constantly monitoring network traffic to identify any signs of suspicious or abnormal activity. As an expert in information security analysis, you'll be responsible for analyzing network packets and patterns to identify any potential security breaches or intrusion attempts. By promptly alerting administrators, you'll ensure that appropriate action is taken to mitigate risks and safeguard the system. Access control mechanisms are responsible for managing user access to network resources and sensitive information. They do this by using predefined authorization levels and authentication credentials. Through the implementation of user authentication, authorization, and accountability measures, access control mechanisms effectively safeguard against unauthorized access to sensitive information and maintain the integrity of the network. There are other conventional security measures that can be implemented, such as antivirus software (Kirubasri, G. et al., 2022), virtual private networks (VPNs), and secure protocols like Secure Sockets Layer (SSL) and Transport Layer Security (TLS). These measures work together to create a secure communication environment, protecting against common cyber threats and vulnerabilities. Similar to an information security analyst, it is important to recognize that traditional security measures alone may not be sufficient to fully protect communication networks from advanced and ever-changing cyber threats. With the rapid advancement of communication technologies and the ever-changing threat landscapes, it is essential to incorporate emerging technologies like artificial intelligence (AI) and machine learning (ML) to strengthen traditional security measures and improve resilience against cyber threats (Huseien, G. F., & Shah, K. W., 2022).

AI's Role in Communication Resilience

The introduction of proactive, adaptive, and intelligent capabilities to mitigate risks and effectively respond to evolving cyber threats. Artificial intelligence (AI) plays a pivotal role in enhancing communication resilience by introducing these capabilities. Utilizing machine learning (Saxena, A. et al., 2024) algorithms to analyze vast amounts of data and identify patterns that are indicative of potential security breaches or vulnerabilities is one of the ways that artificial intelligence (Deepa, R. et al., 2022) enables predictive analysis. Artificial intelligence (Khan, B. et al., 2021) is able to anticipate new threats by continuously monitoring network traffic and proactively implementing countermeasures to deter or lessen the impact of these threats. Through the reduction of the likelihood of successful cyber attacks and the minimization of potential disruptions to network operations, this predictive capability improves the strength of communication resilience. Technology that is driven by artificial intelligence gives communication networks the ability to dynamically adapt to shifting threat landscapes in real time. By automatically modifying security configurations and protocols in response to evolving threat intelligence, artificial intelligence (AI) improves the responsiveness and agility of communication infrastructure (Gohar, A., & Nencioni, G., 2021). The fact that they are confronted with unprecedented challenges, networks are able to effectively defend themselves against sophisticated cyber-attacks and continue to provide uninterrupted service delivery thanks to this adaptable capability. A further benefit of artificial intelligence is that it enhances human capabilities by automating mundane tasks and improving decision-making processes in operations. Artificial intelligence enables professionals in the field to prioritize threats, efficiently allocate resources, and effectively respond to security incidents via the analysis of vast amounts of data and the identification of actionable insights.

Resilience for 6G Connectivity

Resilience is crucial in the world of 6G networks, as it guarantees continuous connectivity and protects data integrity in the face of various challenges. In the ever-changing world of advanced communication technology, with the ambitious goals of 6G networks like massive data transmission and ultra-low latency applications, resilience is not just desirable, but absolutely necessary to maintain seamless service delivery. Resilience is what allows these networks to withstand disruptions and stay connected in different environments, such as urban, remote, or mobile areas. In the era of 6G, where the widespread use of Internet of Things (IoT) devices (Sharma, M. et al., 2022) and the rapid increase in data volumes are the standard, ensuring the security and reliability of data is crucial. Implementing resilience strategies, such as encryption (Pandey, B. K. et al., 2022) and tamper detection, is crucial for protecting data from unauthorized access or manipulation. Ensuring the integrity and dependability of data is crucial for maintaining user trust and ensuring the seamless operation of vital applications. Resilience and security go hand in hand in 6G networks. Implementing strong security measures, such as intrusion detection and access control, is crucial for protecting networks from cyber threats and unauthorized breaches. With a keen eye for detail and a focus on problem-solving, resilient networks are able to proactively address potential risks and safeguard data from unauthorized access or breaches. This ensures that data remains secure and accessible at all times.

Resilience involves the ability of 6G networks to adapt and respond to evolving conditions and demands. Efficient allocation of resources, automatic problem-solving, and adaptive routing guarantee top-notch performance and quick response times, even when dealing with varying network loads or environmental shifts. With its adaptability, 6G networks are able to enhance reliability and effectively meet the evolving needs of users and applications (Tataria, H. et al., 2021).

AI AND COMMUNICATION RESILIENCE

AI Technologies for Networks

AI technologies provide a wide range of tools and techniques that can greatly enhance communication networks, improving their efficiency, security, and overall performance. Machine learning (ML) is a crucial technology that stands out among the rest. ML algorithms enable systems to learn from data and make predictions or decisions without the need for explicit programming. ML is widely used in communication networks for tasks like network optimization, predictive maintenance, and anomaly detection. Through the analysis of network traffic patterns, machine learning algorithms have the ability to detect potential security threats and enhance network performance by predicting and adjusting to fluctuations in traffic loads. Deep learning (Pandey, B. K. et al., 2021), a subset of machine learning, enhances the capabilities of artificial intelligence in communication networks. DL utilizes artificial neural networks (Pandey, D., & Pandey, B. K., 2022) with multiple layers to extract complex features from data. DL techniques such as convolutional neural networks (CNNs) (Anand, R. et al., 2023) and recurrent neural networks (RNNs) are widely used in various applications within communication networks. These applications include image recognition for video surveillance, natural language processing for chatbots, and speech recognition in voice communication systems. Natural Language Processing (NLP) is a crucial AI technology that enables smooth communication between users and machines in communication networks. With NLP, computers gain the ability to comprehend and produce human language. This opens up a world of possibilities, such as voice-controlled interfaces, language translation services, and analyzing the sentiment of social media data. Reinforcement Learning (RL) is becoming a highly promising area in the field of AI. It offers great potential for improving decision-making processes in communication networks. RL algorithms empower agents to acquire knowledge by engaging with their surroundings, strategically making choices to optimize (Pandey, D. et al., 2021b) their overall rewards. With the help of RL, communication networks can achieve dynamic resource allocation, adaptive routing, and automated network management, resulting in improved efficiency and resilience (Murroni, M. et al., 2023).

AI for 6G Security Threats

AI-driven solutions are essential for spotting and reducing security risks in 6G networks because they provide proactive defenses against constantly changing cyberthreats. Challenges in 6G Communication Systems are shown in the figure 1.

Figure 1. Challenges in 6G communication systems

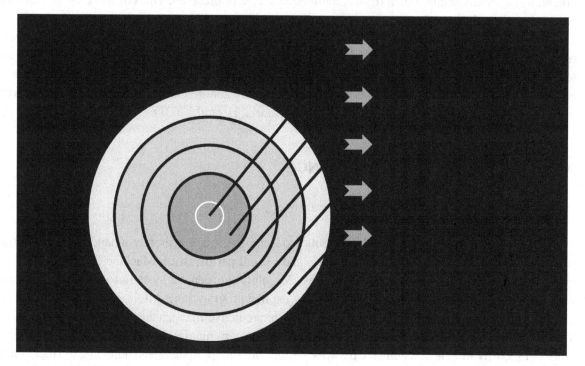

- *Threat Detection and Analysis*: Artificial intelligence-driven solutions for threat detection and analysis in 6G networks make use of complex algorithms to continuously monitor network traffic and identify potential security breaches. In order to differentiate between normal and malicious activities, these algorithms perform real-time analysis of patterns and anomalies, making use of machine learning and deep learning techniques. Artificial intelligence-driven systems are able to detect and categorize suspicious behavior, such as attempts to gain unauthorized access, malware infections, and data exfiltration attempts, by correlating information from a variety of sources, such as network logs, endpoint devices (Du John, H. V. et al., 2022), and threat intelligence feeds. The capabilities of these solutions include advanced data analytics and pattern recognition, which enable them to provide security teams with actionable insights and alerts. This enables security teams to respond promptly to emerging threats and mitigate risks before they become full-blown security incidents.

- *Anomaly Detection:* A key component of AI-driven security solutions for 6G networks is anomaly detection, which enables organizations to identify deviations from normal network behavior that may indicate potential security threats. This is accomplished through the use of artificial intelligence capabilities. Through the examination of historical data and the acquisition of knowledge regarding typical network activity, these solutions establish baseline behavior patterns. Any deviations or anomalies from these baselines that have been established result in the identification of potential security risks, which in turn causes security teams to initiate alerts and conduct investigations. Artificial intelligence algorithms are particularly adept at identifying subtle anomalies that may evade traditional signature-based detection methods. This enables organizations to detect sophisticated threats such as zero-day attacks or insider threats. Anomaly detection solutions that provide early detection of security breaches give organizations the ability to respond proactively and mitigate risks before they cause significant damage to network infrastructure or compromise sensitive data. This is accomplished by providing early detection of security breaches (Singh, P. R. et al., 2023).

- *Predictive Analysis:* The utilization of artificial intelligence (AI) in predictive analysis plays a crucial role in facilitating proactive threat mitigation through the anticipation of potential security risks, drawing upon historical data and prevailing trends. The proposed solutions utilize sophisticated machine learning algorithms to examine extensive quantities of data derived from various sources, such as network logs, endpoint devices, and external threat intelligence feeds. Through the process of pattern recognition and correlation analysis, artificial intelligence algorithms possess the capability to anticipate and forecast potential threats and vulnerabilities prior to their escalation into comprehensive security incidents. Anticipatory analysis solutions offer security teams with practical insights and suggestions, empowering them to proactively implement measures to reduce risks and enhance the overall security stance of 6G networks. Furthermore, through the ongoing acquisition of knowledge from novel data and the ability to adjust to changing threat environments, predictive analysis solutions guarantee the sustained efficacy of security measures in countering emerging cyber threats.

- *Automated Response:* The implementation of AI-driven security solutions allows for the establishment of automated incident response mechanisms, which in turn enables organizations to promptly and effectively address security threats within 6G networks. The aforementioned solutions are designed to seamlessly integrate with network security infrastructure, thereby enabling the automated identification, examination, and resolution of security incidents. Artificial intelligence (AI) algorithms possess the capability to autonomously obstruct malevolent network traffic, isolate compromised devices, or promptly implement security patches on susceptible systems, thereby mitigating the consequences of security breaches and diminishing the duration of response. AI-driven solutions have the capability to automate routine security tasks, such as incident triage and remediation. This allows security teams to allocate their efforts towards more strategic initiatives and intricate threat analysis. Furthermore, the implementation of automated response mechanisms empowers organizations to promptly address security incidents, thereby reducing risks and minimizing the potential harm to network infrastructure and confidential information.

- *Adaptive Security Measures:* Staying ahead of emerging cyber threats is crucial in the ever-evolving realm of 6G networks. That's why adaptive security measures are so important. AI-powered security solutions are highly effective in this field as they constantly learn from fresh data and adapt their threat detection algorithms accordingly. Through careful analysis of security incidents and the integration of up-to-date threat intelligence, these solutions have the ability to adapt security policies and protocols in real-time to effectively combat the ever-evolving landscape of cyber threats. As an illustration, when a novel form of malware is detected in the network, AI algorithms have the ability to swiftly analyze its behavior and generate countermeasures to halt its propagation. This flexibility guarantees that security measures stay efficient in identifying and minimizing various cyber threats, thus protecting 6G networks from potential security breaches and data breaches (Quy, V. K. et al., 2023).

AI-Driven Resilience Enhancements

AI-powered adaptive mechanisms play a crucial role in strengthening communication resilience in 6G networks, providing dynamic responses to the constantly changing threats and challenges. These mechanisms utilize artificial intelligence to constantly monitor network conditions, analyze real-time data, and make necessary adjustments to network configurations to maintain optimal performance and reliability. They possess a remarkable talent for quickly adjusting to ever-changing situations, guaranteeing uninterrupted connectivity even in the face of unexpected challenges and disruptions. Dynamic resource allocation is a crucial aspect of AI-enabled adaptive mechanisms. Through careful analysis of network traffic patterns, device usage, and application demands in real-time, these mechanisms efficiently allocate resources like bandwidth, processing power (Govindaraj, V. et al., 2023), and storage capacity. Ensuring that critical applications receive the necessary resources to operate efficiently is crucial, especially during peak usage periods or instances of network congestion. With its flexibility, 6G networks can easily adapt to different traffic conditions, ensuring that essential services are prioritized and maintaining uninterrupted connectivity for users. Networks can be empowered with AI-driven adaptive mechanisms that enhance communication resilience through self-healing capabilities. By consistently monitoring and analyzing network health metrics, these mechanisms can quickly detect possible problems such as link failures, hardware malfunctions, or malicious activities. They have the ability to reroute traffic, reconfigure network topology, or activate redundant components to restore connectivity and minimize the impact of disruptions. With a network architect's expertise, this proactive approach not only minimizes downtime but also ensures the continuous availability of services, thereby bolstering the network's resilience and user experience. Adaptive mechanisms continuously modify security measures to effectively respond to new threats and vulnerabilities, enhancing the resilience of communication. With a deep understanding of network architecture, these mechanisms are able to constantly analyze network traffic and quickly adapt security policies and protocols to address any potential risks. For example, when a new cyber threat is detected, AI algorithms can automatically update intrusion detection rules, deploy extra security controls, or isolate affected devices to stop the threat from spreading. With the ability to adapt, the network's security posture stays strong and flexible, minimizing the chances of cyber attacks or data breaches (Chataut, R., & Akl, R., 2020).

AI AND COMMUNICATION RESILIENCE

AI Technologies for Networks

AI technologies are transforming communication networks, making them more efficient, secure, and adaptable. These technologies play a crucial role in tackling the ever-growing challenges encountered by modern communication infrastructures. Presented below is a comprehensive overview of AI technologies that can be applied to communication networks:

1. *Machine Learning (ML)*: ML algorithms allow systems to learn from data and make predictions or decisions without the need for explicit programming. ML algorithms are used in communication networks to analyze large datasets and uncover patterns and anomalies. They also help predict network traffic, optimize resource allocation, and improve security by identifying and mitigating threats in real-time.

2. *Deep Learning (DL)*: DL is a branch of Machine Learning that uses artificial neural networks with multiple layers to extract complex features from data. DL techniques, such as convolutional neural networks (CNNs) and recurrent neural networks (RNNs), are widely used in various applications. These include image recognition for video surveillance, natural language processing for chatbots, and speech recognition in voice communication systems within communication networks.

3. *Natural Language Processing (NLP)*: Computers (Bessant, Y. A. et al., 2023) are able to comprehend, interpret, and produce human language thanks to NLP, which allows for smooth communication between users and machines in communication networks. NLP technologies drive voice-controlled interfaces, language translation services, sentiment analysis of social media data, and text-based communication platforms, improving user experience and accessibility.

4. *Reinforcement Learning (RL)*: RL is a field of ML where agents acquire the ability to make decisions through interactions with their environment in order to maximize their overall rewards. Similar to a data scientist, RL algorithms have the potential to enhance communication networks by enabling dynamic resource allocation, adaptive routing, and automated network management. This optimization can greatly improve network performance and efficiency, allowing networks to effectively respond to changing conditions and demands.

5. *Predictive analytics*: Using advanced AI techniques, predictive analytics can accurately forecast future outcomes (Saxena, A. et al., 2021) by analyzing historical data and current trends. Just like a data scientist, predictive analytics can be applied in communication networks to anticipate network congestion, predict equipment failures, and identify potential security threats. This allows for proactive measures to be taken in order to mitigate risks and ensure uninterrupted service delivery.

6. *Autonomous Systems*: Autonomous systems leverage AI technologies to empower communication networks with self-configuring, self-optimizing, and self-healing capabilities. These systems streamline network management tasks, such as configuration management, fault detection, and performance optimization, minimizing operational burden and improving network resilience and reliability.

AI for 6G Security Threats

AI-powered solutions play a crucial role in detecting and addressing security risks in the complex infrastructure of 6G networks. These solutions utilize advanced technology to proactively identify, analyze, and address potential security breaches, ensuring the protection of vital communication infrastructure.

With the use of cutting-edge algorithms and techniques, AI-powered security solutions provide a comprehensive strategy to tackle the ever-changing challenges posed by 6G networks. AI-driven security solutions excel at detecting and analyzing threats, making them an invaluable tool in the field. These solutions are designed to constantly monitor network traffic, analyzing patterns and anomalies in order to quickly detect any potential security breaches. With the power of advanced algorithms, AI systems can differentiate between regular and harmful activities, allowing for the timely identification of security risks like unauthorized access attempts, malware infections, or data breaches. With the mindset of a data scientist, this proactive approach enables security teams to quickly address emerging threats, reducing the impact on network operations and safeguarding the integrity of sensitive data transmitted over 6G networks. When a security incident is detected, AI algorithms can take immediate action to minimize risks and stop the situation from getting worse. As an expert in the field, I can illustrate the effectiveness of AI-driven security solutions during a distributed denial-of-service (DDoS) attack (Serôdio, C. et al., 2023). These solutions have the capability to automatically reroute traffic, block malicious IP addresses, and apply access controls. By doing so, they effectively minimize the impact on network performance. With the ability to automate incident response processes, these solutions empower organizations to swiftly address security threats, saving valuable time and resources while minimizing disruptions to network operations. AI-powered security solutions are essential for improving communication resilience by adjusting security measures to changing threat environments. These solutions are designed to constantly analyze network traffic, detect any unusual behavior, and promptly update security policies and protocols to effectively minimize risks. With the ability to adapt security measures in real-time to counter new threats and vulnerabilities, AI-driven solutions effectively maintain a strong and flexible security posture for the network, minimizing the risk of cyber attacks or data breaches. AI-driven solutions are crucial for detecting and addressing security threats in 6G networks. They offer advanced threat detection, automated response, and adaptive security measures to protect critical communication infrastructure and maintain uninterrupted connectivity for users (Burbank, J. L. et al., 2013).

AI-Driven Resilience Enhancements

AI plays a crucial role in enhancing communication resilience in the intricate realm of 6G networks. These mechanisms utilize advanced AI technology to adapt and address ever-changing challenges, guaranteeing the uninterrupted functioning of vital communication infrastructure. Adaptive mechanisms have the remarkable capability to allocate resources in real-time, adjusting to the ever-changing network conditions. Through constant monitoring of network traffic, user demand, and environmental factors, these mechanisms efficiently allocate bandwidth, processing power, and storage capacity in real-time. With dynamic resource allocation, critical applications can always access the resources they need to operate efficiently, even when there's high demand or network congestion. With its ability to adjust to changing traffic patterns and prioritize essential services, AI-powered adaptive mechanisms improve resilience and guarantee uninterrupted connectivity for users. Communication networks are equipped with AI-enabled adaptive mechanisms that give them the ability to detect and respond to network failures or disruptions on their own, providing self-healing capabilities. By constantly monitoring and analyzing network health metrics, these mechanisms have the ability to detect possible problems like link failures, hardware malfunctions, or cyber attacks. They have the ability to reroute traffic, reconfigure network topology, or activate redundant components to restore connectivity and minimize the impact of disruptions. With an expert understanding of artificial intelligence, this proactive approach effectively reduces downtime and guarantees uninterrupted service availability, ultimately strengthening the resilience of communication networks. AI are essential for adjusting security measures to address new threats and vulnerabilities. Through constant analysis of network traffic, detection of unusual patterns, and immediate updates to security measures, these mechanisms maintain a strong and adaptable security system for the network. For instance, when a new cyber threat is identified, AI algorithms can swiftly update intrusion detection rules, implement extra security measures, or isolate affected devices to halt any further spread. By being highly adaptable, the chances of cyber attacks or data breaches on 6G networks are significantly reduced, thereby improving their overall security and resilience (Garg, V. K. et al., 1999).

Future AI-Driven Developments

Prospects for future advancements and avenues of research in the field of 6G networks offer potential for augmenting communication resilience and enabling novel functionalities. Quantum communication is an area of investigation that aims to create highly secure communication systems. Quantum key distribution (QKD) and quantum-resistant encryption algorithms provide unparalleled levels of security, guaranteeing the confidentiality and integrity of data transmission. Subsequent investigations will prioritize the pragmatic application of quantum communication protocols, effectively incorporating them into 6G networks to enhance resilience against cyber attacks and safeguard data privacy in an ever more linked global landscape. The use of artificial intelligence algorithms at the network edge enables the processing and analysis of data in real-time, resulting in reduced latency and enhanced responsiveness. Future research will focus on exploring innovative edge computing architectures, algorithms, and applications that are specifically designed to meet the distinct requirements of 6G networks. The implementation of these advancements will enhance the capabilities of many applications such as autonomous vehicles, augmented reality, and Internet of Things (IoT) devices, resulting in improved efficiency and dependability. Consequently, this will contribute to the overall resilience of communication networks. 6G networks, future research will place significant emphasis on the imperative task of guaranteeing the security and resilience of IoT infrastructures. The rapid proliferation of Internet of Things (IoT) devices necessitates the development of secure-by-design architectures that integrate strong authentication, encryption, and access control techniques (Pandey, B. K. et al., 2023b). The research will focus on creating robust IoT (Pandey, J. K. et al., 2022) architectures, investigating fault-tolerant designs and adaptive communication protocols to guarantee the dependable functioning of IoT devices in fast-changing and demanding situations. These technological developments will enhance the basic structure of 6G networks, facilitating smooth incorporation of IoT technologies (Iyyanar, P. et al., 2023) while protecting against possible cyber threats and interruptions. The implementation of dynamic spectrum access (DSA) technology shows potential in enhancing spectrum utilization and reducing interference inside 6G networks. The next study will prioritize the advancement of intelligent spectrum management methodologies that effectively distribute spectrum resources in response to real-time demand and environmental factors. The utilization of AI-driven spectrum sensing and allocation algorithms in 6G networks enables the adaptive optimization of spectrum consumption, the maximization of throughput, and the minimization of interference. This, in turn, enhances the resilience and efficiency of communication. These developments will facilitate the development of a communication infrastructure that is more flexible and responsive, capable of well-serving the varied requirements of future applications and services (Dahlman, E. et al., 2016).

AI INTEGRATION IN 6G RESILIENCE

Comparing Traditional and AI Security

An in-depth examination of traditional security measures versus AI-driven approaches in 6G networks offers valuable insights into their respective strengths, limitations, and effectiveness in tackling the ever-changing threat landscape. Conventional security measures in 6G networks usually depend on rule-based approaches, like firewalls, intrusion detection systems (IDS), and encryption protocols, to safeguard against recognized threats and vulnerabilities. These measures have some effectiveness, but they may face challenges in keeping up with rapidly changing cyber threats and advanced attack techniques. In addition, conventional security measures often necessitate manual configuration and management, which hampers their ability to adapt and respond swiftly to evolving threats. Although these measures offer a basic level of security, they might not be sufficient to effectively safeguard 6G networks from sophisticated cyber attacks and unknown vulnerabilities. Utilizing cutting-edge machine learning and artificial intelligence techniques, 6G networks employ AI-driven approaches to bolster security and fortify resilience. These strategies allow for proactive identification of threats, detection of anomalies, and automatic response mechanisms, giving 6G networks the ability to quickly adjust to emerging threats in real-time. AI-powered security solutions have the ability to analyze large volumes of data, recognize patterns, and uncover subtle signs of malicious behavior that might go unnoticed by conventional security measures. With a constant stream of new data and feedback, AI algorithms can enhance their detection capabilities and proactively stay ahead of emerging cyber threats. In addition, AI-driven approaches have the potential to revolutionize security operations by automating tasks and allowing for quicker response times to security incidents, relieving human operators of some of their workload. When examining traditional security measures and AI-driven approaches in 6G networks, a number of notable distinctions become apparent. Conventional security measures offer a solid level of security, but they may face challenges in keeping up with the ever-changing threat landscape and the growing intricacy of 6G networks. On the other hand, AI-driven approaches provide a dynamic and forward-thinking approach to security, utilizing machine learning and AI techniques to identify and address threats in real-time. AI-driven approaches can greatly improve security measures by boosting the security posture, reducing response times, and effectively mitigating risks. AI-driven approaches present certain challenges, including the requirement for extensive datasets for training, the possibility of algorithmic bias, and the vulnerability to adversarial attacks. In addition, incorporating AI-driven security solutions into current infrastructure may necessitate substantial investments in technology, training, and resources. The advantages of utilizing AI-driven methods to improve security resilience and effectiveness are making them more appealing for safeguarding 6G networks against ever-changing cyber threats.

Challenges of AI in 6G Networks

Figure 2. AI challenges in 6G networks

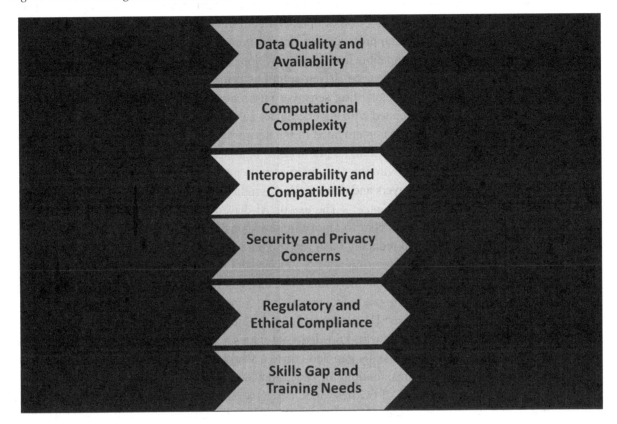

- *Data Quality and Availability*: Acquiring varied and high-quality datasets is essential to training AI models in 6G networks and guaranteeing their efficacy. Privacy considerations provide challenges because proprietary constraints or privacy legislation may apply to sensitive network data. The availability and diversity of training data are hampered by organizational data silos and restricted Network operators, authorities, and industry participants must work together to create data-sharing agreements and create protocols for anonymizing and aggregating data while protecting privacy in order to address these issues.
- *Computational Complexity*: The training and inference processes of AI systems, particularly deep learning models (Anand, R. et al., 2024), require a large amount of processing power. Energy efficiency, latency, and scalability issues arise when deploying these models in 6G networks on devices with limited resources or in edge computing environments. Model compression, quantization, and distributed computing are a few strategies that can reduce computational complexity and maximize resource use. To optimize performance within resource constraints, enterprises must invest in hardware accelerators like GPUs or TPUs and create effective algorithms suited to the limitations of 6G networks.

- *Interoperability and Compatibility*: Ensuring interoperability and compatibility is necessary when integrating AI-driven solutions into legacy systems and network infrastructure. Disparate systems and vendors present challenges due to variations in data formats, communication protocols, and network topologies. By offering similar interfaces and protocols for seamless integration, standardization initiatives and the adoption of open-source frameworks can promote interoperability. Thorough testing and validation protocols are necessary to confirm that AI-driven solutions and current infrastructure are compatible and to spot any dependencies or conflicts that might arise.
- *Security and Privacy Concerns*: - AI-driven solutions bring with them additional privacy and security concerns, such as safeguarding sensitive information and averting malevolent assaults. Significant dangers are associated with adversarial assaults, data breaches, and privacy violations for AI-driven systems that are implemented in 6G networks. Encryption methods, access limits, and strong security measures must all be put in place to protect against cyberattacks and preserve data privacy. In AI-driven environments, organizations need to take a defense-in-depth stance, combining several security layers and continually upgrading defenses to counter evolving threats.
- *Regulatory and Ethical Compliance*: - The use of AI-driven solutions in 6G networks presents ethical and legal issues with regard to algorithmic responsibility, transparency, and data governance. Maintaining trust and reducing legal risks requires strict adherence to industry standards, ethical principles, and data protection laws. To ensure the responsible deployment and usage of AI-driven solutions, ethical factors including bias, fairness, and accountability in AI algorithms require thorough study and mitigation techniques.
- *Skills Gap and Training Needs*: Developing and overseeing AI-driven solutions in 6G networks calls for qualified professionals with knowledge of data science, network engineering, machine learning, and AI. To close the skills gap and create a workforce qualified to design, implement, and oversee AI-driven systems, funding for education and training initiatives is required. In order to satisfy the changing needs of 6G networks, cooperation with academic institutions, business partners, and professional associations can support knowledge transfer and skill development.

The potential for enhancing communication resilience within 6G networks is immense, to the synergies between AI and other emerging technologies. Through the integration of AI with other cutting-edge technologies like blockchain, edge computing, and quantum computing, organizations can develop resilient and flexible communication infrastructures that can effectively handle various challenges which are shown in the figure 2.

Synergies for Resilient Communication

Blockchain technology (David, S. et al., 2023) provides a decentralized and immutable ledger that improves security, transparency, and trust in communication networks. When combined with AI, blockchain offers a reliable and unalterable method for storing essential network data, including authentication credentials, transaction records, and security policies. AI algorithms have the ability to utilize blockchain-based smart contracts to automate and validate security procedures, facilitating secure and dependable communication between network entities. It has the potential to strengthen the durability of AI models. It offers a decentralized and trustworthy platform for training and inference, minimizing the chances of system failures and data tampering. Edge computing allows for the proximity of computation and data storage to the point where data is generated, facilitating immediate processing and analysis of data at the network edge. When AI (Vinodhini, V. et al., 2022) is integrated with edge computing, it can significantly improve communication resilience. This is achieved through faster response times, reduced latency, and decreased dependence on centralized data centers. AI algorithms deployed at the network edge can analyze streaming data in real-time, detect anomalies, and adjust network configurations dynamically to mitigate risks and ensure uninterrupted service delivery. Edge computing allows AI models (Revathi, T. K. et al., 2022) to function in environments with limited resources, like IoT (Pramanik, S. et al., 2023) devices and autonomous vehicles. This improves the scalability and effectiveness of AI-powered solutions in 6G networks. Quantum computing brings forth an incredible level of computational power and capabilities that have the potential to completely transform the resilience of communication in 6G networks. AI algorithms can utilize quantum computing techniques to tackle intricate optimization problems, analyze massive datasets, and simulate complex network scenarios with exceptional speed and precision. Utilizing quantum-enhanced AI algorithms can significantly improve the speed and accuracy of threat detection, anomaly detection, and predictive analytics. This, in turn, boosts the overall security and resilience of 6G networks. In addition, quantum cryptography techniques offer highly secure communication channels, ensuring that sensitive data remains protected from any potential eavesdropping or interception (Shrivastava, U., & Verma, J. K., 2021). This significantly strengthens the overall resilience of communication in 6G networks. This integration allows for a seamless interaction between the physical and digital worlds. AI algorithms have the capability to analyze sensor data from CPS components, accurately predict system failures or anomalies, and autonomously adjust network resources to ensure optimal performance and reliability. By harnessing the power of CPS and AI, organizations can create robust communication infrastructures that can easily adapt to changing and uncertain conditions.

Implications of AI in 6G Resilience

The integration of artificial intelligence (AI) into 6G networks has far-reaching ramifications that go beyond security considerations, significantly influencing multiple aspects of network performance and operational effectiveness. One notable consequence pertains to the domain of improved performance optimization. Through the employment of AI-driven techniques, networks have the ability to dynamically optimize their distribution of resources, utilization of bandwidth, and strategies for routing. The implementation of this proactive management method not only enhances the overall performance of the network, but also guarantees expedited data transmission, decreased latency, and improved availability for users. The enhancement of the quality of service (QoS) results in a smoother and more gratifying user experience in 6G networks. The incorporation of AI enhances the ability to adapt network management, which is a crucial element of communication resilience. Artificial intelligence systems consistently observe network health indicators, allowing them to promptly identify and forecast abnormalities or possible malfunctions. By adopting this proactive strategy, issues may be promptly addressed and maintenance can be carried out, resulting in less downtime and disruptions. Adaptive network management improves the dependability and durability of 6G networks by guaranteeing uninterrupted service availability. This strengthens their capacity to withstand unexpected difficulties and environmental changes. The utilization of AI-driven methodologies enhances the distribution of resources throughout the network infrastructure, resulting in enhanced operational efficiency and cost-effectiveness. AI algorithms have the capability to assign resources, including as bandwidth, processing power, and storage capacity, in a dynamic manner by employing advanced analysis of consumption patterns and demand forecasts. The process of optimization aims to reduce resource wastage, maximize utilization, and improve scalability in order to successfully meet the changing demands of users and applications in 6G networks (Guevara, L., & Auat Cheein, F., 2020). Consequently, firms might attain enhanced adaptability in network operations while concurrently diminishing operating expenses and enhancing cost-effectiveness. The implementation of artificial intelligence (AI) enables the utilization of predictive analytics in the context of capacity planning, hence empowering enterprises to proactively anticipate forthcoming network demands and obstacles. Through the examination of past data and performance metrics, artificial intelligence systems possess the capability to anticipate network traffic patterns, detect potential bottlenecks, and make more precise predictions regarding capacity requirements. This ability to anticipate future events allows firms to take proactive measures such as investing in infrastructure upgrades, increasing capacity, or allocating extra resources to meet expected demand. Organizations may achieve scalability, dependability, and resilience in 6G networks by aligning network capacity with changing requirements.

CONCLUSION

The incorporation of AI into 6G communication networks presents vast possibilities for improving communication resilience and tackling emerging obstacles. Traditional security measures in communication networks lay the groundwork, but the emergence of 6G communication brings forth fresh complexities and requirements. To truly grasp the concept of 6G communication, it's crucial to delve into its definition, characteristics, evolutionary trends, and the unique challenges it presents. One key aspect that cannot be overlooked is the criticality of resilience in ensuring uninterrupted connectivity and safeguarding data integrity. The role of AI in communication resilience is crucial, as AI technologies provide cutting-edge solutions for detecting and reducing security risks and implementing flexible mechanisms. Beyond conventional security measures, incorporating AI into 6G communication resilience requires a thorough examination of the subject, recognizing the connections with emerging technologies, and taking into account wider implications that go beyond just security concerns. Deploying AI-driven solutions in 6G networks comes with its fair share of challenges. These include ethical considerations, technical hurdles, regulatory compliance, and the need to address the skills gap. Addressing these challenges necessitates a collective approach, adherence to ethical principles, and a commitment to education and training. Through addressing these obstacles and adopting ethical guidelines, we can utilize the immense potential of AI to construct robust, protected, and all-encompassing communication systems for the future.

REFERENCES

Anand, R., Khan, B., Nassa, V. K., Pandey, D., Dhabliya, D., Pandey, B. K., & Dadheech, P. (2023). Hybrid convolutional neural network (CNN) for kennedy space center hyperspectral image. *Aerospace Systems*, 6(1), 71–78. 10.1007/s42401-022-00168-4

Anand, R., Lakshmi, S. V., Pandey, D., & Pandey, B. K. (2024). An enhanced ResNet-50 deep learning model for arrhythmia detection using electrocardiogram biomedical indicators. *Evolving Systems*, 15(1), 83–97. 10.1007/s12530-023-09559-0

Bangerter, B., Talwar, S., Arefi, R., & Stewart, K. (2014). Networks and devices for the 5G era. *IEEE Communications Magazine*, 52(2), 90–96. 10.1109/MCOM.2014.6736748

Bessant, Y. A., Jency, J. G., Sagayam, K. M., Jone, A. A. A., Pandey, D., & Pandey, B. K. (2023). Improved parallel matrix multiplication using Strassen and Urdhvatiryagbhyam method. *CCF Transactions on High Performance Computing*, 5(2), 102–115. 10.1007/s42514-023-00149-9

. Bruntha, P. M., Dhanasekar, S., Hepsiba, D., Sagayam, K. M., Neebha, T. M., Pandey, D., & Pandey, B. K. (2023). Application of switching median filter with L 2 norm-based auto-tuning function for removing random valued impulse noise. *Aerospace systems, 6*(1), 53-59.

Burbank, J. L., Andrusenko, J., Everett, J. S., & Kasch, W. T. (2013). Second-generation (2G) cellular communications. In *Wireless Networking: Understanding Internetworking Challenges* (pp. 250–365). IEEE. 10.1002/9781118590775.ch6

Chataut, R., & Akl, R. (2020). Massive MIMO systems for 5G and beyond networks—Overview, recent trends, challenges, and future research direction. *Sensors (Basel)*, 20(10), 2753. 10.3390/s2010275332408531

Dahlman, E., Parkvall, S., & Skold, J. (2016). *4G, LTE-advanced Pro and the Road to 5G*. Academic Press.

David, S., Duraipandian, K., Chandrasekaran, D., Pandey, D., Sindhwani, N., & Pandey, B. K. (2023). Impact of blockchain in healthcare system. In *Unleashing the Potentials of blockchain technology for healthcare industries* (pp. 37–57). Academic Press. 10.1016/B978-0-323-99481-1.00004-3

Deepa, R., Anand, R., Pandey, D., Pandey, B. K., & Karki, B. (2022). Comprehensive performance analysis of classifiers in diagnosis of epilepsy. *Mathematical Problems in Engineering*, 2022, 2022. 10.1155/2022/1559312

Devasenapathy, D., Madhumathy, P., Umamaheshwari, R., Pandey, B. K., & Pandey, D. (2024). Transmission-efficient grid-based synchronized model for routing in wireless sensor networks using Bayesian compressive sensing. *SN Computer Science*, 5(1), 1–11.

Du John, H. V., Jose, T., Jone, A. A. A., Sagayam, K. M., Pandey, B. K., & Pandey, D. (2022). Polarization insensitive circular ring resonator based perfect metamaterial absorber design and simulation on a silicon substrate. *Silicon*, 14(14), 9009–9020. 10.1007/s12633-021-01645-9

Garg, V. K., Halpern, S., & Smolik, K. F. (1999, February). Third generation (3G) mobile communications systems. In *1999 IEEE International Conference on Personal Wireless Communications (Cat. No. 99TH8366)* (pp. 39-43). IEEE.

Gohar, A., & Nencioni, G. (2021). The role of 5G technologies in a smart city: The case for intelligent transportation system. *Sustainability (Basel)*, 13(9), 5188. 10.3390/su13095188

. Govindaraj, V., Dhanasekar, S., Martinsagayam, K., Pandey, D., Pandey, B. K., & Nassa, V. K. (2023). Low-power test pattern generator using modified LFSR. *Aerospace Systems*, 1-8.

Guevara, L., & Auat Cheein, F. (2020). The role of 5G technologies: Challenges in smart cities and intelligent transportation systems. *Sustainability (Basel)*, 12(16), 6469. 10.3390/su12166469

Huseien, G. F., & Shah, K. W. (2022). A review on 5G technology for smart energy management and smart buildings in Singapore. *Energy and AI*, 7, 100116. 10.1016/j.egyai.2021.100116

Iyyanar, P., Anand, R., Shanthi, T., Nassa, V. K., Pandey, B. K., George, A. S., & Pandey, D. (2023). A real-time smart sewage cleaning UAV assistance system using IoT. In *Handbook of Research on Data-Driven Mathematical Modeling in Smart Cities* (pp. 24–39). IGI Global.

Khan, B., Hasan, A., Pandey, D., Ventayen, R. J. M., Pandey, B. K., & Gowwrii, G. (2021). Fusion of datamining and artificial intelligence in prediction of hazardous road accidents. In *Machine learning and iot for intelligent systems and smart applications* (pp. 201–223). CRC Press. 10.1201/9781003194415-12

Kirubasri, G., Sankar, S., Pandey, D., Pandey, B. K., Nassa, V. K., & Dadheech, P. (2022). Software-defined networking-based Ad hoc networks routing protocols. In *Software defined networking for Ad Hoc networks* (pp. 95–123). Springer International Publishing. 10.1007/978-3-030-91149-2_5

Kirubasri, G., Sankar, S., Pandey, D., Pandey, B. K., Singh, H., & Anand, R. (2021, September). A recent survey on 6G vehicular technology, applications and challenges. In *2021 9th International Conference on Reliability, Infocom Technologies and Optimization (Trends and Future Directions)(ICRITO)* (pp. 1-5). IEEE.

Kumar Pandey, B., Pandey, D., Nassa, V. K., Ahmad, T., Singh, C., George, A. S., & Wakchaure, M. A. (2021). Encryption and steganography-based text extraction in IoT using the EWCTS optimizer. *Imaging Science Journal*, 69(1-4), 38–56. 10.1080/13682199.2022.2146885

Murroni, M., Anedda, M., Fadda, M., Ruiu, P., Popescu, V., Zaharia, C., & Giusto, D. (2023). 6G—Enabling the New Smart City: A Survey. *Sensors (Basel)*, 23(17), 7528. 10.3390/s2317752837687986

Pandey, B. K., Mane, D., Nassa, V. K. K., Pandey, D., Dutta, S., Ventayen, R. J. M., & Rastogi, R. (2021). Secure text extraction from complex degraded images by applying steganography and deep learning. In *Multidisciplinary approach to modern digital steganography* (pp. 146–163). IGI Global. 10.4018/978-1-7998-7160-6.ch007

Pandey, B. K., & Pandey, D. (2023). Parametric optimization and prediction of enhanced thermoelectric performance in co-doped CaMnO3 using response surface methodology and neural network. *Journal of Materials Science Materials in Electronics*, 34(21), 1589. 10.1007/s10854-023-10954-1

Pandey, B. K., Pandey, D., & Agarwal, A. (2022). Encrypted information transmission by enhanced steganography and image transformation. [IJDAI]. *International Journal of Distributed Artificial Intelligence*, 14(1), 1–14. 10.4018/IJDAI.297110

Pandey, B. K., Pandey, D., Alkhafaji, M. A., Güneşer, M. T., & Şeker, C. (2023a). A reliable transmission and extraction of textual information using keyless encryption, steganography, and deep algorithm with cuckoo optimization. In *Micro-Electronics and Telecommunication Engineering: Proceedings of 6th ICMETE 2022* (pp. 629–636). Springer Nature Singapore. 10.1007/978-981-19-9512-5_57

Pandey, B. K., Pandey, D., Gupta, A., Nassa, V. K., Dadheech, P., & George, A. S. (2023b). Secret data transmission using advanced morphological component analysis and steganography. In *Role of data-intensive distributed computing systems in designing data solutions* (pp. 21–44). Springer International Publishing. 10.1007/978-3-031-15542-0_2

Pandey, D., Nassa, V. K., Jhamb, A., Mahto, D., Pandey, B. K., George, A. H., & Bandyopadhyay, S. K. (2021a). An integration of keyless encryption, steganography, and artificial intelligence for the secure transmission of stego images. In *Multidisciplinary approach to modern digital steganography* (pp. 211–234). IGI Global. 10.4018/978-1-7998-7160-6.ch010

Pandey, D., & Pandey, B. K. (2022). An efficient deep neural network with adaptive galactic swarm optimization for complex image text extraction. In *Process mining techniques for pattern recognition* (pp. 121–137). CRC Press. 10.1201/9781003169550-10

Pandey, D., Pandey, B. K., & Wairya, S. (2021b). Hybrid deep neural network with adaptive galactic swarm optimization for text extraction from scene images. *Soft Computing*, 25(2), 1563–1580. 10.1007/s00500-020-05245-4

Pandey, J. K., Jain, R., Dilip, R., Kumbhkar, M., Jaiswal, S., Pandey, B. K., & Pandey, D. (2022). Investigating role of iot in the development of smart application for security enhancement. In *IoT Based Smart Applications* (pp. 219–243). Springer International Publishing.

Pramanik, S., Pandey, D., Joardar, S., Niranjanamurthy, M., Pandey, B. K., & Kaur, J. (2023, October). An overview of IoT privacy and security in smart cities. In *AIP Conference Proceedings* (*Vol. 2495*, No. 1). AIP Publishing. 10.1063/5.0123511

Quy, V. K., Chehri, A., Quy, N. M., Han, N. D., & Ban, N. T. (2023). Innovative trends in the 6G era: A comprehensive survey of architecture, applications, technologies, and challenges. *IEEE Access: Practical Innovations, Open Solutions*, 11, 39824–39844. 10.1109/ACCESS.2023.3269297

Revathi, T. K., Sathiyabhama, B., Sankar, S., Pandey, D., Pandey, B. K., & Dadeech, P. (2022). An intelligent model for coronary heart disease diagnosis. In *Networking Technologies in Smart Healthcare* (pp. 309–327). CRC Press. 10.1201/9781003239888-15

Saxena, A., Agarwal, A., Pandey, B. K., & Pandey, D. (2024). Examination of the Criticality of Customer Segmentation Using Unsupervised Learning Methods. *Circular Economy and Sustainability*, 1–14. 10.1007/s43615-023-00336-4

Saxena, A., Sharma, N. K., Pandey, D., & Pandey, B. K. (2021). Influence of tourists satisfaction on future behavioral intentions with special reference to desert triangle of Rajasthan. *Augmented Human Research*, 6(1), 13. 10.1007/s41133-021-00052-4

Serôdio, C., Cunha, J., Candela, G., Rodriguez, S., Sousa, X. R., & Branco, F. (2023). The 6G Ecosystem as Support for IoE and Private Networks: Vision, Requirements, and Challenges. *Future Internet*, 15(11), 348. 10.3390/fi15110348

Sharma, M., Pandey, D., Khosla, D., Goyal, S., Pandey, B. K., & Gupta, A. K. (2022). Design of a GaN-based Flip Chip Light Emitting Diode (FC-LED) with au Bumps & Thermal Analysis with different sizes and adhesive materials for performance considerations. *Silicon*, 14(12), 7109–7120. 10.1007/s12633-021-01457-x

Sharma, M., Pandey, D., Palta, P., & Pandey, B. K. (2022). Design and power dissipation consideration of PFAL CMOS V/S conventional CMOS based 2: 1 multiplexer and full adder. *Silicon*, 14(8), 4401–4410. 10.1007/s12633-021-01221-1

Shrivastava, U., & Verma, J. K. (2021, December). A Study on 5G Technology and Its Applications in Telecommunications. In *2021 International Conference on Computational Performance Evaluation (ComPE)* (pp. 365-371). IEEE.

Singh, P. R., Singh, V. K., Yadav, R., & Chaurasia, S. N. (2023). 6G networks for artificial intelligence-enabled smart cities applications: A scoping review. *Telematics and Informatics Reports*, 9, 100044. 10.1016/j.teler.2023.100044

Tataria, H., Shafi, M., Molisch, A. F., Dohler, M., Sjöland, H., & Tufvesson, F. (2021). 6G wireless systems: Vision, requirements, challenges, insights, and opportunities. *Proceedings of the IEEE*, 109(7), 1166–1199. 10.1109/JPROC.2021.3061701

Tripathi, R. P., Sharma, M., Gupta, A. K., Pandey, D., Pandey, B. K., Shahul, A., & George, A. H. (2023). Timely prediction of diabetes by means of machine learning practices. *Augmented Human Research*, 8(1), 1. 10.1007/s41133-023-00062-4

Vinodhini, V., Kumar, M. S., Sankar, S., Pandey, D., Pandey, B. K., & Nassa, V. K. (2022). IoT-based early forest fire detection using MLP and AROC method. *International Journal of Global Warming*, 27(1), 55–70. 10.1504/IJGW.2022.122794

Chapter 7
Strategic Insights:
Safeguarding 6G Networks in the Era of Intelligent Connectivity

Uchit Kapoor
Arthkalp, India

Sunita Sunil Shinde
https://orcid.org/0000-0003-0979-2043
Nanasaheb Mahadik College of Engineering and Technology, India

Budesh Kanwer
Department of Artificial Intelligence and Data Science, Poornima Institute of Engineering and Technology, India

Sonia Duggal
Manav Rachna International Institute of Research and Studies, Faridabad, India

Lavish Kansal
Lovely Professional University, India

Joshuva Arockia Dhanraj
https://orcid.org/0000-0001-5048-7775
Chandigarh University, India

ABSTRACT

In the age of intelligent connectivity, it is vital to understand how crucial it is to protect these networks as the world prepares for the arrival of 6G networks. Sixth-generation (6G) networks are anticipated to transform a number of sectors, including manufacturing, transportation, and healthcare, by offering previously unheard-of speeds, capacities, and connectivity. These opportunities do, however, come with a number of difficulties, especially in the areas of cybersecurity and data privacy. Understanding the particular threats that 6G networks face is essential to protecting them effectively.

INTRODUCTION

In the age of intelligent connectivity, it is vital to understand how crucial it is to protect these networks as the world prepares for the arrival of 6G (Kirubasri, G. et al., 2021) networks. Sixth-generation (6G) networks are anticipated to transform a number of sectors, including manufacturing, transportation, and healthcare (Tareke, S. A. et al., 2022), by offering previously unheard-of speeds, capacities, and connectivity. These opportunities do, however, come with a number of difficulties, especially in the areas of cybersecurity and data privacy (Sharma, S. et al., 2023). Understanding the particular threats that 6G

DOI: 10.4018/979-8-3693-2931-3.ch007

networks face is essential to protecting them effectively. The threats to cybersecurity are not only ongoing but also dynamic, as hackers are always coming up with new ways to take advantage of weaknesses. These threats could be anything from sophisticated cyberattacks on network infrastructure (Kirubasri, G. et al., 2022) to privacy violations brought on by the massive data exchange that intelligent connectivity technologies enable in the context of 6G networks. A proactive approach to risk mitigation is proactive risk assessment. This means locating possible security holes in 6G networks and filling them in advance by taking preventative action. Furthermore, the confidentiality and integrity of data transferred over these networks depend on strong encryption standards (Kumar Pandey, B. et al., 2021). Organisations can prevent unauthorised access or manipulation of sensitive information by implementing sophisticated encryption algorithms (Pandey, B. K. et al., 2023a) and protocols. Effective protection of 6G networks also requires collaborative security measures. Modern telecommunications systems are interconnected, so no one organization can handle security issues on its own. Collaboration between governments, cybersecurity experts, and industry stakeholders is greatly enhanced by public-private partnerships (Bangerter, B. et al., 2014). These collaborations allow for coordinated action against new threats and vulnerabilities by exchanging threat intelligence and best practices. Building a strong regulatory framework is necessary to guarantee the security and robustness of 6G networks. International standards and guidelines facilitate interoperability and consistency in network security measures by offering a common foundation for cybersecurity practices across national boundaries. Government regulations and oversight procedures, which clearly define requirements for network operators and hold them responsible for compliance, can further strengthen cybersecurity efforts. Effective industry collaboration is another essential component of 6G network security. Through collaborative efforts and combined knowledge, industry participants can create novel approaches to cybersecurity problems. Public-private collaborations enable the development of cutting-edge security practices and technologies by facilitating information exchange and collaborative research projects. Innovations in technology are also essential to improving the security of 6G networks. Large volumes of network traffic can be analyzed in real-time by AI-powered threat detection systems, which can more accurately and efficiently identify potential security threats and anomalies. Blockchain technology (David, S. et al., 2023) lowers the possibility of tampering or unauthorized access by providing a decentralized method of guaranteeing the integrity and transparency of network transactions. Furthermore, sensitive data is protected from quantum computer decryption by quantum-safe cryptography, which offers strong defense against potential threats posed by quantum computing. 6G network security cannot be achieved solely through technological means. Programs for awareness and training are equally crucial to creating an ecosystem that is resistant to cyberattacks. Network operators and end users can benefit from education programs that increase knowledge of cybersecurity best practices and provide them the tools they need to recognize security threats and take appropriate action.

Safeguarding 6G Networks

The protection of 6G networks in this age of intelligent connectivity is of the utmost importance because of the significant role they play in determining the future of communication (Devasenapathy, D. et al., 2024), technology, and society. A wide variety of applications, including augmented reality, autonomous vehicles, smart cities, and the Internet of Things (IoT) (Pandey, J. K. et al., 2022), will be made possible by 6G networks, which are the successor to 5G (Sengupta, R. et al., 2021) networks and promise to deliver unprecedented levels of speed, reliability, and connectivity. On the other hand, these advancements bring with them significant security challenges that need to be addressed in order to prevent potential vulnerabilities and threats. Protecting 6G networks is important for a number of reasons, one of the most important being the critical infrastructure that these networks support. The backbone of digital economies (Swapna, H. R. et al., 2023) will be these networks, which will provide support for essential services across a variety of industries, such as healthcare, finance, transportation (Sahani, K. et al., 2023), and energy. Any compromise to the integrity or availability of 6G networks could have far-reaching consequences, including the disruption of operations, the compromise of sensitive data, and the impact on public safety. On top of that, the proliferation of intelligent connectivity brings about new attack vectors and risks to cybersecurity. As a result of increased interconnectivity and reliance on artificial intelligence (AI) (Pandey, D. et al., 2021a) and machine learning (ML) algorithms (Tripathi, R. P. et al., 2023), 6G networks are more likely to be vulnerable to sophisticated cyber threats. These threats include malware, ransomware, and distributed denial-of-service (DDoS) attacks. These dangers have the potential to exploit vulnerabilities in network infrastructure, software, and Internet of Things devices (Sharma, M. et al., 2022), which could result in widespread disruptions and economic losses (Saxena, A. et al., 2021). It is essential to protect 6G networks in order to protect the privacy of users and the security of their data. It is becoming increasingly difficult to maintain the confidentiality, integrity, and authenticity of data as the number of connected devices (Sharma, M. et al., 2022) continues to increase and the massive amount of data that is generated and transmitted across these networks continues to grow. It is possible that unauthorized access to sensitive information or breaches of privacy could undermine the adoption of emerging technologies and erode trust in the digital ecosystem. In terms of both national security and geopolitical considerations, it is of the utmost importance. The more closely communication networks are intertwined with critical infrastructure and strategic assets, the more appealing they become as targets for cyber espionage, sabotage, or warfare that is sponsored by a state. In order to protect against these dangers and to preserve one's sovereignty in the digital sphere (George, W. K. et al., 2023), it is essential to make certain that 6G networks are resilient and robust. As a means of addressing these challenges, stakeholders are required to adopt a comprehensive and proactive strategy for the protection of 6G networks. This strategy entails the implementation of stringent cybersecurity measures, the utilization of advanced encryption techniques, and the promotion of collaboration between the government, the private sector, and academic institutions. In addition to this, it necessitates investments in research and development in order to develop innovative security solutions that are able to address evolving vulnerabilities and threats. It is absolutely necessary to establish comprehensive regulatory frameworks and international standards in order to guarantee consistency and interoperability across all 6G networks. In the process of establishing cybersecurity policies, enforcing regulations, and holding stakeholders accountable for adhering to security best practices, governments play a crucial role (Sinclair, M. et al., 2023).

Intelligent Connectivity

With the potential to completely transform a variety of sectors, communities, and personal encounters, intelligent connectivity is at the forefront of technological advancement. The convergence of cutting-edge technologies, mainly 5G, artificial intelligence (AI) (Anand, R. et al., 2023), and the Internet of Things (IoT) (Iyyanar, P. et al., 2023), creates this paradigm shift that fundamentally alters how we see and engage with the world. Intelligent connectivity is a shift from static, traditional connectivity models to dynamic, context-aware networks. Intelligent connectivity enables highly personalized and contextualized experiences by leveraging network slicing techniques and integrating AI capabilities at the network edge.

This leads to innovation across multiple domains. The influence of intelligent connectivity on immersive technologies, like virtual reality (VR) and augmented reality (AR), is one of its most obvious applications. These innovations have the potential to completely transform live experiences and entertainment by providing immersive, interactive platforms for events like concerts and sporting events. Users can anticipate smooth, high-fidelity experiences with intelligent connectivity that obfuscate the boundaries between the real and virtual worlds. Through developments like drone delivery services, intelligent connectivity promises to completely transform logistics and transportation. Logistics companies can optimize delivery processes, cut expenses, and improve customer satisfaction by harnessing the potential of 5G networks and AI-powered optimization algorithms (Pandey, D., & Pandey, B. K., 2022). The future of delivery of goods and services will be quicker, safer, and more effective than ever thanks to this convergence of technologies. Intelligent connectivity gives users access to virtual assistants that can handle daily schedules, instantly retrieve information, and automate repetitive tasks, all within the domain of personal assistance. With the help of the enormous amounts of data (Pandey, B. K. et al., 2024) produced by connected devices, these AI-powered assistants can provide individualized recommendations and insights that improve daily convenience and productivity. Intelligent connectivity is changing not just individual experiences but entire economies and industries. Intelligent connectivity is a ubiquitous and transformative force that permeates every aspect of society, with an estimated 9 billion connections made possible by mobile networks worldwide. As businesses use connected data to boost productivity, spur innovation, and open up new revenue opportunities, entire sectors are changing dramatically. A blended environment, where people and things are seamlessly interconnected, arises from the blurring of boundaries between the physical and digital worlds brought about by an increase in digital density. This convergence accelerates the digital transformation of businesses and societies by creating new avenues for innovation, collaboration, and value creation (Ekong, M. O. et al., 2023). Professionals with technical know-how and experience in AI are well-positioned to prosper in this changing environment. People with the ability to fully utilize intelligent connectivity will be the ones driving innovation and influencing the direction of technology as it becomes more and more commonplace.

UNDERSTANDING THREATS

Cybersecurity Risks in the Era of Digital Transformation

Cybersecurity risk has become a major concern for businesses all over the world as they navigate the digital transformation landscape. In addition to spurring corporate innovation, the integration of advanced technologies presents challenging cybersecurity issues. These difficulties are made worse by companies' increasing reliance on third- and fourth-party vendors, which expands the scope of cybersecurity risk outside of their direct control. Comprehending and handling cybersecurity risks have emerged as essential components of a strong cybersecurity risk management plan. Vulnerabilities are increased by adding outside vendors, particularly fourth-party vendors who are one step removed from the supply chain (Malhotra, P. et al., 2021). These organizations may not always have direct control over these entities, but their security procedures have a big influence on the risk profile of the organization. Cybersecurity risk is the possibility of being exposed to or suffering a loss as a result of data breaches or cyberattacks inside an organization. It entails determining possible dangers and weak points in networks and digital systems. This risk goes beyond the possibility of a cyberattack to include possible repercussions like monetary loss, harm to one's reputation, or interruption of business operations. Cyber threats come in many forms and present serious risks to businesses in a variety of industries. For example, ransomware attacks encrypt important data and demand a ransom to unlock it, thereby presenting serious risks to operations and finances. Malware is another common threat that quietly infiltrates systems to either steal or corrupt data, jeopardizing the confidentiality and integrity of sensitive information. Robust access controls and monitoring mechanisms are crucial because insider threats, in which employees misuse their access rights, also pose significant risks to organizations. Furthermore, since attackers use social engineering techniques to fool staff members into disclosing private information, phishing attacks continue to pose a threat. Inadequate compliance management intensifies cybersecurity threats, exposing companies to penalties and legal repercussions. In order to reduce these risks and promote a culture of security within organizations, it is imperative that compliance with industry standards and regulations is maintained. Organizations must prioritize cybersecurity and update their cybersecurity risk management strategies on a regular basis to counter these evolving threats. Implementing strict compliance protocols, deploying strong security systems to efficiently detect and mitigate cyber threats, and providing frequent employee training on threat recognition and response are examples of proactive measures. It is imperative for organizations to implement a multifaceted cybersecurity strategy that includes preventive, investigative, and remedial measures. To prevent unauthorized access and identify suspicious activity in real-time, this involves putting in place firewalls, intrusion detection systems, and endpoint security solutions. To find and fix possible gaps in the organization's cybersecurity posture, regular security audits and vulnerability assessments are essential. In order to mitigate risks related to people, like phishing attacks and insider threats, it is critical to cultivate a culture of cybersecurity awareness among employees. By encouraging a shared responsibility for cybersecurity, companies can increase their overall resistance to cyber threats by giving employees the authority to identify and report suspicious activity.

Potential Vulnerabilities in 6G Networks

The emergence of 6G networks, which are expected to offer unparalleled speed, dependability, and connectivity, presents a challenge that must be identified and resolved. The potential for 6G technology to transform communication and enable game-changing applications is enormous, but it also brings with it new security risks that need to be carefully studied and addressed. Security on the edge of computing is one major area of worry. In order to support high-bandwidth and low-latency applications, 6G networks are anticipated to take advantage of edge computing resources. However, new attack surfaces and potential vulnerabilities are introduced by edge computing's distributed nature. Data confidentiality and integrity are at risk because threat actors may target edge nodes or take advantage of communication channels (Gupta, A. K. et al., 2023) between edge devices. One of the primary characteristics of 6G networks is network slicing, which enables operators to divide up network resources to accommodate different applications with different demands. Concerns concerning security isolation between network slices are raised by network slicing, despite its flexibility and efficiency. Implementations of network slicing may be vulnerable to attacks that compromise the availability and integrity of services through unauthorized access or resource exploitation. The increasing number of Internet of Things (IoT) (Vinodhini, V. et al., 2022) devices in 6G networks presents another vulnerability scenario. Malicious actors can exploit these devices because they frequently lack strong security features. Network infrastructure can be compromised or extensive botnet attacks can be initiated by taking advantage of flaws in IoT devices (Menon, V. et al., 2022), such as unpatched firmware or inadequate authentication methods (Pramanik,

S. et al., 2023). When it comes to 6G networks, privacy issues are also very important. Sensitive and private information shared over these networks may be intercepted or accessed by unauthorized parties due to the exponential increase in data transmission (Gohar, A., & Nencioni, G., 2021). Because of the integration of machine learning (ML) (Pandey, B. K., & Pandey, D., 2023) and artificial intelligence (AI) in 6G networks, there are risks to user privacy and regulatory compliance, as well as concerns regarding algorithmic bias (Bessant, Y. A. et al., 2023) and data privacy. A significant additional vulnerability for 6G networks is supply chain risks. Software and hardware component vulnerabilities are introduced by the intricate supply chain ecosystem that supports network infrastructure. In order to undermine the integrity and security of 6G networks, malicious actors may take advantage of supply chain vulnerabilities such as fake parts or supply chain interruptions. The security and robustness of 6G networks are seriously threatened by cyber-physical assaults. Cyber-physical attacks that aim to compromise interconnected systems can have a domino effect on public safety and infrastructure resilience because these networks allow connectivity to extend beyond conventional communication devices (Raja, D. et al., 2024) and encompass critical infrastructure and cyber-physical systems. To tackle possible weaknesses, cybersecurity must be approached holistically and from multiple angles, including risk analysis, threat reduction, and security-by-design concepts. To guarantee 6G networks' resilience and security in the face of changing cyber threats, cooperation between industry players, government agencies, and cybersecurity specialists is essential. The complete potential of 6G technology can be realized by stakeholders while preserving the availability, confidentiality, and integrity of communication networks by proactively addressing potential vulnerabilities.

Implications of Intelligent Connectivity

Intelligent connectivity, which is being driven by the convergence of 5G, artificial intelligence (AI) (Revathi, T. K. et al., 2022), and the Internet of Things (IoT), has the potential to have significant repercussions for a variety of aspects of society, technology, and the economy. In addition to bringing about a revolution in communication, this paradigm shift also brings about a fundamental change in the way individuals interact with technology and the environment that surrounds them. Due to the improved user experience that it provides, Intelligent connectivity is able to provide users across a wide range of domains with highly personalized and contextualized experiences by utilizing real-time data and artificial intelligence algorithms. Intelligent connectivity enhances user experiences by anticipating and adapting to individual needs and preferences. This improvement can be seen in a variety of ways, including personalized recommendations and seamless integration of services. The disruption of traditional business models and the acceleration of digital transformation across industries come about as a result of intelligent connectivity. Intelligent connectivity enables the development of innovative applications and services that improve operational efficiency, optimize resource allocation, and drive new revenue streams. These applications and services can be found in a variety of industries, including healthcare (Pandey, D. et al., 2023), manufacturing, transportation, and retail. Intelligent connectivity enables the development of smart cities, which are characterized by the presence of interconnected devices and sensors that facilitate the efficient management of resources, further improve public safety, and enhance the quality of life for all residents. The development of urban environments that are both sustainable and resilient can be facilitated by intelligent connectivity, which includes everything from intelligent energy (Govindaraj, V. et al., 2023) grids to intelligent traffic management systems. Intelligent connectivity is a game-changer in the healthcare industry, bringing about revolutionary changes in patient care and healthcare delivery (Singh, S. et al., 2023). Intelligent connectivity helps improve healthcare outcomes while simultaneously lowering costs. This is accomplished through the facilitation of remote patient monitoring, personalized treatment plans, and predictive analytics. Smart connectivity is reshaping the landscape of healthcare (Pandey, B. K. et al., 2011) in a variety of ways, from diagnostic tools powered by artificial intelligence to wearable devices. The development and deployment of autonomous systems, such as robots, drones, and autonomous vehicles, are sped up as a result of this. Through the utilization of real-time data and artificial intelligence algorithms, autonomous systems are able to navigate complex environments, make decisions based on accurate information, and carry out tasks with precision and efficiency, thereby transforming the business sectors of transportation, logistics, and manufacturing. The proliferation of devices that are connected to one another and the enormous amount of data that is generated raises significant concerns regarding the privacy and security of individual data. The introduction of intelligent connectivity results in the creation of new attack vectors and vulnerabilities, which could be exploited by malicious actors in order to compromise sensitive information or disrupt essential services. It is necessary to implement stringent cybersecurity measures, regulatory frameworks, and technologies that protect individuals' privacy in order to address these challenges. Although intelligent connectivity presents opportunities that have the potential to transform the world, it also exacerbates the digital divides and inequalities that already exist. In order to participate in the digital economy and gain access to essential services, having access to high-speed internet and devices that are connected to the internet becomes essential. One of the most important things that can be done to promote social inclusion and economic empowerment is to bridge the digital divide and ensure that everyone has equal access to intelligent connectivity (Tataria et al., 2021).

STRATEGIC APPROACHES

Proactive Risk Assessment

The business world is always changing, and companies work in a risky environment with many different areas of danger. Many different types of risks can happen to businesses, ranging from operational and financial risks to cybersecurity threats and problems with following the rules. You can't protect long-term success and sustainability with reactive risk management, which means making decisions about how to respond only after risks have happened. The more forward-thinking way for modern businesses to find, predict, and reduce risks before they become crises is through proactive risk assessment.

By putting more emphasis on prevention over reaction, proactive risk management requires a major shift in how people think. Companies can protect their operations, assets, and ability to handle uncertainty better by focusing on finding and dealing with risks before they happen. Active risk management lets businesses stop problems before they happen and seize new opportunities, instead of just reacting to risks as they appear. Enterprise risk management (ERM), which tries to lessen the effects of uncertainty and the chance of losing money, is at the heart of proactive risk management. Agencies can improve their ability to deal with unexpected problems and get ready to take advantage of strategic opportunities by using the right risk management strategies. Improving and coming up with new ideas all the time is encouraged by good risk management, which leads to a competitive edge and long-term growth. By making everyone in the company more aware of the things that cause risk and helping people make smart decisions, proactive risk management can be very helpful. Businesses can find new risks, figure out how bad they might be, and take steps to lessen their effects by using up-to-date business intelligence and predictive analyses. In a business world that is always changing, this proactive approach not only lowers the chances of risk events happening, but it also makes the organization more flexible and stronger. The evaluation of possible risks through careful study and measurement is an important part of proactive risk management. In order to find relevant risk drivers and figure out the likelihood and possible effects of risk events, risk managers use historical data, current trends, and predictive analytics. Learning (Saxena, A. et al., 2024) about the factors that cause these risks can help businesses come up with specific plans to lower their chances of happening and lessen the damage they could cause. Closed-loop feedback systems are used in proactive risk management to make sure that things are always getting better and adapting. Professionals who work with risks set limits on what levels of risk are acceptable and test different risk scenarios to see how they might affect things. Companies can increase their overall resilience and proactively manage risk by putting in place plans to get rid of or reduce risks to acceptable levels (Murroni, M. et al., 2023).

Robust Encryption Standards

Strong encryption standards are essential for preserving private and sensitive data in an era of unparalleled digital data volumes and cyber threats that are getting more complex. One essential security feature is encryption, which makes sure that information sent and stored over digital networks is safe and out of the hands of unauthorized individuals. Strong encryption standards are necessary to preserve the confidentiality, integrity, and authenticity of data since both individuals and organizations depend on digital communication and storage for vital tasks. The process of converting plaintext data into ciphertext through the use of cryptographic keys and intricate mathematical algorithms is known as encryption. Only authorized parties with the necessary decryption keys can decrypt and comprehend this encrypted data (Pandey, B. K. et al., 2022). Strong encryption standards protect data from unwanted access and manipulation by making cryptographic algorithms resistant to cryptographic attacks and brute-force decryption attempts. The choice of strong cryptographic algorithms is one of the pillars of robust encryption standards. The security and effectiveness of advanced encryption algorithms, like Elliptic Curve Cryptography (ECC), Rivest-Shamir-Adleman (RSA), and Advanced Encryption Standard (AES), are well known. With the use of sophisticated mathematical operations and key lengths, these algorithms efficiently encrypt data, rendering it computationally impossible for adversaries to decrypt it without the necessary decryption keys. Robust encryption standards stress the significance of safe key management procedures in addition to powerful cryptographic algorithms. The fundamental component of encryption systems, encryption keys control both the encryption and decryption processes. To prevent unauthorized access or compromise, secure key generation (Pandey, B. K. et al., 2023b), distribution, and storage techniques must be used in accordance with strong encryption standards. To guarantee the confidentiality and integrity of encryption keys throughout their lifetime, key management protocols are put into place. These include key rotation, key escrow, and key revocation. Strong encryption standards also include end-to-end encryption protocols to safeguard information while it is being transmitted over digital networks. End-to-end encryption reduces the possibility of malicious actors intercepting or listening in on data as it travels from the sender's device to the recipient's device. Sensitive information is protected from unwanted interception and manipulation by using encryption algorithms to create encrypted communication channels between endpoints in secure communication protocols like Secure Socket Layer (SSL) and Transport Layer Security (TLS). Robust encryption standards are crucial for safeguarding sensitive data in a variety of applications and industries in today's data-driven and networked world. Encryption is a fundamental component of cybersecurity, protecting everything from personal communications and medical records to online (Pandey, D. et al., 2021b) and financial transactions. Strong encryption guidelines promote confidence and trust among users, clients, and stakeholders in addition to reducing the risk of data breaches and cyberattacks. The spread of cutting-edge technologies like blockchain, cloud computing, and the Internet of Things (IoT) highlights how crucial strong encryption standards are to maintaining the security and privacy of digital ecosystems. It is imperative for organizations to prioritize the implementation of robust encryption standards as they embrace digital transformation and innovative technologies. This will help them mitigate the constantly evolving threat landscape and comply with compliance requirements and data protection regulations.

Collaborative Security Measures

Collaborative security is a key defense in modern cybersecurity, where threats evolve quickly and no organization is immune to attack. Collaborative security emphasizes organizations, stakeholders, and cybersecurity professionals sharing intelligence, resources, and expertise to combat cyber threats. This collaboration helps entities strengthen their defenses, identify new threats, and mitigate risks quickly. Information sharing platforms and partnerships underpin collaborative security. Organizations can share threat intelligence, IOCs, and cybersecurity insights on these platforms. Sharing real-time threat and vulnerability data helps organizations improve situational awareness and defend against evolving cyber threats. ISACs and threat intelligence sharing communities help organizations across industries collaborate to fight common enemies. To standardize cybersecurity and boost resilience, collaborative security measures recommend collective defense frameworks and protocols. The NIST Cybersecurity Framework helps organizations manage cybersecurity risk by identifying, protecting, detecting, responding, and recovering. Organisations can improve cyber defences and risk mitigation by following industry standards and best practises. Collective security measures encourage information sharing, collective defense frameworks, and public-private partnerships between government, law enforcement, and the private sector. Partnerships enable cybersecurity initiatives, threat intelligence sharing, and coordinated cyber incident responses. Public-private partnerships strengthen the cybersecurity ecosystem and enable a unified national and international response to cyber threats by leveraging stakeholders' complementary resources and expertise. Collective security emphasizes proactive threat hunting and incident response to detect and neutralize cyber threats before they escalate. Threat hunting uses proactive search for malicious or anomalous behavior in an organization's network to detect and respond to threats. Threat hunting and cyber adversary disruption can be improved by using advanced analytics (Anand, R. et al., 2024), machine learning algorithms (Deepa, R. et al., 2022), and threat intelligence. Collaborative security measures prioritize cybersecurity awareness and training programs to teach employees, stakeholders, and end-users about best practices and emerging threats. Promote cybersecurity awareness and vigilance to empower individuals to identify and report suspicious activity, strengthening the human element of cybersecurity defenses (Quy, V. K. et al., 2023).

REGULATORY FRAMEWORK

International Standards and Guidelines

International standards and guidelines are essential for maintaining consistency, interoperability, and compliance across various industries and jurisdictions in the intricate and globally interconnected realm of cybersecurity, where threats remain transnational and organizations function globally. In an increasingly digital environment, these standards and guidelines provide crucial frameworks for defining security requirements, establishing best practices, and reducing cyber risks. Organizations can improve their cybersecurity posture, build stakeholder trust, and negotiate the challenges of a quickly changing threat landscape by adhering to internationally recognized standards and guidelines. Creating a common framework and language for cybersecurity practices across various industries and geographical areas is one of the main goals of international standards and guidelines. Organizations are given structured methods to recognize, evaluate, and handle cybersecurity risks by standards like the NIST Cybersecurity Framework and ISO/IEC 27001 for information security management systems (ISMS). Organizations can demonstrate their commitment to safeguarding digital assets and sensitive data by adopting globally recognized standards and aligning their cybersecurity efforts with industry best practices. Interoperability and compatibility between various systems, goods, and services are facilitated by international standards and guidelines. To facilitate secure communication over the internet, for instance, standards like Secure Sockets Layer (SSL) and Transport Layer Security (TLS) define protocols that allow data to be transmitted encrypted and seamlessly between various systems and devices. Regardless of the underlying technologies or platforms, organizations can guarantee the authenticity, confidentiality, and integrity of data exchanged across digital networks by adhering to these standards. International standards and guidelines are vital instruments for regulatory compliance and risk management, in addition to fostering uniformity and interoperability. International standards and guidelines are cited by a number of regulatory bodies and industry regulators when creating cybersecurity regulations and requirements. The General Data Protection Regulation (GDPR), for instance, draws its guidelines from international standards like ISO/IEC 27001 and requires the implementation of suitable organizational and technical safeguards to protect personal data. Organizations can demonstrate regulatory compliance and reduce the risk of fines and sanctions by aligning with globally recognized standards. Organizations can evaluate and enhance their cybersecurity maturity and resilience by using international standards and guidelines as a benchmark. Frameworks for assessing cybersecurity practices and pinpointing opportunities for improvement are provided by standards like the NIST Cybersecurity Framework and the Cybersecurity Capability Maturity Model (CMM). Organizations can find gaps in their cybersecurity posture, prioritize investments, and create plans for ongoing improvement by regularly conducting assessments and benchmarking against international standards. The creation and upkeep of global norms and guidelines require cooperation and the ability to reach consensus. Experts from industry, academia, government, and civil society are brought together by international standards bodies like the International Organization for Standardization (ISO), the International Electrotechnical Commission (IEC), and the National Institute of Standards and Technology (NIST) to develop consensus-based standards that represent the combined knowledge and experience of the global cybersecurity community. Organizations can help shape cybersecurity frameworks and standards that address new threats and challenges by taking part in these standards development processes (Shen, F. et al., 2021).

Governmental Policies and Oversight

Government policies and oversight protect cybersecurity, critical infrastructure, sensitive data, and national interests in an era of increasing digitalization and interconnectedness. Governments worldwide must create and enforce cyber threat mitigation, resilience, and digital ecosystem integrity and security policies, regulations, and frameworks. Regulatory frameworks, public-private partnerships, and government policies and oversight are needed to address the complex and evolving cybersecurity challenges of the digital age. Regulations and laws that define cybersecurity requirements, obligations, and responsibilities for organizations, businesses, and individuals are one of the main goals of government policies and oversight. These regulations may include data protection, cybersecurity, incident reporting, and industry-specific regulations to protect critical infrastructure sectors like finance, energy, healthcare, and transportation. Governments can encourage accountability, transparency, and compliance in the private and public sectors by setting cybersecurity standards. To combat cybersecurity threats, government policies and oversight mechanisms help law enforcement, regulatory bodies, and private sector stakeholders work together. Information sharing, threat intelligence dissemination, and joint cyber threat initiatives occur at national cybersecurity strategies, task forces, and coordination centers. ISACs and sector-specific alliances allow government and industry stakeholders to collaborate on cyber resilience, incident response, and cyber defense strategies. Government policies and oversight focus on capacity building, awareness raising, and education to improve cybersecurity awareness and skills among citizens, businesses, and government entities in addition to regulatory frameworks and collaboration mechanisms. Awareness campaigns, training programs, and educational programs teach people how to identify, prevent, and respond to cyber threats. By fostering cybersecurity awareness and resilience, governments can improve cybersecurity and reduce cyber incidents. Regulatory frameworks and strategies are monitored, assessed, and adapted to address emerging cybersecurity challenges and technological advances. As new technologies and cyber threats emerge, governments must be vigilant and proactive in identifying risks, assessing their impact, and updating policies and regulations. To keep up with cyber threats and technological advances, regulatory frameworks must be agile and responsive. Addressing transnational cyber threats, promoting cybersecurity norms, and strengthening global cyber resilience through international cooperation. International treaties, bilateral and multilateral agreements, and cyber defense alliances enable global information sharing, joint exercises, and cyber incident response. Governments can improve collective security, deter bad actors, and build cyberspace trust by cooperating internationally.

Legal Implications and Compliance Requirements

In today's fast-paced digital era, where organizations heavily depend on technology to carry out business operations, navigate complex regulatory landscapes, and protect sensitive information, it is crucial to have a solid grasp of the legal implications and compliance requirements in cybersecurity. Legal frameworks, regulations, and compliance standards provide a set of guidelines and obligations for organizations to uphold data privacy, maintain data security, and effectively manage cyber risks. Not meeting these requirements can have serious repercussions, including legal consequences, financial penalties, damage to your reputation, and a loss of trust from stakeholders. Thus, organizations must actively navigate the intricate legal landscape of cybersecurity to minimize risks and ensure adherence to relevant laws and regulations. One of the key considerations in cybersecurity is the legal aspect, specifically related to data privacy laws and regulations. These laws govern how personal and sensitive information is collected, used, stored, and shared. Regulations like the General Data Protection Regulation (GDPR) in the European Union and the California Consumer Privacy Act (CCPA) in the United States have placed stringent obligations on organizations when it comes to safeguarding data, ensuring transparency, obtaining consent, and upholding individual rights. Failure to comply with these regulations can lead to significant consequences, such as financial penalties, legal action, and regulatory repercussions. Organizations are obligated to protect critical infrastructure, sensitive data, and customer information from cyber threats due to cybersecurity regulations and industry-specific compliance requirements. Regulations in different sectors, like the Health Insurance Portability and Accountability Act (HIPAA) in healthcare and the Payment Card Industry Data Security Standard (PCI DSS) in finance, set cybersecurity standards to protect patient health information 23) and payment card data. Organizations operating in regulated industries must adhere to these requirements in order to prevent any potential legal consequences and safeguard their reputation. Aside from regulatory compliance, organizations need to take into account the legal consequences associated with incident response, data breach notification, and liability in the event of a cyber incident. Incident response plans are a necessary requirement in many jurisdictions to promptly detect, contain, and mitigate cybersecurity incidents. In addition, organizations may be required by law to inform individuals, regulators, and stakeholders if there is a data breach or security incident. Failure to adhere to these notification requirements may lead to legal repercussions and harm the organization's reputation. Legal considerations also encompass liability concerns, whereby organizations can be held responsible for cybersecurity breaches that cause harm to individuals, financial losses, or harm to third parties. Legal actions, lawsuits, and government enforcement can result in substantial financial and reputational consequences for companies that are deemed responsible for not adequately safeguarding sensitive data or neglecting to establish sufficient cybersecurity protocols. Given the nature of the business landscape, it is crucial for organizations to prioritize strong cybersecurity practices, risk management strategies, and compliance frameworks. This will help them effectively address legal risks and liabilities. In order to successfully navigate the intricate legal terrain of cybersecurity, organizations must embrace a proactive mindset towards compliance, risk management, and governance. This requires conducting regular assessments to identify legal requirements, implementing policies and procedures to ensure compliance, and establishing mechanisms for monitoring and enforcement. Furthermore, it is crucial for organizations to stay updated on any shifts in cybersecurity regulations, emerging legal trends, and best practices in order to effectively adjust their cybersecurity strategies.

INDUSTRY COLLABORATION

Public-Private Partnerships

In order to facilitate the development of infrastructure in an effective manner, Public-Private Partnerships (PPPs) make use of the resources and expertise of the private sector. Public-private partnerships (PPPs) contribute to the timely delivery of projects, the guarantee of high-quality services, and the stimulation of economic growth, all of which contribute to the promotion of sustainable development for communities all over the world. Key Benefits of Public-Private Partnerships (PPPs) are as shown in the figure 1.

1. *Efficient Allocation of Financial Resources*: PPPs give governments access to the capital and know-how of the private sector, facilitating the effective distribution of financial resources. Governments can obtain additional funding through private partnerships that may not be possible through traditional budgetary allocations alone. Furthermore, the involvement of the private sector frequently results in innovative and efficient project delivery methods that minimize costs, thereby optimizing the return on public investments.

2. *Risk Sharing*: In contrast to traditional procurement models, PPPs share project risks between the public and private sectors, reducing the financial and operational risks that governments would otherwise bear alone. Private partners take on a portion of the project's risk in exchange for performance-based rewards and revenue-sharing plans, including delays in construction, cost overruns, and revenue fluctuations. This risk-sharing arrangement guarantees efficient risk management and promotes efficiency and innovation in the private sector.

3. *Technology and Innovation*: Participation of the private sector in PPPs encourages innovation and the use of cutting-edge technologies in the construction of infrastructure. Private partners frequently contribute specialized knowledge, creative thinking, and technological innovations to projects, which results in the deployment of more sustainable and effective infrastructure solutions. PPPs stimulate technological innovation and improve the caliber and performance of infrastructure assets by utilizing cutting-edge construction methods, digitization, and smart infrastructure systems.

4. *Timely Project Delivery*: When compared to conventional procurement techniques, PPPs are frequently linked to quicker project delivery schedules. Construction schedule acceleration, expedited approvals, and streamlined decision-making processes are made possible by the efficiency, expertise, and incentives of the private sector. In order to guarantee financial returns and timely project completion and service delivery, private partners are also motivated to meet project milestones and deadlines.

5. *Enhanced Quality of Service*: Strict performance criteria, service level agreements, and quality benchmarks are frequently included in PPP contracts to guarantee that infrastructure projects satisfy user needs and expectations. Contractually, private partners are required to provide superior services and manage infrastructure assets in accordance with predetermined guidelines. By emphasizing service quality and user satisfaction, infrastructure projects guarantee that the general public receives dependable, secure, and efficient services, improving project outcomes overall and stakeholder satisfaction.

6. *Long-Term Value*: By encouraging private partners to invest in asset upkeep, operation, and lifecycle management, PPPs enhance the long-term viability and value of infrastructure assets. Private partners are accountable for long-term upkeep and functionality of infrastructure assets through concession agreements and performance-based contracts. Ensuring that infrastructure assets remain operational, functional, and valuable to communities and stakeholders throughout their entire lifecycle is made possible by this focus on long-term asset management.

7. *Economic Stimulus*: By creating jobs, boosting company activity, and drawing investment, PPP projects promote economic growth and development. Building and running infrastructure projects generates employment in a number of industries, such as engineering, construction, and services, which helps to generate income and jobs. Furthermore, PPP investments frequently serve as a catalyst for more extensive economic development by boosting market accessibility, expanding business opportunities, and enhancing transportation networks—all of which raise productivity, competitiveness, and economic prosperity.

8. *Private Resource Mobilization*: Project-funding and project-delivery partnerships (PPPs) pool private sector resources, including money, know-how, and technology. Governments can launch more expansive and ambitious projects than they could with conventional financing methods alone thanks to private investment, which supplements the limited public funds. By unlocking the value of underutilized or poorly managed assets, privatization of infrastructure assets through PPPs creates opportunities for economic development and private sector investment. This propels innovation throughout the economy, creates jobs, and boosts economic growth.

Figure 1. Key benefits of public-private partnerships (PPPs)

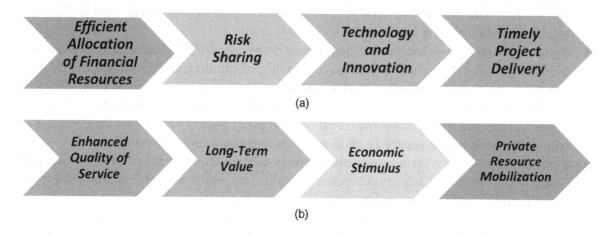

Information Sharing Mechanisms

Efficient information sharing mechanisms are vital in today's society, facilitating the transfer of data, knowledge, and insights between different parties to enhance decision-making, collaboration, and problem-solving. These mechanisms include various protocols, platforms, and frameworks that are intended to promote the secure and efficient exchange of information between different organizations, sectors, and jurisdictions. One of the main goals of information sharing mechanisms is to improve transparency and accountability by providing stakeholders with access to relevant data. By facilitating access to up-to-date and precise information, these mechanisms empower stakeholders to make well-informed decisions, track progress, and ensure individuals and organizations are held responsible for their actions. The sharing of information through various mechanisms helps to enhance interoperability by establishing standardized formats, protocols, and interfaces. Through the adoption of standardized protocols and common systems, information can be easily exchanged between different platforms, minimizing compatibility problems and enhancing the efficiency of data sharing processes. Mechanisms for sharing information promote collaboration and coordination among a wide range of stakeholders, including government agencies, non-profit organizations, private sector entities, and academic institutions. Through the facilitation of communication and knowledge sharing, these mechanisms allow stakeholders to combine resources, utilize expertise, and tackle intricate challenges with greater efficiency. Sharing information improves awareness of the situation and helps manage risks by giving stakeholders access to up-to-date and pertinent data about emerging threats, vulnerabilities, and opportunities. Through the exchange of information on cybersecurity threats, natural disasters, public health crises, and other critical issues, stakeholders can take a proactive approach to identifying and reducing risks, minimizing potential harm, and strengthening resilience. Efficient information sharing mechanisms also foster innovation and knowledge creation by facilitating the exchange of best practices, lessons learned, and research findings. Through the exchange of insights and experiences, stakeholders can uncover emerging trends, delve into fresh ideas, and create inventive solutions to tackle evolving challenges and opportunities. The effective execution of information sharing mechanisms necessitates thoughtful examination of numerous elements, such as data privacy, security, governance, and legal frameworks. It is essential for stakeholders to establish well-defined policies, procedures, and safeguards in order to safeguard sensitive information, adhere to applicable regulations, and uphold trust and confidentiality among participants. The exchange of information is crucial for fostering transparency, interoperability, collaboration, and innovation across various sectors and disciplines. By enabling the flow of information, expertise, and analysis, these mechanisms empower individuals to make well-informed choices, improve collaboration, and tackle intricate problems in a fast-changing and interconnected global landscape.

Joint Research and Development Initiatives

Joint R&D initiatives help advance technology transfer and commercialization by translating research into products and applications. These initiatives foster collaboration between academic researchers, industry partners, and technology transfer offices to seamlessly transfer research findings to the marketplace, boosting economic growth and social impact. Effective governance, communication, and trust are needed for joint R&D projects. Partners must set goals, priorities, and expectations and establish decision-making, conflict resolution, and IP management mechanisms. For transparency, fairness, and regulatory compliance, stakeholders must follow ethical, data sharing, and legal guidelines. Joint R&D initiatives involve government agencies, academic institutions, research organizations, and private companies working together to advance science, technology, and knowledge. These initiatives integrate diverse expertise, resources, and perspectives to solve complex problems, innovate, and benefit all. Coordination and knowledge sharing are key goals of joint R&D projects. Collaboration can solve research and technological problems that individual organizations cannot by sharing resources and expertise. This collaboration gives stakeholders new insights, methods, and technologies, accelerating discovery and innovation. By bringing together researchers and experts from different fields, joint R&D initiatives foster interdisciplinary collaboration. These initiatives promote cross-disciplinary collaboration to integrate diverse perspectives and approaches, resulting in new insights, breakthroughs, and complex problem solutions. It allows participants to share infrastructure, facilities, and resources that may be too expensive or difficult to obtain individually. Collaborations can use cutting-edge equipment, research facilities, and experimental platforms to improve their research and broaden their scope. By helping research findings and innovations become products and applications, joint R&D initiatives aid technology transfer and commercialization. These initiatives foster collaboration between academic researchers, industry partners, and technology transfer offices to seamlessly transfer research findings to the marketplace, boosting economic growth and social impact. Effective governance, communication, and trust are needed for joint R&D projects. Partners must set goals, priorities, and expectations and establish decision-making, conflict resolution, and IP management mechanisms. For transparency, fairness, and regulatory compliance, stakeholders must

CONCLUSION

The protection of 6G networks necessitates the implementation of a comprehensive approach that effectively tackles the dynamic array of cybersecurity risks and susceptibilities. The implementation of proactive risk assessment is of utmost importance in the identification of potential risks and vulnerabilities, enabling stakeholders to proactively implement measures to mitigate their impact. Strong encryption protocols are crucial for safeguarding confidential information and guaranteeing the authenticity of communications across 6G networks. In order to promote cooperation among stakeholders and facilitate collective responses to emerging threats, it is imperative to implement collaborative security measures such as information sharing mechanisms and joint research initiatives.

A strong regulatory framework consisting of global standards, government policies, and legal obligations establishes the essential structure for guaranteeing adherence and responsibility in protecting 6G networks. Public-private partnerships in the industry enable the consolidation of resources, expertise, and insights, thereby promoting innovation and resilience in network security. Advancements in technology, such as the utilization of artificial intelligence for threat detection and the implementation of quantum-safe cryptography, present encouraging prospects for bolstering network resilience and safeguarding against ever-changing threats. To effectively safeguard 6G networks, stakeholders can embrace these innovations and promote collaboration across sectors. Ensuring the security of 6G networks is a collective obligation that necessitates proactive actions, cooperative endeavors, and continuous innovation. Stakeholders can guarantee the integrity, reliability, and resilience of 6G networks by giving priority to security, compliance, and collaboration. This will protect the foundation for the future of intelligent connectivity.

REFERENCES

Anand, R., Khan, B., Nassa, V. K., Pandey, D., Dhabliya, D., Pandey, B. K., & Dadheech, P. (2023). Hybrid convolutional neural network (CNN) for kennedy space center hyperspectral image. *Aerospace Systems*, 6(1), 71–78. 10.1007/s42401-022-00168-4

Anand, R., Lakshmi, S. V., Pandey, D., & Pandey, B. K. (2024). An enhanced ResNet-50 deep learning model for arrhythmia detection using electrocardiogram biomedical indicators. *Evolving Systems*, 15(1), 83–97. 10.1007/s12530-023-09559-0

Bangerter, B., Talwar, S., Arefi, R., & Stewart, K. (2014). Networks and devices for the 5G era. *IEEE Communications Magazine*, 52(2), 90–96. 10.1109/MCOM.2014.6736748

Bessant, Y. A., Jency, J. G., Sagayam, K. M., Jone, A. A. A., Pandey, D., & Pandey, B. K. (2023). Improved parallel matrix multiplication using Strassen and Urdhvatiryagbhyam method. *CCF Transactions on High Performance Computing*, 5(2), 102–115. 10.1007/s42514-023-00149-9

David, S., Duraipandian, K., Chandrasekaran, D., Pandey, D., Sindhwani, N., & Pandey, B. K. (2023). Impact of blockchain in healthcare system. In *Unleashing the Potentials of blockchain technology for healthcare industries* (pp. 37–57). Academic Press. 10.1016/B978-0-323-99481-1.00004-3

Deepa, R., Anand, R., Pandey, D., Pandey, B. K., & Karki, B. (2022). Comprehensive performance analysis of classifiers in diagnosis of epilepsy. *Mathematical Problems in Engineering*, 2022, 2022. 10.1155/2022/1559312

Devasenapathy, D., Madhumathy, P., Umamaheshwari, R., Pandey, B. K., & Pandey, D. (2024). Transmission-efficient grid-based synchronized model for routing in wireless sensor networks using Bayesian compressive sensing. *SN Computer Science*, 5(1), 1–11.

Ekong, M. O., George, W. K., Pandey, B. K., & Pandey, D. (2023). Enhancing the Fundamentals of Industrial Safety Management in TVET for Metaverse Realities. In *Applications of Neuromarketing in the Metaverse* (pp. 19-41). IGI Global. 10.4018/978-1-6684-8150-9.ch002

George, W. K., Ekong, M. O., Pandey, D., & Pandey, B. K. (2023). Pedagogy for Implementation of TVET Curriculum for the Digital World. In *Applications of Neuromarketing in the Metaverse* (pp. 117-136). IGI Global. 10.4018/978-1-6684-8150-9.ch009

Gohar, A., & Nencioni, G. (2021). The role of 5G technologies in a smart city: The case for intelligent transportation system. *Sustainability (Basel)*, 13(9), 5188. 10.3390/su13095188

. Govindaraj, V., Dhanasekar, S., Martinsagayam, K., Pandey, D., Pandey, B. K., & Nassa, V. K. (2023). Low-power test pattern generator using modified LFSR. *Aerospace Systems*, 1-8.

Gupta, A. K., Sharma, R., Pandey, D., Nassa, V. K., Pandey, B. K., George, A. S., & Dadheech, P. (2023). Performance analysis of eight-channel WDM optical network with different optical amplifiers for industry 4.0. In *Innovation and Competitiveness in Industry 4.0 Based on Intelligent Systems* (pp. 197–212). Springer International Publishing. 10.1007/978-3-031-29775-5_9

Iyyanar, P., Anand, R., Shanthi, T., Nassa, V. K., Pandey, B. K., George, A. S., & Pandey, D. (2023). A real-time smart sewage cleaning UAV assistance system using IoT. In *Handbook of Research on Data-Driven Mathematical Modeling in Smart Cities* (pp. 24–39). IGI Global.

Khan, B., Hasan, A., Pandey, D., Ventayen, R. J. M., Pandey, B. K., & Gowwrii, G. (2021). Fusion of datamining and artificial intelligence in prediction of hazardous road accidents. In *Machine learning and iot for intelligent systems and smart applications* (pp. 201–223). CRC Press. 10.1201/9781003194415-12

Kirubasri, G., Sankar, S., Pandey, D., Pandey, B. K., Nassa, V. K., & Dadheech, P. (2022). Software-defined networking-based Ad hoc networks routing protocols. In *Software defined networking for Ad Hoc networks* (pp. 95–123). Springer International Publishing. 10.1007/978-3-030-91149-2_5

Kirubasri, G., Sankar, S., Pandey, D., Pandey, B. K., Singh, H., & Anand, R. (2021, September). A recent survey on 6G vehicular technology, applications and challenges. In *2021 9th International Conference on Reliability, Infocom Technologies and Optimization (Trends and Future Directions)(ICRITO)* (pp. 1-5). IEEE.

Kumar Pandey, B., Pandey, D., Nassa, V. K., Ahmad, T., Singh, C., George, A. S., & Wakchaure, M. A. (2021). Encryption and steganography-based text extraction in IoT using the EWCTS optimizer. *Imaging Science Journal*, 69(1-4), 38–56. 10.1080/13682199.2022.2146885

Malhotra, P., Pandey, D., Pandey, B. K., & Patra, P. M. (2021). Managing agricultural supply chains in COVID-19 lockdown. *International Journal of Quality and Innovation*, 5(2), 109–118. 10.1504/IJQI.2021.117181

Menon, V., Pandey, D., Khosla, D., Kaur, M., Vashishtha, H. K., George, A. S., & Pandey, B. K. (2022). A Study on COVID–19, Its Origin, Phenomenon, Variants, and IoT-Based Framework to Detect the Presence of Coronavirus. In *IoT Based Smart Applications* (pp. 1–13). Springer International Publishing.

Murroni, M., Anedda, M., Fadda, M., Ruiu, P., Popescu, V., Zaharia, C., & Giusto, D. (2023). 6G—Enabling the New Smart City: A Survey. *Sensors (Basel)*, 23(17), 7528. 10.3390/s2317752837687986

Pandey, B. K., & Pandey, D. (2023). Parametric optimization and prediction of enhanced thermoelectric performance in co-doped $CaMnO3$ using response surface methodology and neural network. *Journal of Materials Science Materials in Electronics*, 34(21), 1589. 10.1007/s10854-023-10954-1

Pandey, B. K., Pandey, D., & Agarwal, A. (2022). Encrypted information transmission by enhanced steganography and image transformation. [IJDAI]. *International Journal of Distributed Artificial Intelligence*, 14(1), 1–14. 10.4018/IJDAI.297110

Pandey, B. K., Pandey, D., Alkhafaji, M. A., Güneşer, M. T., & Şeker, C. (2023a). A reliable transmission and extraction of textual information using keyless encryption, steganography, and deep algorithm with cuckoo optimization. In *Micro-Electronics and Telecommunication Engineering: Proceedings of 6th ICMETE 2022* (pp. 629–636). Springer Nature Singapore. 10.1007/978-981-19-9512-5_57

Pandey, B. K., Pandey, D., Gupta, A., Nassa, V. K., Dadheech, P., & George, A. S. (2023b). Secret data transmission using advanced morphological component analysis and steganography. In *Role of data-intensive distributed computing systems in designing data solutions* (pp. 21–44). Springer International Publishing. 10.1007/978-3-031-15542-0_2

Pandey, B. K., Pandey, D., & Sahani, S. K. (2024). Autopilot control unmanned aerial vehicle system for sewage defect detection using deep learning. *Engineering Reports*, 12852. 10.1002/eng2.12852

Pandey, B. K., Pandey, S. K., & Pandey, D. (2011). A survey of bioinformatics applications on parallel architectures. *International Journal of Computer Applications*, 23(4), 21–25. 10.5120/2877-3744

Pandey, D., Hasan, A., Pandey, B. K., Lelisho, M. E., George, A. H., & Shahul, A. (2023). COVID-19 epidemic anxiety, mental stress, and sleep disorders in developing country university students. *CSI Transactions on ICT*, 11(2), 119–127. 10.1007/s40012-023-00383-0

Pandey, D., Nassa, V. K., Jhamb, A., Mahto, D., Pandey, B. K., George, A. H., & Bandyopadhyay, S. K. (2021a). An integration of keyless encryption, steganography, and artificial intelligence for the secure transmission of stego images. In *Multidisciplinary approach to modern digital steganography* (pp. 211–234). IGI Global. 10.4018/978-1-7998-7160-6.ch010

. Pandey, D., Ogunmola, G. A., Enbeyle, W., Abdullahi, M., Pandey, B. K., & Pramanik, S. (2021b). *COVID-19: A framework for effective delivering of online classes during lockdown*. Human Arenas, 1-15.

Pandey, D., & Pandey, B. K. (2022). An efficient deep neural network with adaptive galactic swarm optimization for complex image text extraction. In *Process mining techniques for pattern recognition* (pp. 121–137). CRC Press. 10.1201/9781003169550-10

Pandey, J. K., Jain, R., Dilip, R., Kumbhkar, M., Jaiswal, S., Pandey, B. K., & Pandey, D. (2022). Investigating role of iot in the development of smart application for security enhancement. In *IoT Based Smart Applications* (pp. 219–243). Springer International Publishing.

Pramanik, S., Pandey, D., Joardar, S., Niranjanamurthy, M., Pandey, B. K., & Kaur, J. (2023, October). An overview of IoT privacy and security in smart cities. In *AIP Conference Proceedings (Vol. 2495, No. 1)*. AIP Publishing 10.1063/5.0123511

Quy, V. K., Chehri, A., Quy, N. M., Han, N. D., & Ban, N. T. (2023). Innovative trends in the 6G era: A comprehensive survey of architecture, applications, technologies, and challenges. *IEEE Access: Practical Innovations, Open Solutions*, 11, 39824–39844. 10.1109/ACCESS.2023.3269297

Raja, D., Kumar, D. R., Santhiyakumari, N., Kumarganesh, S., Sagayam, K. M., Thiyaneswaran, B., Pandey, B. K., & Pandey, D. (2024). A compact dual-feed wide-band slotted antenna for future wireless applications. *Analog Integrated Circuits and Signal Processing*, 1–15. 10.1007/s10470-023-02233-0

Revathi, T. K., Sathiyabhama, B., Sankar, S., Pandey, D., Pandey, B. K., & Dadeech, P. (2022). An intelligent model for coronary heart disease diagnosis. In *Networking Technologies in Smart Healthcare* (pp. 309–327). CRC Press. 10.1201/9781003239888-15

Sahani, K., Khadka, S. S., Sahani, S. K., Pandey, B. K., & Pandey, D. (2023). A possible underground roadway for transportation facilities in Kathmandu Valley: A racking deformation of underground rectangular struct es. *Engineering Reports*, 12821. 10.1002/eng2.12821

Saxena, A., Agarwal, A., Pandey, B. K., & Pandey, D. (2024). Examination of the Criticality of Customer Segmentation Using Unsupervised Learning Methods. *Circular Economy and Sustainability*, 1–14. 10.1007/s43615-023-00336-4

Saxena, A., Sharma, N. K., Pandey, D., & Pandey, B. K. (2021). Influence of tourists satisfaction on future behavioral intentions with special reference to desert triangle of Rajasthan. *Augmented Human Research*, 6(1), 13. 10.1007/s41133-021-00052-4

. Sengupta, R., Sengupta, D., Pandey, D., Pandey, B. K., Nassa, V. K., & Dadeech, P. (2021). A Systematic review of 5G opportunities, architecture and challenges. *Future Trends in 5G and 6G*, 247-269. Research Gate.

Sharma, M., Pandey, D., Khosla, D., Goyal, S., Pandey, B. K., & Gupta, A. K. (2022). Design of a GaN-based Flip Chip Light Emitting Diode (FC-LED) with au Bumps & Thermal Analysis with different sizes and adhesive materials for performance considerations. *Silicon*, 14(12), 7109–7120. 10.1007/s12633-021-01457-x

Sharma, M., Pandey, D., Palta, P., & Pandey, B. K. (2022). Design and power dissipation consideration of PFAL CMOS V/S conventional CMOS based 2: 1 multiplexer and full adder. *Silicon*, 14(8), 4401–4410. 10.1007/s12633-021-01221-1

Sharma, S., Pandey, B. K., Pandey, D., Anand, R., Sharma, A., & Saini, S. (2023, March). Character Recognition Technique Implementation for Complicated Deteriorated Scene. In *2023 6th International Conference on Information Systems and Computer Networks (ISCON)* (pp. 1-4). IEEE. 10.1109/ISCON57294.2023.10112185

Shen, F., Shi, H., & Yang, Y. (2021, August). A comprehensive study of 5G and 6G networks. In *2021 international conference on wireless communications and smart grid (ICWCSG)* (pp. 321-326). IEEE.

Sinclair, M., Maadi, S., Zhao, Q., Hong, J., Ghermandi, A., & Bailey, N. (2023). Assessing the socio-demographic representativeness of mobile phone application data. *Applied Geography (Sevenoaks, England)*, 158, 102997. 10.1016/j.apgeog.2023.102997

Singh, S., Madaan, G., Kaur, J., Swapna, H. R., Pandey, D., Singh, A., & Pandey, B. K. (2023). *Bibliometric Review on Healthcare Sustainability. Handbook of Research on Safe Disposal Methods of Municipal Solid Wastes for a Sustainable Environment*, (pp. 142-161). IEEE. 10.4018/978-1-6684-8117-2.ch011

Swapna, H. R., Bigirimana, E., Madaan, G., Hasan, A., Pandey, B. K., & Pandey, D. (2023). Impact of neuromarketing on consumer psychology in digitally connected networks. In *Applications of Neuromarketing in the Metaverse* (pp. 193–205). IGI Global. 10.4018/978-1-6684-8150-9.ch015

Tareke, S. A., Lelisho, M. E., Hassen, S. S., Seid, A. A., Jemal, S. S., Teshale, B. M., & Pandey, B. K. (2022). The prevalence and predictors of depressive, anxiety, and stress symptoms among Tepi town residents during the COVID-19 pandemic lockdown in Ethiopia. *Journal of Racial and Ethnic Health Disparities*, 1–13.35028903

Tataria, H., Shafi, M., Molisch, A. F., Dohler, M., Sjöland, H., & Tufvesson, F. (2021). 6G Wireless Systems: Vision, Requirements, Challenges, Insights, and Opportunities. *Proceedings of the IEEE*, 109(7), 1166–1199. 10.1109/JPROC.2021.3061701

Tripathi, R. P., Sharma, M., Gupta, A. K., Pandey, D., Pandey, B. K., Shahul, A., & George, A. H. (2023). Timely prediction of diabetes by means of machine learning practices. *Augmented Human Research*, 8(1), 1. 10.1007/s41133-023-00062-4

Vinodhini, V., Kumar, M. S., Sankar, S., Pandey, D., Pandey, B. K., & Nassa, V. K. (2022). IoT-based early forest fire detection using MLP and AROC method. *International Journal of Global Warming*, 27(1), 55–70. 10.1504/IJGW.2022.122794

Chapter 8
Leadership in the Age of Holographic Connectivity:
Securing the Future of 6G

M. Beulah Viji Christiana

*Department of Master of Business Administration,
Panimalar Engineering College, Chennai, India*

Thaya Madhavi

 https://orcid.org/0000-0002-9635-3841

Mohan Babu University, Tirupati, India

S. Bathrinath

 https://orcid.org/0000-0002-5502-6203

*Kalasalingam Academy of Research and
Education, Krishnankoil, India*

Vellayan Srinivasan

S.A. Engineering College, Chennai, India

Mano Ashish Tripathi

*Motilal Nehru National Institute of Technology,
Allahabad, India*

B. Uma Maheswari

 https://orcid.org/0000-0001-9707-285X

St. Joseph's College of Engineering, India

Pankaj Dadheech

 https://orcid.org/0000-0001-5783-1989

*Swami Keshvanand Institute of Technology,
Management, and Gramothan, India*

ABSTRACT

The forthcoming transition to 6G brings with it the promise of holographic connectivity, which has the potential to revolutionize communication in a variety of fields, including marketing, education, medicine, business, and entertainment, among others. Users are able to interact with one another through the use of high-quality 3D representations while overcoming geographical barriers thanks to this technology. On the other hand, in order to successfully navigate this transition, effective leadership is required. This leadership must be able to anticipate technological advancements, encourage innovation, and strike a balance between risks and opportunities within the 6G ecosystem.

DOI: 10.4018/979-8-3693-2931-3.ch008

INTRODUCTION

Countries all over the world are actively pursuing initiatives to push the development of 6G technology as the global market for 6G technology (Kirubasri, G. et al., 2021) continues to pick up steam. Nations competing to become leaders in the emerging 6G space are spending a lot of money and effort in this endeavor. This coordinated effort demonstrates the understanding of 6G's enormous potential to influence wireless communication (Kirubasri, G. et al., 2022) in the future and propel global economic growth. Every new generation of wireless communication has brought about a major advancement in terms of speed, connectivity, and capabilities within the ever-changing landscape of technological advancement (Devasenapathy, D. et al., 2024). As the next development in this line of advancement, 6G appears to be building on the groundwork set by its predecessors. With 5G (Sengupta, R. et al., 2021) already completely changing the way we connect and communicate, there is a lot of excitement about 6G, which is expected to usher in an even more revolutionary period of seamless connectivity and ground-breaking innovation. But as we move closer to 6G, it's critical that we put these cutting-edge networks' security (Pandey, B. K. et al., 2021) and integrity first. Strong cybersecurity measures and proactive risk management strategies are necessary to secure (Pandey, B. K. et al., 2023a) the future of 6G given the growth of connected devices (Vinodhini, V. et al., 2022) and the sophistication of cyber threats. This means strengthening network infrastructure, putting strict encryption protocols in place, and encouraging stakeholders to adopt a cybersecurity-aware mindset (Pandey, J. K. et al., 2022). Mitigating potential risks of downtime and disruption is crucial, as 6G technologies are expected to support an unprecedented volume of data (Iyyanar, P. et al., 2023) and facilitate a wide range of applications, including infrastructure and critical services. To maintain service continuity in the face of unforeseen obstacles, robust network architectures, redundancy measures, and disaster recovery plans must be developed. As cutting-edge features like holographic communication and ubiquitous connectivity become more common, protecting user privacy (Pramanik, S. et al., 2023) rights and making sure data is stewarded responsibly become crucial. Strict privacy regulations, open data governance structures, and procedures for user consent and control over personal data must all be put in place in order to achieve this (Dogra, A. et al., 2020).

Holographic Connectivity

Innovative communication technology like holographic connectivity has a lot of potential, especially when it comes to 6G, the next generation of wireless communication (Gupta, A. K. et al., 2023). This cutting-edge method simulates face-to-face communication by allowing users to interact in real-time with holograms or three-dimensional representations, regardless of geographic distance. The following technological developments and their ramifications highlight the importance of holographic connectivity in the context of 6G (Attanasio, B. et al., 2021).

- *Bandwidth and Data Transfer Rates*: First off, 6G is expected to provide unmatched data transfer speeds—higher than those of 5G. In order to smoothly transfer high-resolution 3D holographic images in real-time, holographic communication requires a significant amount of bandwidth. Large data volumes needed for holographic communication can be transmitted seamlessly thanks to 6G networks' improved data transfer rates.

- *Low Latency*: for holographic projections to be displayed without discernible lag, holographic communication needs to have incredibly low latency. It is anticipated that 6G networks will achieve ultra-low latency levels, possibly measured in microseconds, by drastically reducing latency. This makes it possible to interact in real time with little delay, which improves the immersive experience of communicating through holography.

- *High Throughput and Massive Connectivity*: Massive connectivity and extraordinarily high throughput are anticipated from 6G networks. These networks support many devices (Sasidevi S et al., 2024) connecting at once without experiencing congestion and enable the simultaneous transmission of multiple high-definition holographic streams (Sahani, S. K. et al., 2024). This opens the door for holographic communication to be used widely.

- *Advanced Signal Processing and Beamforming*: 6G networks are probably going to use beamforming and advanced signal processing (Govindaraj, V. et al., 2023) techniques to maximize holographic data transmission and reception. Reliable and excellent holographic projections are guaranteed by intelligent algorithms (Bessant, Y. A. et al., 2023) and adaptive antenna system, which also improve signal quality (Du John, H. V. et al., 2022) and minimize interference (Bruntha, P. M. et al., 2023).

- *Edge Computing and AI Integration*: The fusion of artificial intelligence (Pandey, B. K., & Pandey, D., 2023) and edge computing powers improves the performance of holographic communication in 6G networks. The production and transmission of holographic content require massive volumes of data, which edge computing handles, while AI algorithms maximize network capacity and enhance user experience overall.

- *Enhanced Security Measures*: In order to protect sensitive holographic data from cyber threats and unauthorized access, strong security measures (Kumar Pandey, B. et al., 2021) are essential. Advanced encryption methods (Pandey, B. K. et al., 2023b), authentication procedures, and privacy-improving technologies are anticipated to be incorporated into 6G networks to enable secure holographic communication and promote user confidence.

Impending Transition to 6G

Countries all over the world are getting ready for this transformative shift by allocating substantial resources towards the development of 6G technology. The impending transition to 6G marks a significant milestone in the evolution of wireless communication technology. This change is a continuation of the pattern that has been observed with previous generations, in which each successive generation has brought about significant advancements in terms of speed, connectivity, and capabilities. The transition to 6G holds immense promise for further pushing the boundaries of seamless connectivity and transformative innovation. It is the successor to 5G, which has already revolutionized the way in which we interact and communicate with one another. Countries are actively investing time and money in order to position themselves as leaders in this emerging space. This is because it is anticipated that the global market for 6G will experience exponential growth. As a result of the transition to 6G, it is anticipated that a multitude of new opportunities will become available across a variety of industries, including healthcare, education, transportation (Sahani, K. et al., 2023), and entertainment, amongst others. With the help of cutting-edge technologies like terahertz frequency bands, advanced antenna systems, and artificial intelligence (Anand, R. et al., 2024), the sixth-generation or sixth-generation wireless network (also known as 6G) intends to provide unprecedented data transfer rates, extremely low latency, and widespread connectivity (Raja,

D. et al., 2024). This set of capabilities has the potential to bring about a revolution in various industries (Pandey, B. K. et al., 2024), to propel economic growth, and to improve the quality of life for people all over the world. It is important to note that the transition to 6G is not without its difficulties. In order to ensure a smooth transition to 6G networks, it will be essential to address issues such as the allocation of spectrum, the deployment of infrastructure, and interoperability (Giordani, M. et al., 2020). The future of 6G will be significantly influenced by the aspects of security, privacy, and regulatory frameworks that are taken into consideration. The transition presents an opportunity for stakeholders to collaborate, innovate, and shape the future of connectivity. Despite these challenges, the potential benefits of 6G are vast, and the transition presents another opportunity for stakeholders to collaborate. In light of the fact that we are on the verge of this transformative transition, it is of the utmost importance to acknowledge the significance of proactive planning, investment, and collaboration in order to fully realize the potential of 6G and ensure a more promising future for future generations.

HOLOGRAPHIC CONNECTIVITY AND 6G

Holographic Communication Technology

Driven by growing demand for immersive communication experiences and technological advancements, the holographic communication market saw significant growth and innovation in 2023. The potential for holographic communication to become commonplace with the upcoming switch to 6G networks has drawn interest from both consumers and businesses. The market consequently witnessed a spike in funding, R&D, and other initiatives targeted at improving the functionality and accessibility of holographic communication technology. Leading companies in the industry (Saxena, A. et al., 2021) kept launching state-of-the-art products and services to satisfy the expanding need for holographic communication in a variety of industries. These solutions included immersive holographic displays for gaming and entertainment as well as holographic telepresence systems for remote collaboration. Furthermore, the potential for holographic communication was increased by developments in augmented reality (AR) and virtual reality (VR), which made it possible for more engaging and realistic experiences. Additionally, holographic communication systems have significantly improved due to the integration of machine learning algorithms (Saxena, A. et al., 2024) and artificial intelligence (AI) (Anand, R. et al., 2023), which has decreased latency, improved image quality, and allowed for more natural interactions. In industries like telemedicine, education, retail, and automotive, new applications and use cases have been made possible by this convergence of technologies. Industries like healthcare, education, and entertainment have been early adopters of holographic communication technology in terms of market adoption.

Figure 1. Holographic communication market

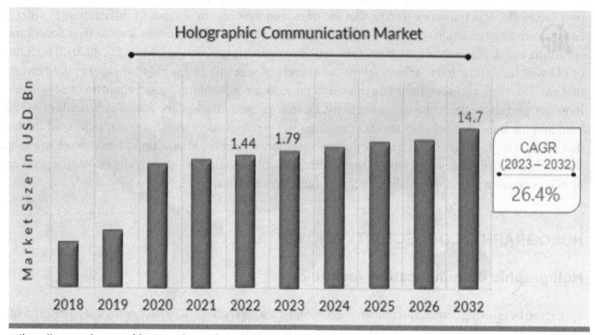

While educational institutions integrated holographic displays into their classrooms to create more engaging learning environments, healthcare providers used holographic telepresence systems for remote consultations and surgical training. Entertainment companies have also experimented with using holographic technology for virtual experiences, concerts, and live events. Looking ahead, as 6G networks spread and technological advancements continue to accelerate, the holographic communication market is expected to see sustained expansion and innovation. Holographic communication is anticipated to become more and more important in determining the direction of connectivity and immersive experiences, as it has the potential to completely transform the ways in which we interact, communicate, and collaborate. Because of this, it is expected that stakeholders from a variety of industries will keep funding holographic communication solutions, which will spur future innovation and market growth. Holographic communication technology creates immersive, lifelike interactions across distances, revolutionizing telecommunications. Holographic communication uses three-dimensional holograms to convey information, giving it a sense of presence and realism like face-to-face interactions. Holographic communication technology uses holography to capture and reproduce an object's light field as a three-dimensional image. A laser beam is split into a reference beam and an object beam. Object beam interacts with captured object or scene, while reference beam remains unchanged. Two beams converge on a holographic plate or medium, creating an interference pattern that is recorded as a hologram (Bhattarai, S. et al., 2016). Holographic communication technology allows real-time transmission and reception of holographic images, allowing users to see and interact with remote participants. High-speed networks and advanced display systems enable seamless, high-definition holographic experiences with this technology. Holographic communication technology overcomes the limitations of video conferencing and telepresence.

Holographic communication improves collaboration, creativity, and productivity across industries and applications by making communication more immersive and engaging. Holographic communication allows healthcare professionals to collaborate and share expertise in real time during remote consultations, surgical planning, and medical training. Holographic displays can make classrooms interactive, allowing students to explore complex concepts and simulations in three dimensions. Holographic communication technology can also create interactive, virtual experiences that blur the lines between the physical and digital worlds (Swapna, H. R. et al., 2023), revolutionizing entertainment, gaming, and live events. Holographic communication has endless possibilities, from holographic concerts and immersive gaming to virtual meetings and telepresence robots. Holographic communication will change how we communicate, collaborate, and interact in the digital age (George, W. K. et al., 2023) as technology and networks advance. Holographic communication technology, which creates lifelike, immersive experiences, is a groundbreaking innovation that will change telecommunications and beyond.

Potential Implications and Capabilities

With implications and capabilities spanning multiple industries and applications, holographic communication technology has the potential to revolutionize many aspects of interaction, collaboration, and communication. Improving teamwork is one of the main areas where holographic communication technology could have a big impact. Holographic communication facilitates more organic and engaging teamwork by allowing users to engage with distant participants as though they were in person. The healthcare industry, where medical professionals can consult, work together to diagnose patients, and conduct virtual surgeries regardless of their physical locations, stands to benefit greatly from this capability. Furthermore, by establishing immersive learning environments, holographic communication technology holds the potential to revolutionize the field of education. Students can study difficult ideas, interact with holographic simulations, and take part in virtual experiments to improve their comprehension and memory of the material. This creative approach to teaching not only increases student engagement but also creates new opportunities for inaccessible hands-on learning experiences.

Holographic communication technology presents creative opportunities for producing immersive and interactive experiences in the entertainment sector. This technology blurs the boundaries between the real and virtual worlds, enabling audiences to interact with content in completely new ways. Examples of its applications include interactive gaming experiences and holographic concerts and live events. When it comes to providing audiences with realistic and captivating experiences, holographic communication holds great potential for the entertainment sector. Furthermore, virtual collaboration spaces are made possible by holographic communication technology, enabling users to connect, communicate, and collaborate in real-time from anywhere in the world. These virtual spaces can be tailored to meet a range of requirements, including social gatherings, networking events, and business conferences and meetings. Working together virtually not only increases output but also strengthens relationships between people, no matter where they live. Additionally, remote workers can engage in meetings, presentations, and discussions to the same extent as their in-person counterparts thanks to holographic communication technology. With the growing prevalence of remote collaboration in today's decentralized and globalized workforce, this capability is especially important. Holographic communication technology adds to the effectiveness and success of remote work arrangements by enabling efficient communication and collaboration among remote teams (Chowdhury, M. Z. et al., 2020).

LEADERSHIP CHALLENGES IN 6G DEVELOPMENT

Adapting to Technological Advancements

Technological developments are transforming businesses, societies, and industries at a rate never seen before in today's quickly changing global environment. These developments, which range from blockchain technology and virtual reality to artificial intelligence and machine learning (Pandey, D. et al., 2022), are not only changing our everyday lives but also bringing with them new opportunities and challenges. Both people and organizations must accept and adjust to these technological changes in order to survive in this changing environment and use them to spur innovation and expansion. Individuals have many opportunities for both professional and personal growth when they adjust to technological advancements. Gaining new expertise in cutting-edge technologies can improve employability and open up exciting career opportunities. For instance, becoming proficient in coding languages can lead to profitable opportunities in software development. Knowledge of virtual reality and blockchain technology, on the other hand, can position people as leaders in these quickly developing fields, spurring innovation and opening up new avenues. Technological developments are transforming businesses, societies, and industries at a rate never seen before in today's quickly changing global environment. These developments, which range from blockchain technology and virtual reality to artificial intelligence (Revathi, T. K. et al., 2022) and machine learning, are not only changing our everyday lives but also bringing with them new opportunities and challenges (Tripathi, R. P. et al., 2023). Both people and organizations must accept and adjust to these technological changes in order to survive in this changing environment and use them to spur innovation and expansion (Deepa, R. et al.,(2022).). Individuals have many opportunities for both professional and personal growth when they adjust to technological advancements (Mahmoud, H. H. H. et al., 2021). Gaining new expertise in cutting-edge technologies can improve employability and open up exciting career opportunities. For instance, becoming proficient in coding languages can lead to profitable opportunities in software development. Knowledge of virtual reality and blockchain technology, on the other hand, can position people as leaders in these quickly developing fields, spurring innovation and opening up new avenues. In a similar vein, companies need to keep up with technology developments in order to be competitive and relevant in today's market. Adopting technology can result in higher productivity, better client interactions, and better ability to make decisions. Businesses can automate tedious tasks and free up resources for more strategic initiatives by incorporating artificial intelligence into their operations (Pandey, B. K. et al., 2024). Furthermore, big data analytics and other technologies offer insightful information about consumer behavior, allowing companies to effectively modify their product offerings to meet changing needs. Individuals and organizations alike must take a proactive approach to learning and innovation in order to adapt to technological advancements. People should look for opportunities to learn new skills and knowledge through online courses, workshops, and industry conferences as they are vital to lifelong learning. Maintaining ties with communities and professional networks can also give you important insider knowledge about the newest innovations and trends in technology. In addition, it is critical to cultivate an innovative and experimenting culture within organizations. Businesses can create ground-breaking solutions and gain a competitive edge by pushing employees to experiment with new technologies and step outside of their comfort zones. In a technology landscape that is changing quickly, cooperation and partnerships with other people or organizations can also spur innovation and success (Elmeadawy, S., & Shubair, R. M., 2019).

Fostering Innovation and Collaboration

In today's quickly changing business environment, an organization's ability to stay competitive and drive growth depends on its ability to foster innovation and collaboration are as shown in the figure 1, within the organization. Multiple important tactics can be used to foster an innovative culture. First and foremost, a common vision must be established. When the company's objectives and direction are made clear, staff members are motivated to contribute creative ideas that support the company's vision and can comprehend the meaning behind their work. For instance, a technology startup with the goal of revolutionizing communication can motivate its staff by outlining a clear vision for changing the way individuals connect and engage. Facilitating risk-taking is an additional crucial facet of cultivating innovation. Innovation frequently entails venturing beyond one's comfort zone and investigating novel concepts, even if they appear unorthodox or dangerous. Supervisors have the ability to foster a culture where workers are encouraged to take chances and learn from mistakes. Implementing guidelines such as Google's "20% time" policy, for example, encourages staff members to take on creative side projects without worrying about failing. Fostering innovation also requires collaboration.

Figure 2. Fostering collaboration and vision

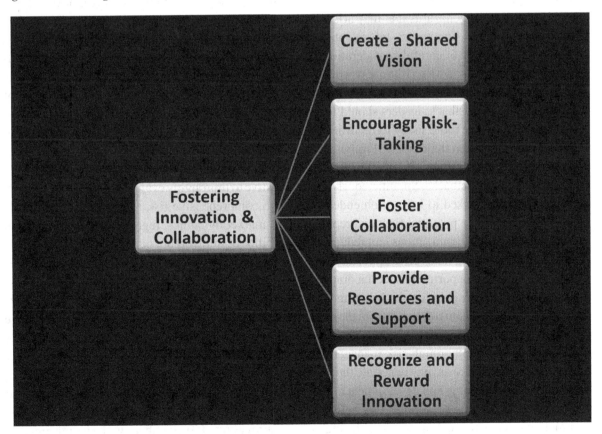

Organizations can harness the collective insights and creativity of individuals with diverse backgrounds and levels of expertise to effectively tackle intricate challenges. Opportunities for cross-functional collaboration, like brainstorming sessions or workshops, allow staff members to exchange ideas, dismantle organizational silos, and promote an innovative culture. Giving workers the tools and encouragement they need to innovate is crucial. This entails putting money into research and development, giving access to resources and technologies that encourage creative thinking, and providing opportunities for training and development. Employees can test out new concepts and technologies in dedicated innovation labs at companies like Apple and Microsoft. The last step in enhancing innovation's significance within the company is to acknowledge and reward it. A culture that values and encourages creativity is fostered by publicly praising individuals or groups for their creative solutions and by celebrating their accomplishments. Recognition initiatives, like those carried out by 3M, can encourage staff members to keep coming up with new ideas and promote constructive change within the company.

Balancing Risks and Opportunities

One of the most important things in encouraging innovation and teamwork in organizations is striking a balance between opportunities and risks. Even though innovation frequently entails taking chances and investigating novel concepts, it's critical to establish a balance between welcoming innovation and skillfully handling any hazards that may arise. Encouraging a culture of prudent risk-taking is one way to strike this balance. Encourage staff members to experiment with novel concepts and methods while highlighting the value of careful consideration, analysis, and risk assessment. Employees can make educated decisions about which risks are worthwhile to pursue and which may require additional analysis or mitigating measures if they receive the necessary direction and assistance. Additionally, instead of viewing failure as a setback, leaders should foster an environment where it is seen as a teaching opportunity. Employees are more likely to innovate and try out new ideas when they feel empowered to take chances and learn from their errors. Promote a culture of continuous improvement by promoting open communication and transparency, enabling staff members to share their experiences and insights—whether they are triumphs or setbacks. Simultaneously, it is imperative to thoroughly evaluate the possible hazards and consequences linked to innovation endeavors. Make comprehensive risk assessments, taking into account variables like potential financial repercussions, market dynamics, regulatory compliance, and stakeholder impact. Organizations can reduce the possibility of unfavorable outcomes and make better decisions about seizing innovative opportunities by anticipating and addressing potential risks. In order to balance risks and opportunities, collaboration is also very important. Organizations can better assess risks and find creative solutions by bringing together people with different viewpoints and areas of expertise. This allows them to benefit from the collective insights and experiences. Teams that collaborate across functional boundaries are better able to take into account different points of view, foresee possible problems, and create thorough risk management plans. Leaders should also give staff members clear direction and assistance as they negotiate the challenges of risk and innovation. Provide employees with opportunities for training and development to improve their risk management abilities. They should also have access to tools and resources that help them make well-informed decisions. Organizations may promote an innovative culture while reducing risks by providing staff members with the information and tools they need to manage risks (Lu, Y., & Zheng, X., 2020.

SECURITY IN HOLOGRAPHIC 6G

Identifying and Addressing Threats

One of the most important aspects of encouraging innovation and teamwork in organizations is recognizing and responding to threats. When companies pursue innovation, they have to anticipate and address possible roadblocks that might get in the way of their progress or compromise their ability to meet their goals. A thorough approach incorporates a number of important tactics. First and foremost, it is critical to perform in-depth risk assessments. Companies must carefully examine all external and internal factors that may affect their capacity for effective collaboration and innovation. These variables include changes in regulations, market dynamics, cybersecurity vulnerabilities, and internal operating difficulties, among other things. In order to guard against possible threats, organizations should also put proactive monitoring systems in place. This entails monitoring market trends, keeping an eye on rivalry, and keeping up with new threats. Organizations can identify possible threats early and take preventative action to address them before they become serious problems by continuously monitoring the business environment. In addition, cultivating an environment of candid communication is crucial. Employers need to foster a culture where staff members are encouraged to raise issues and provide information about possible dangers. Encouragement of cross-functional collaboration can facilitate the consideration of diverse perspectives. Organizations can identify threats that may have gone unnoticed and develop effective strategies to mitigate them by welcoming input from a variety of perspectives.

Collaborative problem-solving is also very important. Putting workers in interdisciplinary teams to address threats that have been identified promotes shared knowledge and understanding. By working together, organizations can create comprehensive solutions that address the underlying causes of risks and lessen their influence on initiatives involving innovation and collaboration. Effective threat mitigation also requires agility and adaptability. Businesses need to recognize that the business environment is dynamic and be ready to quickly adapt by reallocating resources, adjusting strategies, and taking action when new risks arise. Organizations that adopt a flexible approach are better equipped to quickly address threats and seize new opportunities. Putting money into resilience is also essential. Businesses should diversify their sources of income, invest in strong cybersecurity measures, and create backup plans in order to protect themselves from possible threats and disruptions. By incorporating resilience into the organizational structure, businesses can weather unforeseen obstacles and keep up their innovative and cooperative efforts. Finally, companies ought to embrace a continuous improvement mentality. Organizations can remain ahead of potential threats by regularly reviewing and reevaluating risk factors and updating mitigation strategies. Through iterative improvement of procedures and methods, organizations can eventually improve their capacity to recognize and neutralize threats (Ullah, Y. et al., 2023).

Implementing Encryption and Protocols

Encryption and protocol implementation are essential for protecting data and communication routes inside enterprises. Sensitive information is shielded from unwanted access or interception by encryption, which acts as a basic security measure by encoding data in a way that only authorized parties can access and decode. Ensuring confidentiality, integrity, and authenticity of data while it's in use, in transit, and at rest requires the use of strong encryption algorithms. This entails encrypting data before it is transferred over networks or stored on servers in order to prevent breaches even in the event that data is accessed by unauthorized parties. One of the most important functions of secure communication protocols (David, S. et al., 2023) is to protect data transfer across networks. Data exchanged between clients and servers is encrypted by protocols like Secure Sockets Layer (SSL) and Transport Layer Security (TLS), which guard against data manipulation, man-in-the-middle attacks, and eavesdropping. Organizations can strengthen their overall cybersecurity posture by putting these protocols into place, which guarantee the confidentiality and integrity of data while it is being transmitted. Encrypting data from beginning to end is essential for maintaining its confidentiality from sender to recipient. By limiting access to the encryption keys needed to decrypt the data (Pandey, B. K. et al., 2022), this method of encryption makes sure that only the sender and the intended recipient have access to the plaintext. For sensitive communications, like messaging apps or email exchanges, end-to-end encryption is especially important to prevent illegal access or interception. For encrypted data to remain secure, efficient key management procedures are necessary. To reduce the risk of key compromise, organizations should set up robust key management procedures. These procedures should include creating strong encryption keys, distributing and storing keys securely (Sharma, S. et al., 2023), and rotating keys on a regular basis. Furthermore, adding an additional layer of security to the encryption key access process through the use of multi-factor authentication helps to further protect sensitive data. It is crucial to guarantee adherence to industry norms and laws that regulate encryption and data security procedures. Strict encryption requirements are mandated by laws like the Health Insurance Portability and Accountability Act (HIPAA) and the General Data Protection Regulation (GDPR) in order to protect sensitive data. Following these guidelines strengthens an organization's defenses against possible data breaches and helps it avoid legal ramifications. Frequent security assessments and audits are necessary to determine possible vulnerabilities and gauge how well encryption implementations are working. Organizations can strengthen their cybersecurity defenses by promptly addressing weaknesses found in encryption protocols or key management practices through the use of penetration testing, vulnerability scanning, and security audits.Ultimately, thorough training and awareness campaigns are essential for teaching staff members the value of protocols and encryption in preserving data security. Best practices for protecting sensitive data, spotting phishing scams, and sending data securely should be taught to staff members. Organizations can reduce the human element in cybersecurity risks and strengthen their defenses against illegal access or data breaches by cultivating a culture of security awareness (Wu, W. et al., 2022).

Establishing Regulatory Frameworks

Regulatory frameworks must be established in order to regulate different facets of operations in businesses and sectors. In order to guarantee adherence to regulatory requirements and industry standards, these frameworks function as standards and guidelines that specify appropriate practices, procedures, and behaviors. Regulatory frameworks safeguard the interests of stakeholders, such as customers, workers, and the general public, while simultaneously advancing accountability, transparency, and fairness. Financial regulations, environmental standards, data privacy laws, and occupational health and safety regulations are just a few of the many domains they cover (Singh, S. et al., 2023). Reduced risk and potential harm are two of the main goals of regulatory frameworks. Regulatory frameworks assist in identifying and addressing potential risks connected to particular activities or industries by establishing explicit rules and regulations. Financial regulatory frameworks, for instance, set limits on capital requirements, risk management procedures, and disclosure requirements for financial institutions in an effort to preserve stability and integrity in the financial markets. In a similar vein, environmental regulations establish guidelines for reducing pollution, conserving resources, and implementing sustainable practices in order to prevent environmental degradation and safeguard ecosystems. Furthermore, regulatory frameworks are essential for guaranteeing responsibility and encouraging moral behavior. Regulatory frameworks hold people and organizations responsible for their choices and actions by outlining rights, obligations, and expectations. By establishing procedures for enforcement and sanctions for non-compliance, they serve to discourage wrongdoing and unethical behavior. Furthermore, openness and disclosure clauses are frequently included in regulatory frameworks, allowing stakeholders to hold companies responsible for their actions and performance as well as to make educated decisions. Regulatory frameworks also help to promote confidence and trust in markets and industries. Regulatory frameworks contribute to the development of trust among stakeholders, including investors and consumers, by establishing minimum standards and requirements. For instance, by setting guidelines for medical procedures, medication approvals, and healthcare facilities, regulatory frameworks in the healthcare sector guarantee patient safety, privacy, and high-quality treatment. Similar to this, data privacy laws set limits on data collection, storage, and use in order to safeguard people's private information and promote confidence in digital services and platforms (Habibi, M. A. et al., 2023).

ETHICAL AND SOCIETAL IMPLICATIONS

Safeguarding Privacy and Data

With so much sensitive and personal data being collected, stored, and sent electronically in the modern digital age, protecting privacy and data has become more and more important. To safeguard people's right to privacy and guarantee the security and confidentiality of their data, appropriate measures must be put in place. This entails building strong frameworks, guidelines, and technology to protect data and privacy in a variety of fields and businesses. Enacting thorough data protection laws and regulations is one of the essential components of protecting privacy and data. These laws establish requirements for businesses that gather, use, or store personal information as well as the rights of individuals with regard to that information. Regulations such as the California Consumer Privacy Act (CCPA) and the General Data Protection Regulation (GDPR) of the European Union, for instance, set stringent guidelines for data security and privacy. These guidelines include provisions for consenting to data collection, data minimization, purpose limitation, and notification of data breaches. Organizations must also implement security measures and privacy-enhancing technologies to guard against misuse, disclosure, and illegal access to data. Techniques like tokenization, anonymization, and encryption can be used to protect data while it's in transit and at rest, making sure that only people with the proper authorization can access and decrypt sensitive data. Strong authentication and access control measures are also helpful in preventing unauthorized users from accessing data, and frequent vulnerability assessments and security audits aid in identifying and reducing potential security risks. Furthermore, it's critical to promote a privacy- and data-protective culture within businesses. This entails educating staff members about the value of protecting privacy and data as well as offering training on security procedures, compliance requirements, and best practices for data protection. Workers should be alert in spotting and reporting possible security incidents or breaches, as well as aware of their roles and responsibilities when handling sensitive information. Organizations should build alliances and collaborate on projects to improve privacy and data protection in addition to internal initiatives. Organizations can exchange best practices, resources, and threat intelligence with government agencies, regulatory bodies, and other industry peers through collaboration. This promotes a team approach to tackling privacy and data security issues. In addition, interacting with consumer advocacy groups and privacy advocacy groups shows a dedication to accountability and transparency, which promotes confidence and trust among stakeholders and customers.

Promoting Inclusivity and Accessibility

The removal of barriers and the cultivation of environments in which everyone, regardless of their abilities or backgrounds, is able to fully participate are both necessary steps in the promotion of equal access and inclusivity are as shown in the figure 2. In order to accomplish this, it is necessary to establish settings that are easily accessible, to encourage diverse representation, and to embrace inclusive practices in every facet of society. Art can bring people together, inspire them, and help them think. It can also break down barriers and touch people's hearts from all walks of life. But people often can't fully access and engage with artistic experiences because of physical, sensory (Du John, H. V. et al., 2021), or mental limitations. To encourage a real appreciation of culture and creativity, it is important to make sure that everyone, no matter what their abilities or limitations are, has the chance to interact with and enjoy art. Physical accessibility is very important for making the art world more welcoming to everyone. Museums, art galleries, and exhibition spaces should make an effort to be welcoming by including things like ramps, elevators, and bathrooms that people with disabilities can use. Tactile maps and clear signs can help people who are blind or have low vision get around the space on their own, making sure they can fully enjoy the artistic experience. Art is an experience that involves more than one sense, and being inclusive means making accommodations for people who have trouble with certain senses. Giving people different ways to experience art, like tactile displays, audio descriptions, or multisensory (Dhanasekar, S. et al., 2023) installations, makes sure that everyone can feel comfortable expressing themselves through art. Businesses can help make the world a more welcoming place by providing quiet rooms or sensory (Abdulkarim, Y. I. et al., 2023) friendly events for people who need a calmer environment to fully enjoy art. In this digital age, it is important to make things more accessible outside of physical spaces. People who can't get to art shows in person or because of where they live can still enjoy art through virtual exhibitions and online platforms. Making digital resources accessible by adding features like closed captions for videos, alt text for photos, and screen reader compatibility lets disabled people enjoy art from the comfort of their own homes. Another important way to make the arts more accessible is to put together exhibitions that include works from a wide range of artists and backgrounds. By displaying a wide range of artistic styles, museums and galleries make the space feel welcoming and connect with more people. Working together with local groups and disabled artists is a great way to show how important it is to include everyone and build community in the art world. Education is one of the most important ways to make the arts more open and accessible to everyone. Institutions can make their educational programs more accessible by making art therapy sessions, workshops, and classes more accessible to people with disabilities. Working together with community centers, schools, and disability groups makes sure that everyone can get art instruction, giving them the tools to discover their creative side and express themselves through art.

Figure 3. Advancing accessibility and inclusivity

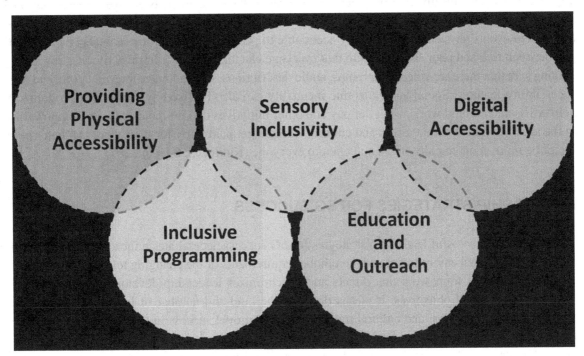

Addressing Socio-Economic Disparities

It is important to deal with differences in socioeconomic status when it comes to art and culture in order to make society fairer and more open so that everyone can enjoy and benefit from artistic experiences. People's ability to take part in cultural activities and get artistic opportunities is often based on their socioeconomic status, such as their income, education, and access to resources. Multiple strategies can be used to make the art world more welcoming and equal for everyone. For starters, making art and culture events cheaper or even free can help get more people to attend. Low-income people can get cheaper tickets, community events can be held for free, or programs can be funded so that people from underserved areas can get into cultural institutions. Targeted outreach programs that work with underrepresented groups can also help close the gap between rich and poor people's access to art and culture. Partnering up with schools, community groups, and social service agencies can make it easier to reach out to people and make sure that artistic opportunities reach people who may be having trouble with money. Another important way to deal with socioeconomic differences is to put money into arts education programs in schools and communities. Schools can give all students, no matter how much money they have, the chance to express themselves through art by including arts education in the regular school day and offering extracurricular activities. It's also important to support new artists from a range of socioeconomic backgrounds. New artists can pursue their art and contribute to culture through grants, scholarships, residencies, and mentorship programs. This way, talented people don't miss out on opportunities because they can't afford them. To fix social and economic problems, we need to support cultural diversity and representation in the arts. Cultural institutions can create a more inclusive and fair art scene that reflects the diversity of society by showcasing the work of artists from different backgrounds and giving a voice to those who aren't heard as much. Accessibility programs that make cultural institutions and events physically and digitally accessible to people with disabilities can also help close the gap between rich and poor. To make sure that everyone can take part in cultural activities, this includes making sure that there are wheelchair ramps, audio descriptions, sign language interpretation, and online accessibility features. Social and economic inequality can also be fixed by supporting programs that help artists and cultural workers get fair pay and other initiatives that promote economic empowerment in the arts. Making sure that artists and cultural workers get paid fairly for their work can help the arts sector be more economically stable and open to everyone (Kukliński, S. et al., 2021).

LEADERSHIP STRATEGIES FOR 6G SUCCESS

Developing successful leadership strategies for 6G success necessitates a thorough comprehension of the organizational environment and the unique requirements of the changing telecom sector. A leadership strategy is a framework that directs and synchronizes leadership development programs with the business's overall objectives. It means defining the kind and number of leaders required as well as the behaviors, abilities, and cultural traits required to propel success in the 6G era. Understanding how leadership capabilities and business strategies are intertwined is essential to creating a successful leadership strategy. This entails evaluating the formal and informal leadership positions that are currently held and those that are anticipated within the company, as well as determining the competencies and conduct that are necessary for putting the 6G business plan into practice. It takes skill for leaders to guide their teams through the intricacies of new technologies, predict market trends, and encourage creativity. Furthermore, cultivating a culture of collective leadership is critical in the context of 6G, as cooperation and cross-border thinking are essential for advancing innovation and effectively addressing changing obstacles. Effective leaders are able to work across functional lines, take advantage of different viewpoints, and inspire groups of people to achieve shared objectives. Fostering a culture of leadership that prioritizes employee engagement, accountability, mentorship, and ongoing learning is imperative. Creating a strong leadership plan is essential for 6G success, and this calls for a deliberate effort to match industry changes in demand with talent development programs. It involves developing a culture of adaptability and resilience in the face of rapid change, as well as identifying the core competencies and knowledge base necessary to propel technological advancements. Businesses need to spend money on focused leadership development initiatives that give executives at all levels the abilities and perspective they need to succeed in the 6G ecosystem. Organizations can improve their ability to manage the intricacies of the 6G environment and seize new opportunities by outlining a clear leadership strategy. This entails developing a culture of leadership that places a high value on cooperation, creativity, and ongoing learning in addition to coordinating talent development programs with the business's strategic goals. In the end, businesses can set themselves up for success in the competitive and dynamic 6G market by investing in strong leadership strategies.

CONCLUSION

In the upcoming 6G era, the revolutionary development of holographic communications technologies has the potential to completely transform digital communication. Transmitting three-dimensional images of users over long distances is extremely important since it can create immersive and interactive experiences in a variety of industries. But to successfully navigate these complex challenges, effective leadership is required. It is imperative for leaders to anticipate and adjust to technological progress, promote creativity and teamwork, and weigh potential advantages and disadvantages within the dynamic 6G network. It is imperative that ethical and societal considerations be given careful thought to security, privacy protection, inclusivity, and socioeconomic inequality. Stakeholders can take advantage of the revolutionary potential of holographic connectivity in the 6G landscape by placing a high priority on cooperation, innovation, and adaptability. Strategic initiatives should be in line with the organization's overall objectives and make use of resources like the Health Technologies Knowledge Transfer Network to encourage creativity and teamwork. Organizations must embrace a continuous improvement and learning culture while they set out on this transformative journey. They also need to stay forward-thinking and agile. Through strategic leadership approaches, stakeholders can effectively manage the intricacies of the 6G shift and usher in an era enhanced by the potential of holographic communication.

REFERENCES

Abdulkarim, Y. I., Awl, H. N., Muhammadsharif, F. F., Saeed, S. R., Sidiq, K. R., Khasraw, S. S., & Pandey, D. (2023). Metamaterial-based sensors loaded corona-shaped resonator for COVID-19 detection by using microwave techniques. *Plasmonics*, 1–16.

Anand, R., Khan, B., Nassa, V. K., Pandey, D., Dhabliya, D., Pandey, B. K., & Dadheech, P. (2023). Hybrid convolutional neural network (CNN) for kennedy space center hyperspectral image. *Aerospace Systems*, 6(1), 71–78. 10.1007/s42401-022-00168-4

Anand, R., Lakshmi, S. V., Pandey, D., & Pandey, B. K. (2024). An enhanced ResNet-50 deep learning model for arrhythmia detection using electrocardiogram biomedical indicators. *Evolving Systems*, 15(1), 83–97. 10.1007/s12530-023-09559-0

Attanasio, B., Mazayev, A., du Plessis, S., & Correia, N. (2021). Cognitive load balancing approach for 6G MEC serving iot mashups. *Mathematics*, 10(1), 101. 10.3390/math10010101

Bessant, Y. A., Jency, J. G., Sagayam, K. M., Jone, A. A. A., Pandey, D., & Pandey, B. K. (2023). Improved parallel matrix multiplication using Strassen and Urdhvatiryagbhyam method. *CCF Transactions on High Performance Computing*, 5(2), 102–115. 10.1007/s42514-023-00149-9

Bhattarai, S., Park, J. M. J., Gao, B., Bian, K., & Lehr, W. (2016). An overview of dynamic spectrum sharing: Ongoing initiatives, challenges, and a roadmap for future research. *IEEE Transactions on Cognitive Communications and Networking*, 2(2), 110–128. 10.1109/TCCN.2016.2592921

. Bruntha, P. M., Dhanasekar, S., Hepsiba, D., Sagayam, K. M., Neebha, T. M., Pandey, D., & Pandey, B. K. (2023). Application of switching median filter with L 2 norm-based auto-tuning function for removing random valued impulse noise. *Aerospace systems*, 6(1), 53-59.

Chowdhury, M. Z., Shahjalal, M., Ahmed, S., & Jang, Y. M. (2020). 6G wireless communication systems: Applications, requirements, technologies, challenges, and research directions. *IEEE Open Journal of the Communications Society*, 1, 957–975. 10.1109/OJCOMS.2020.3010270

David, S., Duraipandian, K., Chandrasekaran, D., Pandey, D., Sindhwani, N., & Pandey, B. K. (2023). Impact of blockchain in healthcare system. In *Unleashing the Potentials of blockchain technology for healthcare industries* (pp. 37–57). Academic Press. 10.1016/B978-0-323-99481-1.00004-3

Deepa, R., Anand, R., Pandey, D., Pandey, B. K., & Karki, B. (2022). Comprehensive performance analysis of classifiers in diagnosis of epilepsy. *Mathematical Problems in Engineering*, 2022, 2022. 10.1155/2022/1559312

Devasenapathy, D., Madhumathy, P., Umamaheshwari, R., Pandey, B. K., & Pandey, D. (2024). Transmission-efficient grid-based synchronized model for routing in wireless sensor networks using Bayesian compressive sensing. *SN Computer Science*, 5(1), 1–11.

Dhanasekar, S., Martin Sagayam, K., Pandey, B. K., & Pandey, D. (2023). Refractive Index Sensing Using Metamaterial Absorbing Augmentation in Elliptical Graphene Arrays. *Plasmonics*, 1–11. 10.1007/s11468-023-02152-w

Dogra, A., Jha, R. K., & Jain, S. (2020). A survey on beyond 5G network with the advent of 6G: Architecture and emerging technologies. *IEEE Access : Practical Innovations, Open Solutions*, 9, 67512–67547. 10.1109/ACCESS.2020.3031234

Du John, H. V., Jose, T., Jone, A. A. A., Sagayam, K. M., Pandey, B. K., & Pandey, D. (2022). Polarization insensitive circular ring resonator based perfect metamaterial absorber design and simulation on a silicon substrate. *Silicon*, 14(14), 9009–9020. 10.1007/s12633-021-01645-9

. Du John, H. V., Moni, D. J., Ponraj, D. N., Sagayam, K. M., Pandey, D., & Pandey, B. K. (2021). Design of Si based nano strip resonator with polarization-insensitive metamaterial (MTM) absorber on a glass substrate. *Silicon*, 1-10.

Ekong, M. O., George, W. K., Pandey, B. K., & Pandey, D. (2023). Enhancing the Fundamentals of Industrial Safety Management in TVET for Metaverse Realities. In *Applications of Neuromarketing in the Metaverse* (pp. 19-41). IGI Global. 10.4018/978-1-6684-8150-9.ch002

. Elmeadawy, S., & Shubair, R. M. (2019, November). 6G wireless communications: Future technologies and research challenges. In *2019 international conference on electrical and computing technologies and applications (ICECTA)* (pp. 1-5). IEEE.

George, W. K., Ekong, M. O., Pandey, D., & Pandey, B. K. (2023). Pedagogy for Implementation of TVET Curriculum for the Digital World. In *Applications of Neuromarketing in the Metaverse* (pp. 117-136). IGI Global. 10.4018/978-1-6684-8150-9.ch009

Giordani, M., Polese, M., Mezzavilla, M., Rangan, S., & Zorzi, M. (2020). Toward 6G networks: Use cases and technologies. *IEEE Communications Magazine*, 58(3), 55–61. 10.1109/MCOM.001.1900411

. Govindaraj, V., Dhanasekar, S., Martinsagayam, K., Pandey, D., Pandey, B. K., & Nassa, V. K. (2023). Low-power test pattern generator using modified LFSR. *Aerospace Systems*, 1-8.

Gupta, A. K., Sharma, R., Pandey, D., Nassa, V. K., Pandey, B. K., George, A. S., & Dadheech, P. (2023). Performance analysis of eight-channel WDM optical network with different optical amplifiers for industry 4.0. In *Innovation and Competitiveness in Industry 4.0 Based on Intelligent Systems* (pp. 197–212). Springer International Publishing. 10.1007/978-3-031-29775-5_9

Habibi, M. A., Han, B., Fellan, A., Jiang, W., Sánchez, A. G., Pavón, I. L., Boubendir, A., & Schotten, H. D. (2023). Towards an open, intelligent, and end-to-end architectural framework for network slicing in 6G communication systems. *IEEE Open Journal of the Communications Society*, 4, 1615–1658. 10.1109/OJCOMS.2023.3294445

Iyyanar, P., Anand, R., Shanthi, T., Nassa, V. K., Pandey, B. K., George, A. S., & Pandey, D. (2023). A real-time smart sewage cleaning UAV assistance system using IoT. In *Handbook of Research on Data-Driven Mathematical Modeling in Smart Cities* (pp. 24–39). IGI Global.

Kirubasri, G., Sankar, S., Pandey, D., Pandey, B. K., Nassa, V. K., & Dadheech, P. (2022). Software-defined networking-based Ad hoc networks routing protocols. In *Software defined networking for Ad Hoc networks* (pp. 95–123). Springer International Publishing. 10.1007/978-3-030-91149-2_5

Kirubasri, G., Sankar, S., Pandey, D., Pandey, B. K., Singh, H., & Anand, R. (2021, September). A recent survey on 6G vehicular technology, applications and challenges. In *2021 9th International Conference on Reliability, Infocom Technologies and Optimization (Trends and Future Directions)(ICRITO)* (pp. 1-5). IEEE.

Kukliński, S., Tomaszewski, L., Kołakowski, R., & Chemouil, P. (2021). 6G-LEGO: A framework for 6G network slices. *Journal of Communications and Networks (Seoul)*, 23(6), 442–453. 10.23919/JCN.2021.000025

Kumar Pandey, B., Pandey, D., Nassa, V. K., Ahmad, T., Singh, C., George, A. S., & Wakchaure, M. A. (2021). Encryption and steganography-based text extraction in IoT using the EWCTS optimizer. *Imaging Science Journal*, 69(1-4), 38–56. 10.1080/13682199.2022.2146885

Lu, Y., & Zheng, X. (2020). 6G: A survey on technologies, scenarios, challenges, and the related issues. *Journal of Industrial Information Integration*, 19, 100158. 10.1016/j.jii.2020.100158

Lu, Y., & Zheng, X. (2020). 6G: A survey on technologies, scenarios, challenges, and the related issues. *Journal of Industrial Information Integration*, 19, 100158. 10.1016/j.jii.2020.100158

Mahmoud, H. H. H., Amer, A. A., & Ismail, T. (2021). 6G: A comprehensive survey on technologies, applications, challenges, and research problems. *Transactions on Emerging Telecommunications Technologies*, 32(4), e4233. 10.1002/ett.4233

Pandey, B. K., & Pandey, D. (2023). Parametric optimization and prediction of enhanced thermoelectric performance in co-doped CaMnO3 using response surface methodology and neural network. *Journal of Materials Science Materials in Electronics*, 34(21), 1589. 10.1007/s10854-023-10954-1

Pandey, B. K., Pandey, D., & Agarwal, A. (2022). Encrypted information transmission by enhanced steganography and image transformation. [IJDAI]. *International Journal of Distributed Artificial Intelligence*, 14(1), 1–14. 10.4018/IJDAI.297110

Pandey, B. K., Pandey, D., Alkhafaji, M. A., Güneşer, M. T., & Şeker, C. (2023a). A reliable transmission and extraction of textual information using keyless encryption, steganography, and deep algorithm with cuckoo optimization. In *Micro-Electronics and Telecommunication Engineering: Proceedings of 6th ICMETE 2022* (pp. 629–636). Springer Nature Singapore. 10.1007/978-981-19-9512-5_57

Pandey, B. K., Pandey, D., Gupta, A., Nassa, V. K., Dadheech, P., & George, A. S. (2023b). Secret data transmission using advanced morphological component analysis and steganography. In *Role of data-intensive distributed computing systems in designing data solutions* (pp. 21–44). Springer International Publishing. 10.1007/978-3-031-15542-0_2

Pandey, B. K., Pandey, D., & Sahani, S. K. (2024). Autopilot control unmanned aerial vehicle system for sewage defect detection using deep learning. *Engineering Reports*, 12852. 10.1002/eng2.12852

Pandey, B. K., Pandey, D., Wairya, S., & Agarwal, G. (2021). An advanced morphological component analysis, steganography, and deep learning-based system to transmit secure textual data. [IJDAI]. *International Journal of Distributed Artificial Intelligence*, 13(2), 40–62. 10.4018/IJDAI.2021070104

Pandey, D., Wairya, S., Sharma, M., Gupta, A. K., Kakkar, R., & Pandey, B. K. (2022). An approach for object tracking, categorization, and autopilot guidance for passive homing missiles. *Aerospace Systems*, 5(4), 553–566. 10.1007/s42401-022-00150-0

Pandey, J. K., Jain, R., Dilip, R., Kumbhkar, M., Jaiswal, S., Pandey, B. K., & Pandey, D. (2022). Investigating role of iot in the development of smart application for security enhancement. In *IoT Based Smart Applications* (pp. 219–243). Springer International Publishing.

Pramanik, S., Pandey, D., Joardar, S., Niranjanamurthy, M., Pandey, B. K., & Kaur, J. (2023, October). An overview of IoT privacy and security in smart cities. In *AIP Conference Proceedings* (*Vol. 2495*, No. 1). AIP Publishing. 10.1063/5.0123511

Raja, D., Kumar, D. R., Santhiyakumari, N., Kumarganesh, S., Sagayam, K. M., Thiyaneswaran, B., Pandey, B. K., & Pandey, D. (2024). A compact dual-feed wide-band slotted antenna for future wireless applications. *Analog Integrated Circuits and Signal Processing*, 1–15. 10.1007/s10470-023-02233-0

Revathi, T. K., Sathiyabhama, B., Sankar, S., Pandey, D., Pandey, B. K., & Dadeech, P. (2022). An intelligent model for coronary heart disease diagnosis. In *Networking Technologies in Smart Healthcare* (pp. 309–327). CRC Press. 10.1201/9781003239888-15

Sahani, K., Khadka, S. S., Sahani, S. K., Pandey, B. K., & Pandey, D. (2023). A possible underground roadway for transportation facilities in Kathmandu Valley: A racking deformation of underground rectangular structures. *Engineering Reports*, 12821. 10.1002/eng2.12821

Sahani, S. K., Pandey, B. K., & Pandey, D. (2024). *Single-valued Signals, Multi-valued Signals and Fixed-Point of Contractive Signals*. Mathematics Open. 10.1142/S2811007224500020

Sasidevi, S., Kumarganesh, S., Saranya, S., Thiyaneswaran, B., Shree, K. V. M., & Martin Sagayam, K. (2024, May 15). Binay Kumar Pandey & Digvijay Pandey. (2024). Design of Surface Plasmon Resonance (SPR) Sensors for Highly Sensitive Biomolecular Detection in Cancer Diagnostics. *Plasmonics*. 10.1007/s11468-024-02343-z

Saxena, A., Agarwal, A., Pandey, B. K., & Pandey, D. (2024). Examination of the Criticality of Customer Segmentation Using Unsupervised Learning Methods. *Circular Economy and Sustainability*, 1–14. 10.1007/s43615-023-00336-4

Saxena, A., Sharma, N. K., Pandey, D., & Pandey, B. K. (2021). Influence of tourists satisfaction on future behavioral intentions with special reference to desert triangle of Rajasthan. *Augmented Human Research*, 6(1), 13. 10.1007/s41133-021-00052-4

. Sengupta, R., Sengupta, D., Pandey, D., Pandey, B. K., Nassa, V. K., & Dadeech, P. (2021). A Systematic review of 5G opportunities, architecture and challenges. *Future Trends in 5G and 6G*, 247-269.

Sharma, S., Pandey, B. K., Pandey, D., Anand, R., Sharma, A., & Saini, S. (2023, March). Character Recognition Technique Implementation for Complicated Deteriorated Scene. In *2023 6th International Conference on Information Systems and Computer Networks (ISCON)* (pp. 1-4). IEEE. 10.1109/ISCON57294.2023.10112185

Singh, S., Madaan, G., Kaur, J., Swapna, H. R., Pandey, D., Singh, A., & Pandey, B. K. (2023). Bibliometric Review on Healthcare Sustainability. *Handbook of Research on Safe Disposal Methods of Municipal Solid Wastes for a Sustainable Environment,* (pp. 142-161). IGI Global. 10.4018/978-1-6684-8117-2.ch011

Swapna, H. R., Bigirimana, E., Madaan, G., Hasan, A., Pandey, B. K., & Pandey, D. (2023). Impact of neuromarketing on consumer psychology in digitally connected networks. In *Applications of Neuromarketing in the Metaverse* (pp. 193–205). IGI Global. 10.4018/978-1-6684-8150-9.ch015

Tripathi, R. P., Sharma, M., Gupta, A. K., Pandey, D., Pandey, B. K., Shahul, A., & George, A. H. (2023). Timely prediction of diabetes by means of machine learning practices. *Augmented Human Research*, 8(1), 1. 10.1007/s41133-023-00062-4

Ullah, Y., Roslee, M. B., Mitani, S. M., Khan, S. A., & Jusoh, M. H. (2023). a survey on handover and mobility management in 5G HetNets: Current state, challenges, and future directions. *Sensors (Basel)*, 23(11), 5081. 10.3390/s2311508137299808

Vinodhini, V., Kumar, M. S., Sankar, S., Pandey, D., Pandey, B. K., & Nassa, V. K. (2022). IoT-based early forest fire detection using MLP and AROC method. *International Journal of Global Warming*, 27(1), 55–70. 10.1504/IJGW.2022.122794

Wu, W., Zhou, C., Li, M., Wu, H., Zhou, H., Zhang, N., Shen, X. S., & Zhuang, W. (2022). AI-native network slicing for 6G networks. *IEEE Wireless Communications*, 29(1), 96–103. 10.1109/MWC.001.2100338

Chapter 9
Navigating 6G Challenges:
A Managerial Perspective on Security Solutions – Strategic Insights

Shefali
Department of Management, Institute of Innovation in Technology and Management, Janakpuri, India

S. Prema
Department of Information Technology, Mahendra Engineering College, Namakkal, India

Vishal Ashok Ingole
P.R. Pote Patil College of Engineering and Management, India

G. Vikram
Karunya Institute of Technology and Sciences, Coimbatore, India

Smita M. Gaikwad
CMS B-School, Jain University (Deemed), India

H. Mickle Aancy
Department of Master of Business Administration, Panimalar Engineering College, Chennai, India

Pankaj Dadheech
https://orcid.org/0000-0001-5783-1989
Swami Keshvanand Institute of Technology, Management, and Gramothan, India

ABSTRACT

This chapter delves into the challenges posed by the advent of 6G technology from a managerial standpoint, particularly focusing on security solutions. As the telecommunications landscape evolves rapidly, it becomes imperative for managers to navigate the intricacies of ensuring robust security measures amidst technological advancements. Through strategic insights, this chapter explores the complexities associated with 6G security and provides managerial perspectives aimed at fostering proactive and effective security strategies.

INTRODUCTION TO 6G TECHNOLOGY

In the fast-paced landscape of technological innovation, the advent of each new generation of wireless communication brings with it transformative capabilities and unprecedented opportunities. As we stand on the cusp of the next evolutionary leap, the emergence of 6G (Kirubasri, G. et al., 2021) technology represents a seminal moment in the trajectory of connectivity. At its core, technology builds

DOI: 10.4018/979-8-3693-2931-3.ch009

upon the fundamental principles of wireless communication established by its predecessors, including 1G through 5G (Sengupta, R. et al., 2021). The adoption of terahertz (THz) frequencies, which provide unmatched bandwidth and data transmission (Iyyanar, P. et al., 2023) capabilities, is one of the distinguishing characteristics of 6G technology. Through the utilization of THz waves, which span frequencies from 0.1 to 10 THz, 6G networks are able to attain data speeds that are multiple orders of magnitude faster than those of 5G. The way we interact with digital content (George, W. K. et al., 2023) is being revolutionized by the seamless integration of bandwidth-intensive applications like augmented reality (AR), virtual reality (VR), and holographic communication made possible by this exponential increase in bandwidth (Raja, D. et al., 2024). Furthermore, to improve spectral efficiency and network capacity (Devasenapathy, D. et al., 2024), 6G technology integrates cutting-edge antenna (Bruntha, P. M. et al., 2023) technologies like massive MIMO (Multiple Input Multiple Output) and beamforming. By enabling precise beam steering and spatial multiplexing, these antenna technologies (Pandey, B. K. et al., 2024) maximize spectrum utilization and boost communication link dependability (Govindaraj, V. et al., 2023). Furthermore, autonomous network management, intelligent resource allocation, and predictive analytics are made possible by the integration of artificial intelligence (AI) (Pandey, B. K., & Pandey, D., 2023) and machine learning algorithms (Anand, R. et al., 2023) at the core of 6G networks, which improves user experience and network performance (Tripathi, R. P. et al., 2023). 6G technology is poised to usher in a new era of innovation and connectivity by revolutionizing numerous industries and domains (Jayapoorani, S. et al., 2023). 6G-enabled telemedicine platforms, for instance, have the potential to transform surgical techniques, diagnostic imaging, and remote patient monitoring in the healthcare industry (Singh, S. et al., 2023). This would allow healthcare providers to provide prompt, individualized care to patients regardless of their location. Similar to this, 6G connectivity in the automotive industry makes it possible for autonomous cars to seamlessly communicate with their environment (Swapna, H. R. et al., 2023), improving road safety, streamlining traffic, and completely changing the logistics of transportation. In addition, the introduction of 6G networks has the potential to close the digital gap (Ekong, M. O. et al., 2023) by giving underprivileged areas and marginalized communities ubiquitous connectivity. With applications ranging from precision agriculture and environmental monitoring to smart cities and sustainable energy grids, 6G technology is both diverse and transformative, providing answers to some of the most important problems (Pandey, D. et al., 2023) facing modern society (Apriliyanti, M., 2022).

CHALLENGES IN 6G TECHNOLOGY

Increased Complexity

The complex interweave of today's society has reshaped our way of navigating life, creating a more complicated landscape. Globalization, environmental changes, socio-political dynamics, and technological advancements are some of the interconnected factors that contribute to this complexity (Vinodhini, V. et al., 2022). These components combine to create a variety of opportunities and problems that call for careful consideration, strategic thinking, and tactical action. The speed at which technology is developing is one of the main causes of growing complexity. Technological advancements in areas like automation, biotechnology, and artificial intelligence (Anand, R. et al., 2024) are changing employment markets, upending industries, and fundamentally changing how people interact and communicate. Although these developments present unmatched chances for efficiency and advancement, they also bring with them concerns about cybersecurity (Pandey, B. K. et al., 2021), privacy, and the moral implications of developing technologies. Furthermore, the digital divide makes already-existing disparities worse, resulting in a complicated web of exclusion and access. On an unprecedented scale, the process of globalization has linked economies, cultures, and societies. Because of this interconnection, trade, capital, and information can move more easily across national boundaries, promoting both cultural and economic development. It also makes people more susceptible to systemic risks like pandemics, financial crises, and climate change. To effectively address shared challenges, the complex web of global supply chains and interdependent networks necessitates strong governance mechanisms and international cooperation. The complexity of contemporary society is greatly influenced by social and political dynamics. Urbanization, cultural diversity, and changing demographics upend established institutions and reshape communities. Concerns about identity politics, polarization, and inequality highlight the necessity of inclusive governance frameworks and social cohesion mechanisms. Furthermore, the emergence of populism and authoritarianism in different regions of the world complicates geopolitical relations and global governance, necessitating thoughtful responses to protect democratic values and human rights. A crucial component of societal complexity is the growing environmental crisis, which is typified by resource depletion, biodiversity loss, and climate change. Because ecological systems are interrelated, perturbations in one place can have a significant global impact. In order to effectively address environmental challenges, a comprehensive strategy that takes into account the complex interactions between human activity and the natural world is needed. Shifting to sustainable practices and reducing the effects of climate change require concerted efforts involving a variety of stakeholders and sectors at the local, national, and international levels. To navigate uncertainty and take advantage of opportunities for innovation and collaboration, people, organizations, and governments must adopt flexible and agile strategies in the face of growing complexity. Developing interdisciplinary viewpoints, building resilience, and encouraging ongoing learning and experimentation are all components of embracing complexity. Creating strong networks and collaborations can improve group problem-solving abilities and make knowledge and best practices easier to share. Moreover, making investments in infrastructure, technology, and education can enable people and communities to prosper in a world that is changing quickly.

Pervasive Connectivity

Modern life is characterized by pervasive connectivity, where technology has seamlessly integrated into our interactions, work, and lives. Pervasive connectivity is the presence of interconnected digital networks that transcend geographical and time boundaries, enabled by advances in telecommunications, the internet, and semiconductor devices (Du John, H. V. et al., 2022). Today, people carry powerful computing devices (Dhanasekar, S. et al., 2023) in their pockets to stay connected to the internet (Menon, V. et al., 2022) and communicate instantly from anywhere on Earth. Today, social media, messaging apps, and video conferencing enable real-time, remote communication and collaboration. Pervasive connectivity has transformed commerce, education, healthcare (KVM, S. et al., 2024), and entertainment. E-commerce platforms have made shopping and paying online easier for consumers. Distance learning programs use digital technology to reach students worldwide and encourage lifelong learning. Telemedicine platforms improve healthcare outcomes and access to medical expertise, especially in underserved areas. Pervasive connectivity has also changed media consumption, with streaming services providing on-demand content across devices. Data flows across networks, exposing individuals to privacy and security threats (Pandey, J. K. et al., 2022). Strong measures are needed to protect personal information and digital assets. Lack of reliable internet infrastructure worsens connectivity gaps between urban and rural areas and affluent and marginalized communities (Castelo-Branco, I. et al., 2022).

Quantum Computing Threats

With its promise of revolutionary computational powers (Bessant, Y. A. et al., 2023) and exponentially faster processing speeds, quantum computing offers enormous opportunities as well as serious threats to a number of industries, including data privacy (Pandey, B. K. et al. 2023a), cybersecurity, and cryptography (Pramanik, S. et al., 2023). The possibility that quantum computing will make current cryptographic techniques outdated is one of the biggest threats it poses. Conventional encryption techniques, like RSA and ECC, rely on the difficulty of discrete logarithm problems and factoring big prime numbers, which are computationally demanding tasks for computers running on the classical era. However, Shor's algorithm and other algorithms like it could be used by quantum computers to factor large numbers quickly, cracking encryption schemes and jeopardizing sensitive data that has been stored in encrypted form (Pandey, B. K. et al., 2022). Sensitive data kept in databases, safe communication routes (Pandey, D. et al., 2022), and financial transactions may all be significantly impacted by this vulnerability. Digital signatures are generated by mathematical (Deepa, R. et al., 2022) algorithms like RSA and ECDSA, which produce distinct signatures that are verifiable using matching public keys. Nevertheless, Grover's algorithm may be used by quantum computers to crack these algorithms, jeopardizing the security (Sharma, S. et al., 2023) of digital signatures and undermining confidence in electronic transactions, contracts, and documentation. Apart from weaknesses in cryptography, quantum computing also presents threats to the security of blockchain (David, S. et al., 2023), the technology that powers cryptocurrencies and decentralized applications. Cryptographic hash functions are essential to blockchain networks in order to secure transactions and preserve distributed ledger integrity. However, by taking advantage of flaws in hash-based algorithms, quantum computers may be able to undermine the immutability of blockchain networks and allow for the manipulation of transaction histories or the compromise of the consensus mechanism. Although it is thought that these cryptographic primitives will not be compromised by quantum computer attacks, further research and development into quantum algorithms may eventually make them susceptible, emphasizing the necessity of ongoing innovation and adaptation in cybersecurity practices.

Privacy Concerns

The proliferation of online platforms and services that collect user data for a variety of purposes, including targeted advertising, behavioral analysis, and algorithmic decision-making, is one of the primary sources of privacy concerns about the internet. Users frequently give their consent to the collection and processing of their data without even realizing it. This consent is typically given through lengthy terms of service agreements and privacy policies, which are frequently written in difficult legal language and may not disclose the full extent of the data collection and sharing practices that are undertaken. This is because these devices continuously collect data about the behaviors, preferences, and physical environments of users. There are numerous opportunities for data collection within the Internet of Things ecosystem, which includes connected cars and medical devices, as well as smart home devices and wearable fitness trackers. However, there are also significant privacy risks associated with this ecosystem. Vulnerabilities in Internet of Things devices have the potential to expose sensitive information to unauthorized access or exploitation by malicious actors, which poses a threat to the privacy and security of individuals. It is becoming increasingly common for government agencies, law enforcement agencies, and private companies to increase their use of surveillance technologies in order to monitor the movements, activities, and interactions of individuals in both physical and digital environments. In spite of the fact that these technologies have the potential to be beneficial in terms of public safety and security, they also raise ethical and legal questions concerning privacy, consent, and the possibility of abuse or misuse of surveillance data (Akundi, A. et al., 2022).

Interoperability Issues

Interoperability means that different systems, devices, or applications can easily share and understand data. It is important for making sure that everything works together and is connected in today's digital world. But interoperability problems still exist, making it hard for information to flow freely and limiting how different technologies and platforms can be used. One of the main problems with interoperability is that there are so many proprietary standards and closed ecosystems that are run by tech companies. A lot of the time, businesses make their own protocols, formats, or interfaces to keep users inside their own ecosystems and stop them from using competing products or services. This lack of interoperability can cause vendor lock-in, which is when users can't switch between products or platforms without a lot of trouble or losing functionality. This makes the market less open to choose and new ideas. Interoperability problems also happen a lot when the data formats, protocols, or APIs (Application Programming Interfaces) that different systems or apps use are not the same. Without standard formats or protocols for exchanging data, developers may run into problems when trying to connect different systems or make solutions that work with other systems. Different industries and domains may not have the same standards and protocols. This can make it harder to share data, work together, and integrate systems. This can slow down innovation and efficiency in areas like healthcare, finance, and transportation. Problems with interoperability can also happen with older systems and infrastructure that don't support the latest standards or protocols that are needed for them to work with newer technologies. Moving data and programs from old systems to new ones can be hard and cost a lot of money. It needs careful planning, testing for compatibility, and maybe even custom integration solutions to make sure that everything works together and the data stays the same. There may also be problems with interoperability when old systems are combined with cloud-based or SaaS (Software as a Service) solutions. This is because different architectures or data models can make integration harder and negatively impact performance or dependability. Regulatory and compliance requirements can also make it harder for systems to work together, especially in industries with a lot of rules, like healthcare or finance. Industry-specific rules or standards may make it harder to share data, connect systems, or integrate them. To make sure they follow the rules while still being able to connect with other systems or stakeholders, businesses may need to use custom solutions or workflows (Akundi et al., 2022).

Regulatory Compliance

The term "regulatory compliance" describes following the rules, laws, and guidelines set forth by the government, business associations, or regulatory agencies. These rules are intended to safeguard moral conduct, safeguard consumers, encourage fair competition, and lessen risks across a range of sectors and industries. Understanding and abiding by the applicable laws and regulations that control particular businesses or activities is one of the most important components of regulatory compliance. Regulations like the Sarbanes-Oxley Act (SOX), the Dodd-Frank Act, and the Payment Card Industry Data Security Standard (PCI DSS), for instance, must be followed by businesses in the financial sector. Respecting patient privacy and maintaining the security of medical data in the healthcare industry requires adherence to laws like the Health Insurance Portability and Accountability Act (HIPAA). Regulatory compliance also entails putting policies, procedures, and controls in place to guarantee that businesses follow the law and industry norms. This could entail creating compliance programs, risk assessments, and internal controls to keep an eye on and enforce regulatory compliance. In many cases, compliance officers or departments are in charge of managing compliance initiatives, carrying out audits, and resolving any issues related to non-compliance. As technology develops, new risks arise, and corporate practices change, regulations must adapt accordingly. To handle new requirements and successfully reduce compliance risks, organizations need to be aware of regulatory developments and update their compliance programs proactively. Serious repercussions, such as monetary fines, legal ramifications, harm to one's reputation, and lost commercial opportunities, may arise from breaking regulatory requirements. Regulatory agencies are empowered to conduct compliance investigations and enforce compliance through audits, enforcement actions, and inspections. The organization's reputation and long-term viability can also be negatively impacted by noncompliance with regulatory requirements, which can erode trust and credibility with stakeholders, partners, and customers (Gladysz, B. et al., 2023).

MANAGERIAL PERSPECTIVES ON SECURITY SOLUTIONS

Proactive Security Measures

Given the pervasive and evolving nature of cybersecurity threats, proactive security measures are essential to modern organizational strategies. Managers must implement proactive security solutions to protect an organization's assets, reputation, and sensitive data. Managers must first conduct thorough risk assessments to identify threats to their systems, networks, and data. These assessments help managers prioritize security investments and allocate resources to mitigate the biggest risks. Once risks are identified, strong security policies are needed. Such policies outline employees' security duties. These policies must be reviewed and updated to address new threats and business changes. Along with policies, employee training and awareness programs are essential. Continuous training is crucial to defending against threats because employees are often a company's weakest link. Phishing emails, strong passwords, and data handling protocols should be covered in training. Access control and authentication are key to proactive security. Strong access controls like multi-factor authentication and least privilege access protect sensitive data and systems. User access rights are reviewed regularly to ensure authorized personnel have privileges. Both continuous monitoring and incident response are crucial. Organizations must constantly monitor networks and systems for unusual activity or breaches to practice proactive security. Security incidents must be detected, contained, and mitigated quickly with strong incident response plans. These audits ensure an organization's security controls are effective and meet industry standards. Close collaboration with internal and external auditors helps identify and fix weaknesses. A holistic approach to proactive security includes people, processes, and technology. Organizations can reduce risk and improve security by being proactive and vigilant against evolving threats. Cybersecurity is an ongoing process, not a one-time effort. Thus, organizations must constantly evaluate their security, adapt to new threats, and update their proactive security measures. In an increasingly complex threat landscape, managers can mitigate risks and protect assets by prioritizing security at all levels (Adel, A., 2022).

Machine Learning in Security

ML algorithms are great at analyzing large amounts of data, finding patterns, and making predictions, which helps prevent security breaches. Anomaly detection is a major ML security application. ML algorithms can learn baseline system, network, and user behavior and flag any deviations as security incidents. This proactive approach lets organizations respond quickly to emerging threats before they become attacks By analyzing historical data and trends, ML models can predict cyber threats and vulnerabilities. This lets organizations patch vulnerabilities, strengthen defenses, and mitigate risks before attackers do. ML-based behavior analysis also helps identify insider threats. ML algorithms can identify abnormal user behavior that may indicate insider threats or compromised accounts. Training data quality and diversity are crucial to ML algorithm performance. Training data biases can skew predictions, compromising organization security. Adversarial attacks can also fool ML algorithms into making incorrect predictions or classifications. As organizations use more digital technologies and face more sophisticated cyber threats, ML must be integrated into security frameworks to stay ahead of adversaries and protect sensitive data and assets. Organizations can improve security, threat detection, and cyber threat mitigation by using ML (Wang, B. et al., 2024)

Collaboration with Stakeholders

Developing effective security solutions that satisfy the various needs and concerns of all parties involved requires collaboration with stakeholders. Stakeholders in the cybersecurity space can include a broad spectrum of organizations, including executives, staff members, clients, partners, regulators, and even the general public. Working together effectively makes sure that security measures meet legal requirements, support stakeholder trust, and are in line with organizational objectives. Including executive leadership in security initiatives is one of the most important parts of working with stakeholders. In an organization, executives are critical in establishing the security culture, assigning resources, and making strategic choices about risk management. Organizations can guarantee that security is a top priority by including executives in security discussions and giving them regular updates on security posture and incidents. Employee awareness of and adherence to security policies are essential in reducing risks, as they are frequently the first line of defense against cyberattacks. Building a strong security culture where everyone feels accountable for safeguarding sensitive data and systems can be achieved by regularly offering security training, sharing security best practices, and asking staff members for their opinions. Maintaining the security of shared data and systems also requires cooperation with clients and partners (Gupta, A. K. et al., 2023). When sharing information with outside parties, organizations need to set clear policies and procedures for security and data protection. Strong connections and reducing security risks in cooperative endeavors require open communication, transparency, and mutual trust. Keeping up with changing security standards and compliance requirements also depends on working together with industry associations and regulatory bodies (Paschek, D. et al., 2019).

Investment in Research and Development

Investing in research and development is one of the best ways to maximize excess cash flow for capital expansion (R&D). This enables businesses to improve their goods and services while also innovating and staying one step ahead of the competition. Although R&D can be expensive, it is necessary for success and growth over the long run. This section will cover the advantages of funding research and development (R&D), the various forms of R&D, and how to calculate the right amount of investment.

A company that invests in R&D can reap many benefits, including, better goods and services: Research and development (R&D) can result in the creation of new, better goods and services that adapt to the changing needs and preferences of consumers. Businesses that make significant R&D investments can outperform their rivals by providing distinctive and cutting-edge goods and services. Over time, research and development can result in lower costs, more productivity, and larger profit margins. R&D can produce priceless intellectual property, like trademarks and patents, which can be used to safeguard the business's inventions and bring in more money. Basic, applied, and development research are the three primary categories of research and development. This kind of study aims to comprehend underlying ideas and principles. The goal of this kind of study is to create useful applications by applying the knowledge gathered from basic research. Developing new products and services or testing and improving already-existing ones are examples of applied research projects. This is the procedure used to transform applied research findings into goods and services that can be sold. Prototypes must be created, tested, and improved upon before they are prepared for market release. Determining the right amount of investment is crucial because investing in R&D involves a substantial financial commitment (Longo, F. et al., 2020). To find out if they are investing more or less than average, businesses should examine the R&D expenditures of their peers and competitors. Businesses should take into account the potential for new and emerging markets as well as the degree of demand for their goods and services. To establish themselves and get a foothold in the market, businesses that are just starting out may need to make larger investments in R&D. Businesses that are more willing to take chances and explore new opportunities may be more inclined to invest in R&D. An intelligent strategy to maximize excess cash flow for capital expansion is to invest in research and development. Better goods and services, a competitive edge, higher profitability, and priceless intellectual property are all possible outcomes. Businesses should think about the various forms of R&D and assess the right amount of investment depending on market demand, industry standards, development stage, and risk tolerance. Businesses can ensure their long-term success and growth by making R&D investments. Investing in research and development are as shown in Figure 1.

Figure 1. Investing in research and development

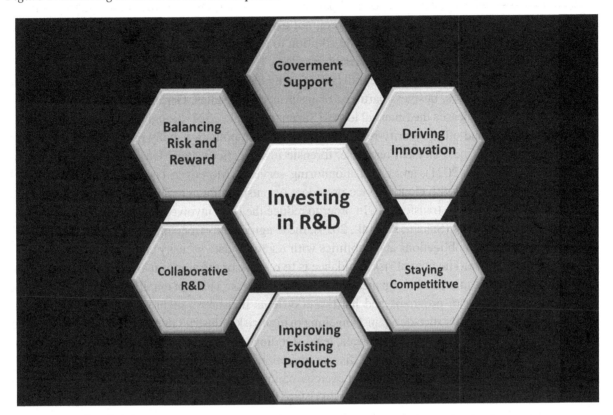

STRATEGIC INSIGHTS FOR SECURITY SOLUTIONS

Risk Assessment and Mitigation Strategies

Identification of possible risks and assessment of their impact and likelihood on a project or organization comprise risk assessment. Mitigation strategies are as shown in the figure 2, which include risk transfer, contingency planning, and the implementation of safeguards in an effort to reduce these risks as much as possible. To guarantee efficacy, regular observation and modification are essential.

1. *Installation of Controls*: Controls must be put in place in order to mitigate risks that are found during the assessment process. Technological controls monitor and filter network traffic, identify malicious activity, and stop unauthorized access. Examples of these controls include firewalls, antivirus software, and intrusion detection systems. These measures help safeguard against cyber threats. The purpose of these controls is to improve the security posture of the organization and reduce the probability of successful attacks. Employees must adhere to procedures and guidelines established by procedural controls, such as incident response plans, data encryption protocols, and access control policies, in order to guarantee the security and integrity of organizational assets. In order to fix vulnerabilities

that have been identified and lower the likelihood of exploitation, software (Kirubasri, G. et al., 2022) and systems must be updated and patched on a regular basis. Furthermore, putting in place a thorough security awareness training program gives staff members the knowledge and skills to identify and report suspicious activity in addition to teaching them about common security threats like phishing scams and social engineering attacks.

2. *Transfer of Risk*: Transferring risk entails transferring the cost or liability of particular risks to other parties, like suppliers, business partners, or insurance companies. Getting insurance coverage for cybersecurity can lessen the financial losses (Saxena, A. et al., 2024) brought on by network failures, data breaches, and other security-related incidents. These policies usually pay for costs associated with regulatory fines, incident response, forensic investigations, legal fees, customer notification (Saxena, A. et al., 2021), and credit monitoring services. Moreover, contracts like service-level agreements (SLAs) with cloud service providers or outsourcing agreements with outside vendors allow organizations to transfer risk. In order to reduce the risks involved with depending on outside services or suppliers (Sahani, K. et al., 2023), these agreements frequently contain provisions outlining each party's obligations and liabilities with regard to data security and privacy.

3. *Avoidance of Risk*: The goal of risk avoidance is to remove or refrain from any actions, procedures, or circumstances that put the organization at intolerable risk. This can entail getting rid of insecure or outdated software programs and hardware systems that aren't secure or maintained enough. Organizations may also decide against engaging in high-risk industries or commercial endeavors that put them at serious risk to their finances, legal standing, or reputation. Organizations can mitigate the risk of adverse events that could disrupt operations or harm the business by taking proactive measures to evaluate potential risks and exercise caution in decision-making processes. Risk avoidance tactics may involve trade-offs between risk reduction and innovation or business growth, and they necessitate a thorough analysis of the possible costs and benefits of particular investments or actions. Risk avoidance is the best course of action for safeguarding the organization's interests, though, in certain situations where the possible consequences of a given risk may outweigh the potential benefits.

4. *Acceptance of Risk*: Accepting risk entails accepting and living with some risks that are considered necessary or acceptable in light of the organization's resources, strategic goals, and risk tolerance. It may not be feasible or cost-effective to try to completely eliminate or mitigate every risk. As an alternative, businesses may decide to take on specific risks and concentrate their efforts on controlling and reducing their possible effects through different channels, like business continuity plans, crisis management techniques, or contingency planning. Accepting risk necessitates a thorough analysis of the possibility and possible outcomes of hazards that have been identified, as well as a consideration of the organization's capacity to absorb or recover from any unfavorable events that may arise. Accepting risks entails making informed decisions about which risks to prioritize and allocating resources towards addressing, while also acknowledging that some residual risk may always exist. Accepting risks does not mean ignoring or neglecting them.

5. *Ongoing Observation and Evaluation*: Effective risk management requires ongoing monitoring and review to make sure mitigation techniques are still applicable and efficient in dealing with new threats and vulnerabilities. Companies should periodically evaluate and reevaluate their risk posture, accounting for modifications to the threat landscape, technological advancements, legal requirements, operational procedures, and other elements that could affect the risk profile of the company. For the purpose of identifying new risks and evaluating the efficacy of current controls, periodic risk

assessments, vulnerability scans, penetration tests, and security audits must be carried out. In order to identify and address possible security incidents as soon as they arise, organizations should also set up procedures for continuous monitoring and surveillance of network activity, system logs, and security alerts. Organizations can guarantee the continuous protection of their assets, reputation, and operations by being proactive and alert in monitoring and assessing their security posture. This allows them to promptly detect new threats and vulnerabilities, plug any holes in their defenses, and modify their risk management plans as needed.

Figure 2. Mitigation strategies

End-to-End Encryption Implementation

Modern security solutions must include end-to-end encryption, which provides strong protection for sensitive data while it's in transit. From a managerial standpoint, deploying requires strategic choices and operational judgments to guarantee successful deployment while striking a balance between security requirements and usability and compliance specifications. First and foremost, managers need to comprehend the tenets and advantages. Data is encrypted at the source and decrypted only at the intended destination with E2EE, rendering it unreadable by intermediaries or unauthorized parties, in contrast to traditional encryption techniques, which only secure data while it's in transit between endpoints. This protects against listening in on someone else's conversations, intercepting them, and tampering with them by guaranteeing confidentiality and integrity throughout. Managers must strategically determine whether it is appropriate for the unique use cases and communication channels of their company. E2EE offers robust security, but it can also be complicated and have drawbacks, especially in group settings where access control and data sharing are crucial. In order to choose the best strategy for implementing, managers must weigh the trade-offs between security, usability, and functionality. Technologies, incorporating them into current workflows and systems, and creating guidelines and protocols for user authentication, key management, and incident response are all operational considerations. Additionally, managers have to make sure that solutions meet all applicable industry standards, legal requirements, and contractual obligations. This is especially important in industries like healthcare, finance, and government that deal with sensitive or regulated data. Efficient training and communication are essential for a successful implementation. It is the responsibility of managers to inform staff members and other interested parties about the significance and, its effects on privacy and security (Revathi, T. K. et al., 2022), and the safest ways to use tools and applications that support E2EE (Ghobakhloo, M. et al., 2022).

Robust Authentication Mechanisms

Today's digital landscape, with its proliferation of online services and growing cyber threats, requires strong authentication mechanisms to protect user identities and sensitive data. Development of effective authentication mechanisms requires technical innovation, user experience considerations, and security best practices. Multi-factor authentication (MFA) or two-factor authentication (2FA) is essential to strong authentication. Compared to single-factor authentication, these methods require users to provide multiple verification methods like passwords, biometrics, or mobile one-time codes, improving security. MFA protects against unauthorized access by combining a password with a smartphone or fingerprint.

Cryptographic (Pandey, B. K. et al., 2023b) methods like public-key cryptography and digital signatures are also necessary for authentication integrity and confidentiality. Cryptographic protocols like TLS and SSL secure client-server communication, preventing eavesdropping and man-in-the-middle attacks during authentication.

User-centric design principles and technical measures are essential to strong authentication mechanisms. User experience factors like ease of use, accessibility, and seamless workflow integration are crucial for end-user adoption and compliance. Complex authentication processes can frustrate users and allow security bypasses, reducing authentication effectiveness. Continuous authentication monitors user behavior and contextual factors to verify identity throughout a system or service interaction. User behavioral patterns, such as keystroke dynamics, mouse movements, and touchscreen gestures, can be used to passively authenticate users for a frictionless and adaptive authentication experience while maintaining security. Adaptive authentication mechanisms that dynamically adjust authentication requirements based on risk factors like the user's location, device type, and behavior can help organizations balance security and user convenience. By using risk-based authentication policies, organizations can strengthen authentication for high-risk or suspicious behavior while minimizing disruptions for legitimate users. Regular security assessments, penetration testing, and vulnerability scanning help identify and fix authentication vulnerabilities. Developers can protect authentication mechanisms from new threats and attack vectors by proactively identifying and fixing security flaws (Ghobakhloo et al., 2022).

Leveraging Zero-Trust Architecture

The idea behind zero-trust architecture (ZTA) is that you should never trust something and should always check it. It goes against the usual perimeter-based security model by assuming that threats may already be inside the network and that all users and devices should be verified and given permission every time they try to connect, no matter where they are or what access rights they have had before. Using ZTA means putting in place a set of rules and technologies that are meant to improve security and lower the risk of data breaches and unauthorized access. Using strict access controls based on the principle of least privilege is an important part of using ZTA. This is done by giving users and devices only the minimal access they need to do their jobs. This lowers the damage that a security breach or insider threat could do. Solutions for identity and access management (IAM), like role-based access control (RBAC), attribute-based access control (ABAC), and continuous authentication, can be used to control who can see and change what. In ZTA, it is also important to use strong authentication and authorization systems. Strong encryption protocols, digital certificates, and multi-factor authentication (MFA) help make sure that users and devices are who they say they are before they can access sensitive data or resources. Authorization policies should be enforced dynamically based on things like the user's location, the type of device they are using, the network environment, and their past behavior. This will make sure that access privileges are flexible and in line with security policies. By keeping an eye on system logs, network traffic, and user behavior all the time, companies can quickly spot strange behavior or possible security problems. Advanced technologies for finding threats, like intrusion detection systems (IDS), intrusion prevention systems (IPS), and security information and event management (SIEM) solutions, help connect and analyze security events across the network and make the process of responding to incidents more automated. Organizations can reduce the attack surface and keep hackers from getting to sensitive data and critical assets by separating the network into smaller, separate segments and controlling access between them (Moro, S. R. et al., 2023).

Enhancing Incident Response Capabilities

Today's complex cybersecurity landscape requires organizations to improve incident response. A comprehensive incident response plan is essential to this process. This plan describes how the company will detect, assess, contain, and mitigate security incidents. It defines key stakeholders' roles and responsibilities, escalation paths and communication protocols, and incident response steps. To address changing threats and the organization's environment, this plan is reviewed and updated regularly. Strong detection and alerting systems are needed for incident response. Companies use intrusion detection systems (IDS), security information and event management (SIEM) solutions, and endpoint detection and response (EDR) platforms to monitor their networks, systems, and endpoints for suspicious activity. Responders can quickly address potential security incidents with these tools' real-time alerts and notifications. Incident response playbooks help responders handle specific incidents. These playbooks provide detailed instructions, tools, and resources for incident containment and mitigation, ensuring a coordinated and consistent organization-wide response. Incident response team members need regular training and exercises to learn their roles and follow the plan and playbooks. Tabletop exercises and simulated cyberattack scenarios help teams assess their security incident response readiness and identify areas for improvement. Creating a centralized incident response team with diverse skills and expertise is essential for effective coordination and execution. This team works with law enforcement, regulatory bodies, cybersecurity vendors, and incident response firms to share threat intelligence, coordinate response efforts, and access specialized expertise and resources. Continuous evaluation and improvement of incident response processes keep the organization agile and responsive to new threats and challenges. By proactively improving incident response, organizations can mitigate security incidents, protect critical assets, and maintain business continuity in the face of evolving cyber threats.

CONCLUSION

Taking into consideration the analysis of 6G technology and the challenges that are associated with it, it is clear that preventative security measures and strategic solutions are of significant importance. Organizations need to navigate a complex landscape in order to ensure the secure deployment and operation of 6G networks. This landscape includes the complexities of implementation, the looming threats of quantum computing, and concerns regarding privacy. When it comes to proactive security, managerial perspectives advocate for the implementation of machine learning, collaboration with stakeholders, and investment in research and development as essential pillars. There is a need for comprehensive risk assessment, end-to-end encryption, robust authentication mechanisms, zero-trust architecture, and enhanced incident response capabilities, according to the strategic insights that were gleaned from this examination. It is possible for organizations to not only mitigate the challenges that are associated with 6G technology but also to harness the transformative potential of this technology if they adopt these measures and solutions. The realization of a future of seamlessly connected and secure 6G networks is possible through concerted efforts in the areas of security, collaboration, and innovation. This will bring about new opportunities for the advancement of society as well as the expansion of the economy.

REFERENCES

Adel, A. (2022). Future of Industry 5.0 in society: Human-centric solutions, challenges and prospective research areas. *Journal of Cloud Computing (Heidelberg, Germany)*, 11(1), 40. 10.1186/s13677-022-00314-536101900

Akundi, A., Euresti, D., Luna, S., Ankobiah, W., Lopes, A., & Edinbarough, I. (2022). State of Industry 5.0—Analysis and identification of current research trends. *Applied System Innovation*, 5(1), 27. 10.3390/asi5010027

Anand, R., Khan, B., Nassa, V. K., Pandey, D., Dhabliya, D., Pandey, B. K., & Dadheech, P. (2023). Hybrid convolutional neural network (CNN) for kennedy space center hyperspectral image. *Aerospace Systems*, 6(1), 71–78. 10.1007/s42401-022-00168-4

Anand, R., Lakshmi, S. V., Pandey, D., & Pandey, B. K. (2024). An enhanced ResNet-50 deep learning model for arrhythmia detection using electrocardiogram biomedical indicators. *Evolving Systems*, 15(1), 83–97. 10.1007/s12530-023-09559-0

Apriliyanti, M. (2022). Challenges of The Industrial Revolution Era 1.0 to 5.0: University Digital Library In Indoensia. *Library Philosophy and Practice*, 1-17.

Bessant, Y. A., Jency, J. G., Sagayam, K. M., Jone, A. A. A., Pandey, D., & Pandey, B. K. (2023). Improved parallel matrix multiplication using Strassen and Urdhvatiryagbhyam method. *CCF Transactions on High Performance Computing*, 5(2), 102–115. 10.1007/s42514-023-00149-9

. Bruntha, P. M., Dhanasekar, S., Hepsiba, D., Sagayam, K. M., Neebha, T. M., Pandey, D., & Pandey, B. K. (2023). Application of switching median filter with L 2 norm-based auto-tuning function for removing random valued impulse noise. *Aerospace systems*, 6(1), 53-59.

Castelo-Branco, I., Oliveira, T., Simões-Coelho, P., Portugal, J., & Filipe, I. (2022). Measuring the fourth industrial revolution through the Industry 4.0 lens: The relevance of resources, capabilities and the value chain. *Computers in Industry*, 138, 103639. 10.1016/j.compind.2022.103639

David, S., Duraipandian, K., Chandrasekaran, D., Pandey, D., Sindhwani, N., & Pandey, B. K. (2023). Impact of blockchain in healthcare system. In *Unleashing the Potentials of blockchain technology for healthcare industries* (pp. 37–57). Academic Press. 10.1016/B978-0-323-99481-1.00004-3

Deepa, R., Anand, R., Pandey, D., Pandey, B. K., & Karki, B. (2022). Comprehensive performance analysis of classifiers in diagnosis of epilepsy. *Mathematical Problems in Engineering*, 2022, 2022. 10.1155/2022/1559312

Devasenapathy, D., Madhumathy, P., Umamaheshwari, R., Pandey, B. K., & Pandey, D. (2024). Transmission-efficient grid-based synchronized model for routing in wireless sensor networks using Bayesian compressive sensing. *SN Computer Science*, 5(1), 1–11.

Dhanasekar, S., Martin Sagayam, K., Pandey, B. K., & Pandey, D. (2023). Refractive Index Sensing Using Metamaterial Absorbing Augmentation in Elliptical Graphene Arrays. *Plasmonics*, 1–11. 10.1007/s11468-023-02152-w

Du John, H. V., Jose, T., Jone, A. A. A., Sagayam, K. M., Pandey, B. K., & Pandey, D. (2022). Polarization insensitive circular ring resonator based perfect metamaterial absorber design and simulation on a silicon substrate. *Silicon*, 14(14), 9009–9020. 10.1007/s12633-021-01645-9

Ekong, M. O., George, W. K., Pandey, B. K., & Pandey, D. (2023). Enhancing the Fundamentals of Industrial Safety Management in TVET for Metaverse Realities. In *Applications of Neuromarketing in the Metaverse* (pp. 19-41). IGI Global. 10.4018/978-1-6684-8150-9.ch002

George, W. K., Ekong, M. O., Pandey, D., & Pandey, B. K. (2023). Pedagogy for Implementation of TVET Curriculum for the Digital World. In *Applications of Neuromarketing in the Metaverse* (pp. 117-136). IGI Global. 10.4018/978-1-6684-8150-9.ch009

Ghobakhloo, M., Iranmanesh, M., Mubarak, M. F., Mubarik, M., Rejeb, A., & Nilashi, M. (2022). Identifying industry 5.0 contributions to sustainable development: A strategy roadmap for delivering sustainability values. *Sustainable Production and Consumption*, 33, 716–737. 10.1016/j.spc.2022.08.003

Gladysz, B., Tran, T. A., Romero, D., van Erp, T., Abonyi, J., & Ruppert, T. (2023). Current development on the Operator 4.0 and transition towards the Operator 5.0: A systematic literature review in light of Industry 5.0. *Journal of Manufacturing Systems*, 70, 160–185. 10.1016/j.jmsy.2023.07.008

. Govindaraj, V., Dhanasekar, S., Martinsagayam, K., Pandey, D., Pandey, B. K., & Nassa, V. K. (2023). Low-power test pattern generator using modified LFSR. *Aerospace Systems*, 1-8.

Gupta, A. K., Sharma, R., Pandey, D., Nassa, V. K., Pandey, B. K., George, A. S., & Dadheech, P. (2023). Performance analysis of eight-channel WDM optical network with different optical amplifiers for industry 4.0. In *Innovation and Competitiveness in Industry 4.0 Based on Intelligent Systems* (pp. 197–212). Springer International Publishing. 10.1007/978-3-031-29775-5_9

Iyyanar, P., Anand, R., Shanthi, T., Nassa, V. K., Pandey, B. K., George, A. S., & Pandey, D. (2023). A real-time smart sewage cleaning UAV assistance system using IoT. In *Handbook of Research on Data-Driven Mathematical Modeling in Smart Cities* (pp. 24–39). IGI Global.

Jayapoorani, S., Pandey, D., Sasirekha, N. S., Anand, R., & Pandey, B. K. (2023). Systolic optimized adaptive filter architecture designs for ECG noise cancellation by Vertex-5. *Aerospace Systems*, 6(1), 163–173. 10.1007/s42401-022-00177-3

Kirubasri, G., Sankar, S., Pandey, D., Pandey, B. K., Nassa, V. K., & Dadheech, P. (2022). Software-defined networking-based Ad hoc networks routing protocols. In *Software defined networking for Ad Hoc networks* (pp. 95–123). Springer International Publishing. 10.1007/978-3-030-91149-2_5

Kirubasri, G., Sankar, S., Pandey, D., Pandey, B. K., Singh, H., & Anand, R. (2021, September). A recent survey on 6G vehicular technology, applications and challenges. In *2021 9th International Conference on Reliability, Infocom Technologies and Optimization (Trends and Future Directions)(ICRITO)* (pp. 1-5). IEEE.

KVM, S., Pandey, B. K., & Pandey, D. (2024). Design of Surface Plasmon Resonance (SPR) Sensors for Highly Sensitive Biomolecular Detection in Cancer Diagnostics. *Plasmonics*, 1-13.

Longo, F., Padovano, A., & Umbrello, S. (2020). Value-oriented and ethical technology engineering in industry 5.0: A human-centric perspective for the design of the factory of the future. *Applied Sciences (Basel, Switzerland)*, 10(12), 4182. 10.3390/app10124182

Menon, V., Pandey, D., Khosla, D., Kaur, M., Vashishtha, H. K., George, A. S., & Pandey, B. K. (2022). A Study on COVID–19, Its Origin, Phenomenon, Variants, and IoT-Based Framework to Detect the Presence of Coronavirus. In *IoT Based Smart Applications* (pp. 1–13). Springer International Publishing.

Moro, S. R., Cauchick-Miguel, P. A., de Sousa-Zomer, T. T., & de Sousa Mendes, G. H. (2023). Design of a sustainable electric vehicle sharing business model in the Brazilian context. *International Journal of Industrial Engineering and Management*, 14(2), 147–161. 10.24867/IJIEM-2023-2-330

Pandey, B. K., & Pandey, D. (2023). Parametric optimization and prediction of enhanced thermoelectric performance in co-doped CaMnO3 using response surface methodology and neural network. *Journal of Materials Science Materials in Electronics*, 34(21), 1589. 10.1007/s10854-023-10954-1

Pandey, B. K., Pandey, D., & Agarwal, A. (2022). Encrypted information transmission by enhanced steganography and image transformation. [IJDAI]. *International Journal of Distributed Artificial Intelligence*, 14(1), 1–14. 10.4018/IJDAI.297110

Pandey, B. K., Pandey, D., Alkhafaji, M. A., Güneşer, M. T., & Şeker, C. (2023a). A reliable transmission and extraction of textual information using keyless encryption, steganography, and deep algorithm with cuckoo optimization. In *Micro-Electronics and Telecommunication Engineering: Proceedings of 6th ICMETE 2022* (pp. 629–636). Springer Nature Singapore. 10.1007/978-981-19-9512-5_57

Pandey, B. K., Pandey, D., Gupta, A., Nassa, V. K., Dadheech, P., & George, A. S. (2023b). Secret data transmission using advanced morphological component analysis and steganography. In *Role of data-intensive distributed computing systems in designing data solutions* (pp. 21–44). Springer International Publishing. 10.1007/978-3-031-15542-0_2

Pandey, B. K., Pandey, D., & Sahani, S. K. (2024). Autopilot control unmanned aerial vehicle system for sewage defect detection using deep learning. *Engineering Reports*, 12852. 10.1002/eng2.12852

Pandey, B. K., Pandey, D., Wairya, S., & Agarwal, G. (2021). An advanced morphological component analysis, steganography, and deep learning-based system to transmit secure textual data. [IJDAI]. *International Journal of Distributed Artificial Intelligence*, 13(2), 40–62. 10.4018/IJDAI.2021070104

Pandey, D., Hasan, A., Pandey, B. K., Lelisho, M. E., George, A. H., & Shahul, A. (2023). COVID-19 epidemic anxiety, mental stress, and sleep disorders in developing country university students. *CSI Transactions on ICT*, 11(2), 119–127. 10.1007/s40012-023-00383-0

Pandey, D., Wairya, S., Sharma, M., Gupta, A. K., Kakkar, R., & Pandey, B. K. (2022). An approach for object tracking, categorization, and autopilot guidance for passive homing missiles. *Aerospace Systems*, 5(4), 553–566. 10.1007/s42401-022-00150-0

Pandey, J. K., Jain, R., Dilip, R., Kumbhkar, M., Jaiswal, S., Pandey, B. K., & Pandey, D. (2022). Investigating role of iot in the development of smart application for security enhancement. In *IoT Based Smart Applications* (pp. 219–243). Springer International Publishing.

Paschek, D., Mocan, A., & Draghici, A. (2019, May). *Industry 5.0—The expected impact of next industrial revolution. In Thriving on future education, industry, business, and Society.* Proceedings of the MakeLearn and TIIM International Conference, Piran, Slovenia.

Pramanik, S., Pandey, D., Joardar, S., Niranjanamurthy, M., Pandey, B. K., & Kaur, J. (2023, October). An overview of IoT privacy and security in smart cities. In *AIP Conference Proceedings* (*Vol. 2495*, No. 1). AIP Publishing. 10.1063/5.0123511

Raja, D., Kumar, D. R., Santhiyakumari, N., Kumarganesh, S., Sagayam, K. M., Thiyaneswaran, B., Pandey, B. K., & Pandey, D. (2024). A compact dual-feed wide-band slotted antenna for future wireless applications. *Analog Integrated Circuits and Signal Processing*, 118(2), 1–15. 10.1007/s10470-023-02233-0

Revathi, T. K., Sathiyabhama, B., Sankar, S., Pandey, D., Pandey, B. K., & Dadeech, P. (2022). An intelligent model for coronary heart disease diagnosis. In *Networking Technologies in Smart Healthcare* (pp. 309–327). CRC Press. 10.1201/9781003239888-15

Sahani, K., Khadka, S. S., Sahani, S. K., Pandey, B. K., & Pandey, D. (2023). A possible underground roadway for transportation facilities in Kathmandu Valley: A racking deformation of underground rectangular structures. *Engineering Reports*, 12821. 10.1002/eng2.12821

Saxena, A., Agarwal, A., Pandey, B. K., & Pandey, D. (2024). Examination of the Criticality of Customer Segmentation Using Unsupervised Learning Methods. *Circular Economy and Sustainability*, 1–14. 10.1007/s43615-023-00336-4

Saxena, A., Sharma, N. K., Pandey, D., & Pandey, B. K. (2021). Influence of tourists satisfaction on future behavioral intentions with special reference to desert triangle of Rajasthan. *Augmented Human Research*, 6(1), 13. 10.1007/s41133-021-00052-4

. Sengupta, R., Sengupta, D., Pandey, D., Pandey, B. K., Nassa, V. K., & Dadeech, P. (2021). A Systematic review of 5G opportunities, architecture and challenges. *Future Trends in 5G and 6G*, 247-269.

Sharma, S., Pandey, B. K., Pandey, D., Anand, R., Sharma, A., & Saini, S. (2023, March). Character Recognition Technique Implementation for Complicated Deteriorated Scene. In 2023 6th International Conference on Information Systems and Computer Networks (ISCON) (pp. 1-4). IEEE. 10.1109/ISCON57294.2023.10112185

Singh, S., Madaan, G., Kaur, J., Swapna, H. R., Pandey, D., Singh, A., & Pandey, B. K. (2023). *Bibliometric Review on Healthcare Sustainability. Handbook of Research on Safe Disposal Methods of Municipal Solid Wastes for a Sustainable Environment*, 142-161. IGI Global. 10.4018/978-1-6684-8117-2.ch011

Swapna, H. R., Bigirimana, E., Madaan, G., Hasan, A., Pandey, B. K., & Pandey, D. (2023). Impact of neuromarketing on consumer psychology in digitally connected networks. In *Applications of Neuromarketing in the Metaverse* (pp. 193–205). IGI Global. 10.4018/978-1-6684-8150-9.ch015

Tripathi, R. P., Sharma, M., Gupta, A. K., Pandey, D., Pandey, B. K., Shahul, A., & George, A. H. (2023). Timely prediction of diabetes by means of machine learning practices. *Augmented Human Research*, 8(1), 1. 10.1007/s41133-023-00062-4

Vinodhini, V., Kumar, M. S., Sankar, S., Pandey, D., Pandey, B. K., & Nassa, V. K. (2022). IoT-based early forest fire detection using MLP and AROC method. *International Journal of Global Warming*, 27(1), 55–70. 10.1504/IJGW.2022.122794

Wang, B., Zhou, H., Li, X., Yang, G., Zheng, P., Song, C., Yuan, Y., Wuest, T., Yang, H., & Wang, L. (2024). Human Digital Twin in the context of Industry 5.0. *Robotics and Computer-integrated Manufacturing*, 85, 102626. 10.1016/j.rcim.2023.102626

Chapter 10
Distributed Ledger Technology in 6G Management Strategies for Threat Mitigation

Chetan Thakar

Savitribai Phule Pune University, India

Rashi Saxena

Department of AIMLE, Gokaraju Rangaraju Institute of Engineering and Technology, Hyderabad, India

Modi Himabindu

Institute of Aeronautical Engineering, Dundigal, India

C. Rakesh

Department of Mechanical Engineering, New Horizon College of Engineering, Bangalore, India

Amit Dutt

Lovely Professional University, India

Joshuva Arockia Dhanraj

https://orcid.org/0000-0001-5048-7775

Dayananda Sagar University, India

ABSTRACT

Establishing strong security protocols is crucial in the quickly changing internet of things (IoT) environment to reduce potential risks and weaknesses. It is possible to manage security risks more effectively, but there are also challenges because of the interconnected nature of IoT devices, the introduction of 6G networks, and the incorporation of distributed ledger technology (DLT). The focus of this note is on proactive methods of protecting infrastructure and sensitive data. It explores different management strategies that are intended to mitigate threats in IoT environments. Using a security-by-design methodology is a fundamental tactic for threat mitigation in internet of things settings. Every phase of the lifecycle of an IoT device, from design and development to deployment and operation, must incorporate security measures.

INTRODUCTION

Establishing strong security protocols is crucial in the quickly changing Internet of Things (IoT) environment to reduce potential risks and weaknesses. It is possible to manage security risks more effectively, but there are also challenges because of the interconnected nature of IoT semiconductor devices (Du John,

DOI: 10.4018/979-8-3693-2931-3.ch010

H. V. et al., 2021), the introduction of 6G (Kirubasri, G. et al., 2021) networks, and the incorporation of Distributed Ledger Technology (DLT). The focus of this note is on proactive methods of protecting infrastructure and sensitive data. It explores different management strategies that are intended to mitigate threats in IoT environments. Using a Security-by-Design methodology is a fundamental tactic for threat mitigation in Internet of Things settings. Every phase of the lifecycle of an IoT device (Du John, H. V. et al., 2023a), from design and development to deployment and operation, must incorporate security measures. Organizations can strengthen their defences against potential cyberattacks by integrating security features like encryption, access controls, and secure boot mechanisms into Internet of Things devices. Putting strong authentication measures in place, like multi-factor authentication or biometric recognition, can also aid in preventing unwanted access to IoT networks and devices. It is impossible to overestimate the contribution that Distributed Ledger Technology (DLT) makes to improving security and resilience in IoT environments. Data tampering (Sharma, S. et al., 2023) and manipulation is prevented by DLT, which is often linked to blockchain technology (David, S. et al., 2023). It provides transactional data with tamper-resistant, decentralized storage. Decentralized ledger technology (DLT) enables Internet of Things (IoT) devices (Du John, H. V. et al., 2024) to safely log and validate transactions, build confidence between parties, and improve data consistency over dispersed networks. Particularly in situations where sensitive data exchange and financial transactions are involved, implementing DLT-based solutions can greatly improve the security posture of IoT ecosystems. In IoT environments, implementing encryption and authentication mechanisms is a critical component of threat mitigation. Through the encoding of data transferred between IoT devices and networks, encryption plays a critical role in protecting data integrity and confidentiality (Letaief, K. B. et al., 2019). Organizations can guarantee the privacy and security of sensitive information by encrypting data using strong cryptographic algorithms (Kumar Pandey, B. et al., 2021), which can thwart unauthorized interception and logging in. Deploying robust authentication techniques, like public-key infrastructure (PKI) or digital certificates, can also help confirm the identity of users and devices connecting to IoT networks, reducing the possibility of impersonation attacks and illegal access. Effective threat mitigation in IoT environments requires proactive monitoring in addition to incident response techniques and preventive measures. Organizations can minimize the potential impact of cyberattacks by promptly detecting and responding to security incidents through the continuous monitoring of devices and networks. Anomalies, suspicious activity, and possible security breaches within ecosystems can be found by putting Intrusion Detection Systems (IDS) and Security Information and Event Management (SIEM) solutions into place. Additionally, the overall resilience of infrastructures can be improved and timely remediation of security incidents can be ensured by implementing incident response protocols and regularly performing security audits and assessments (Sheth, K. et al., 2020).

6G IoT and DLT

The Internet of Things (IoT) is a network of sensors, software, and connectivity-enabled devices, vehicles, appliances, and other objects that collect, exchange, and analyze data autonomously or with minimal human intervention. IoT allows devices and systems to communicate and interact, enabling data-driven decision-making, automation, and efficiency improvements across domains. Smartphones, wearable, home appliances, industrial machinery, environmental sensors, and infrastructure components are IoT devices. Sensors (Du John, H. V. et al., 2023b) collect data about their surroundings, processors process and analyze it, and communication interfaces send it to other devices or centralized systems. Wireless or wired IoT devices can communicate (Gupta, A. K. et al., 2023), exchange data, and respond to commands, creating interconnected ecosystems called the Internet of Things. A revolutionary method for organizing data and enabling transactions over decentralized networks is Distributed Ledger Technology (DLT). DLT functions as the network infrastructure that permits concurrent record access, validation, and updating in a networked database. The blockchain systems that have gained a lot of attention and traction in recent years are built on this technology. DLT offers a safe, transparent, and unchangeable method for recording and verifying transactions. DLT divides data and processing duties among a network of linked nodes, in contrast to traditional centralized databases, where data management is managed by a single entity. To ensure consensus and integrity throughout the network, every node keeps a copy of the ledger and works with other nodes to verify the correctness of transactions. Although distributed ledger technology has been around for many years, Bitcoin and other cryptocurrencies have brought it more notoriety (Yang, H. et al., 2020). The decentralized recording and verification of transactions for these digital currencies is made possible by blockchain technology, a particular kind of DLT. Since then, developments in data science, computing, software (Kirubasri, G. et al., 2022), hardware, and networking technologies have led to a significant evolution in DLT. Transparency and accountability in data management are two of DLT's primary characteristics. DLT improves the reliability of data stored on the ledger and lessens the need for laborious auditing procedures by enabling users to view any changes made to the ledger and identify the parties responsible for those changes. DLT also makes it possible to implement access control systems, which guarantee that only individuals with permission can access or alter particular data. To secure and authenticate transactions within the network, DLT uses cryptography. The ledgers data is safeguarded against unauthorized access and manipulation by means of cryptographic keys and signatures. Information entered into the ledger is considered immutable once it is recorded and cannot be removed or changed without the approval of all network users. By doing this, the accuracy and integrity of the data kept on the ledger are guaranteed. DLT systems rely heavily on consensus mechanisms because they allow network users to decide whether a transaction is valid without the need for a central authority. By guaranteeing that every transaction is authenticated and documented precisely, these systems forestall fraudulent and duplicate spending practices. A few instances of consensus mechanisms are Practical Byzantine Fault Tolerance (PBFT), Proof of Stake (PoS), and Proof of Work (PoW). DLT is useful in a wide range of sectors where security, transparency, and data management are critical. It has applications in supply chain management, identity verification, healthcare (Tareke, S. A. et al., 2022), and finance, among other areas. Although distributed ledgers are a component of every blockchain, not every distributed ledger is a blockchain. With DLT, you can create decentralized applications and completely transform the way data is managed and transactions are carried out (Huang, T. et al., 2019).

IoT Threat Management

The significance of effectively managing threats in IoT (Internet of Things) environments is of utmost importance in the contemporary interconnected global landscape. The proliferation of Internet of Things (IoT) devices in diverse sectors and fields, including healthcare (Du John, H. V. et al., 2023c), manufacturing, transportation, and smart homes, presents a multitude of security risks and vulnerabilities that necessitate proactive measures for mitigation. The safeguarding of confidential information. Internet of Things (IoT) devices gather and transmit extensive quantities of sensitive data, encompassing personal information, financial data, and proprietary business data. Insufficient security protocols may result in the unauthorized access of this data, thereby giving rise to privacy infringements (Pandey, D. et al., 2021a), identity theft, financial deception, and the theft of intellectual property. Ensuring the confidentiality and integrity of this data requires the implementation of efficient threat management strategies (Pandey, B. K. et al., 2024). Ensuring the prevention of cyberattacks is of utmost importance in IoT environments. The susceptibility of IoT devices to cyberattacks is heightened by their frequently insufficient security protocols and uninterrupted connectivity (Rupprecht, D. et al., 2018). Cybercriminals leverage vulnerabilities present in Internet of Things (IoT) devices to initiate a range of attacks, including Distributed Denial of Service attacks, ransomware attacks, and breach of data security. The effective management of threats plays a crucial role in mitigating the risk of cyberattacks and minimizing their potentially detrimental effects on both organizations and individuals. It is imperative for businesses and critical infrastructure that depend on IoT devices to uphold operational continuity. Numerous Internet of Things (IoT) devices play a crucial role in several essential operations, including industrial automation systems, healthcare monitoring systems (Singh, S. et al., 2023), and smart city infrastructure. Cyberattacks or security breaches have the potential to cause substantial downtime, financial losses, and safety hazards. Efficient threat management is crucial for maintaining the continuous functioning of Internet of Things systems and mitigating the risk of disruptions. Effectively mitigating risks in IoT environments is crucial for safeguarding the authenticity and dependability of data and services. Internet of Things devices produce substantial quantities of data that are utilized to facilitate decision-making, optimize (Pandey, D. et al., 2021b). Hybrid deep) operational procedures, and improve user satisfaction. Any compromise in the integrity of this data, whether due to unauthorized access, manipulation, or tampering, can result in inaccurate insights, flawed decisions, and compromised outcomes. Organizations can ensure the integrity and reliability of their data and services by implementing comprehensive security measures and employing effective threat management strategies (Kato, N. et al., 2020).

INTERSECTION OF IOT AND 6G NETWORKS

IoT Evolution in 6G

The Internet of Things (IoT) is a network of sensors, software, and connectivity-enabled devices, vehicles, appliances, and other objects that collect, exchange, and analyze data autonomously or with minimal human intervention. IoT's evolution in 6G wireless communication technology (Raja, D. et al., 2024) has major implications for connectivity, data transmission, and smart device proliferation. Initial IoT devices used 4G LTE and Wi-Fi for connectivity. These technologies faced bandwidth, latency, and scalability issues as the number of connected devices grew exponentially. 5G (Sengupta, R. et al., 2021) networks addressed these issues by providing faster data speeds, lower latency, and more capacity to support a massive number of IoT devices. 5G was the predecessor to 6G, which will push wireless communication speed, latency, and connectivity. 6G is expected to use terahertz (THz) frequencies to achieve unprecedented data rates and ultra-low latency, unlike 5G, which uses sub-6 GHz and millimeter-wave frequencies. In IoT, 5G to 6G communication promises to enable new applications and use cases that require ultra-reliable, low-latency communication. IoT devices can communicate more efficiently and seamlessly with 6G, enabling real-time data transmission, mission-critical applications, and immersive experiences. IoT devices will benefit from 6G communication paradigms like holographic networking and intelligent beamforming. These advances make IoT devices smarter, autonomous, and secure, enabling transformative applications in healthcare (KVM, S. et al., 2024), transportation (Sahani, K. et al.,

2023), manufacturing, agriculture, and more. IoT evolution in 6G is closely linked to edge computing, AI (Tripathi, R. P. et al., 2023), and quantum computing. IoT devices can process data faster, have lower latency, and use less cloud infrastructure with edge computing. AI algorithms let IoT devices analyze and respond to data in real time, optimizing performance (Pandey, D., & Pandey, B. K., 2022) and making smart decisions. Quantum computing can improve IoT network security and efficiency with advanced encryption and optimization algorithms (Karjaluoto, H., 2006).

6G Advancements for IoT

The Internet of Things (IoT) is made possible in large part by the features and advancements of 6G networks, which promise revolutionary capabilities that will eclipse those of earlier generations. 6G networks have adopted terahertz frequency bands (Anand, R. et al., 2023), which extend from 95 GHz to 3 THz. This is a major improvement over the spectrum that 5G networks were using. These higher frequency bands open the door for cutting-edge Internet of Things applications and allow 6G networks to reach previously unheard-of data transfer speeds. 6G networks can handle large amounts of data transmission because they operate in the terahertz range, which makes it easier for IoT devices to connect and communicate with one another. By providing additional channels for fast data exchange, this widened spectrum enables intricate IoT ecosystems with a wide range of devices and uses. 6G networks are notable for their exceptionally fast data transfer rates, which can potentially reach 1 TB/s, or 8,000 GB/s. This promises to transform IoT connectivity and is a major advancement over the capabilities of 5G networks. Sixth generation (6G) networks enable Internet of Things (IoT) devices to exchange massive volumes of data in real time, with transfer speeds 100 times faster than 5G. With this high-speed connectivity, Internet of Things applications become more responsive and efficient, allowing them to support cutting-edge technologies like industrial automation, smart cities, and autonomous vehicles with unmatched performance. 6G networks are anticipated to dramatically lower latency, from 1 ms to as low as 1 µs, in addition to improved data transfer speeds. IoT devices operating in dynamic environments require quick data processing and decision-making, which is ensured by this minimal latency. 6G networks reduce latency, allowing for responsiveness and real-time communication, which improves the capabilities of IoT applications in a variety of fields. In applications such as autonomous manufacturing, smart grid management, and remote healthcare monitoring, low latency in 6G networks guarantees smooth communication between IoT devices, enhancing productivity and dependability. 6G networks utilize submillimeter waves and decreased packet transmission rates to enhance wireless sensing technology, real-time computing, and network dependability (Saxena & Chaudhari, 2014). These technological advancements enhance the communication capabilities of IoT devices, allowing for faster exchange and analysis of data. 6G networks improve latency and data transmission reliability by optimizing packet transmission rates, ensuring uninterrupted connectivity for IoT devices in various environments. The incorporation of submillimeter waves facilitates precise and accurate communication among Internet of Things (IoT) devices, thereby enhancing the overall performance of the network. It support in 6G networks facilitates the implementation of intelligent and autonomous computing capabilities at end terminals. Predictive analytics enables real-time network decisions, enhancing efficiency and reducing processing time. 6G networks enhance the computational capabilities of IoT applications by incorporating High-Performance Computing (HPC) into the network infrastructure. This integration allows these applications to efficiently handle substantial amounts of data. The autonomous performance of IoT devices facilitates the execution of intricate tasks, thereby enhancing the efficiency and responsiveness of IoT ecosystems. The integration of artificial intelligence (AI) (Pandey, B. K., & Pandey, D., 2023) capabilities with edge computing facilitates the improvement of data storage and processing capabilities in proximity to connected devices, while simultaneously ensuring minimal latency. By reducing data transmission times and enhancing real-time decision-making, this approach enhances the efficiency of IoT applications. Integrating AI-based technology into edge computing infrastructure allows for efficient computation of big data analytics and device processing at the edge. As a result, there is an emergence of more intelligent devices and applications that possess dynamic capabilities, thereby augmenting the overall efficacy of Internet of Things (IoT) systems. Energy efficiency and battery-limited devices, given the increased frequency bands of 6G networks, energy efficiency becomes essential, particularly for IoT devices with limited battery capacity. The utilization of energy harvesting and wireless energy transfer technologies enhances the capabilities of these devices, thereby guaranteeing their extended operational lifespan and the reduction of maintenance needs. 6G networks facilitate off-grid operations and longevity of IoT devices and sensors by utilizing energy harvesting technologies. By decreasing dependence on conventional power sources, this approach improves the sustainability of Internet of Things (IoT) implementations, thereby promoting the development of energy-efficient IoT systems. Beamforming techniques, such as holographic beamforming (HBF), improve the efficiency, coverage, and throughput of 6G networks. The implementation of software-defined antennas (SDAs) in high-bandwidth band (HBF) techniques enables 6G networks to achieve signal transmission and reception with exceptional flexibility and efficiency. The developments in beamforming technology have enhanced the overall efficiency of 6G networks, allowing them to effectively address the growing requirements of Internet of Things (IoT) applications for dependable and fast connectivity. The integration of blockchain technology into 6G communication systems serves to augment security, transparency, and reliability. Blockchain technology offers a secure and verifiable method for spectrum management, verified transactions, and the prevention of unauthorized access due to its decentralized attributes of tamper-resistance and secrecy (Pandey, B. K. et al., 2023). By harnessing the capabilities of blockchain technology, 6G networks facilitate the establishment of trust among interconnected applications and obviate the necessity for trusted intermediaries. The implementation of this technology guarantees the protection of data during transmission and improves the overall security stance of 6G networks, rendering them well-suited for facilitating Internet of Things (IoT) applications that necessitate secure and transparent communication channels (Jefferies, N., 1995).

IoT-6G Synergies

A new era of connectivity and functionality is being ushered in by the convergence of 6G and Internet of Things (IoT) technologies, which is revolutionizing how we use and interact with data. The 6G networks' ultra-reliable low-latency communication (URLLC) capability, which guarantees low latency and high reliability essential for real-time Internet of Things applications, is at the core of this synergy. The smooth integration of IoT devices with 6G networks facilitates fast and effective data exchange, improving operational efficiency and decision-making processes in a variety of sectors, from autonomous vehicles navigating city streets to remote healthcare monitoring systems. 6G networks enable edge computing, giving IoT devices AI-driven intelligence and localized processing power. Edge computing lowers latency and bandwidth requirements by enabling data processing closer to the point of collection, allowing for faster response times and better scalability for Internet of Things deployments. In addition to improving IoT applications responsiveness, this decentralized approach to data processing also makes them more resilient and adaptable in changing settings. 6G networks surpass previous generations' limitations in data transfer rates by utilizing terahertz frequency bands. By enabling IoT devices to send and receive enormous volumes of data with ease, this high-speed connectivity opens up new opportunities for creativity and teamwork. IoT deployments within 6G networks can function with increased trust, reliability, and integrity when combined with the integration of blockchain technology, which guarantees safe and transparent data transactions. This opens the door for revolutionary applications in smart cities, healthcare, agriculture, and other areas.

DISTRIBUTED LEDGER TECHNOLOGY (DLT)

DLT Types for IoT in 6G

Distributed Ledger Technology (DLT) signifies a significant change in the method of recording and verifying transactions in digital systems (Swapna, H. R. et al., 2023). DLT is a system that allows for the recording and validation of transactions across multiple nodes within a network in a decentralized and distributed manner (George, W. K. et al., 2023). Pedagogy for Imple). In contrast to conventional centralized systems, DLT operates on a peer-to-peer network, enabling all participants to possess a copy of the ledger. Similar to a network architect, this decentralization removes the necessity for intermediaries, minimizing the risk of a single point of failure and improving the overall resilience of the system. Transparency is a core principle of DLT. In a distributed ledger technology system, transactions are visible to all participants in the network. Every transaction (Ekong, M. O. et al., 2023) is meticulously logged in a sequential manner and securely connected to the preceding one, creating a series of blocks, commonly referred to as a blockchain. Ensuring transparency allows all participants to have equal access to information, fostering trust and accountability within the network. In addition, the transactions are immutable, meaning that once they are recorded, they cannot be changed or removed. This feature improves the integrity and security of the ledger, making it highly resistant to any fraudulent activity or tampering. DLT depends on a consensus mechanism to verify and reach an agreement on the status of the ledger among all nodes in the network. Consensus mechanisms differ based on the particular DLT implementation, but typically involve a process where nodes reach a collective agreement on the legitimacy of transactions before they are included in the ledger. Ensuring synchronization and consistency across all copies of the ledger, even in the face of malicious actors or network failures, is crucial. Various forms of Distributed Ledger Technology (DLT) are applicable to the integration of Internet of Things (IoT) within the framework of 6G networks. These DLT types have unique features and capabilities that are shown in the figure 1, which can effectively meet the specific needs and challenges of IoT applications in the 6G ecosystem:

1. *Blockchain*: Blockchain is widely recognized as a popular form of DLT, known for its decentralized and unchangeable record of transactions. In the realm of IoT and 6G, blockchain offers the ability to maintain transparent and tamper-proof records for a wide range of IoT devices and sensors. Smart contracts, a feature of blockchain technology, allow for the automated and secure execution of agreements between IoT devices, ensuring trustless interactions and transactions.

2. *Directed Acyclic Graphs*: It provide an alternative to blockchain, emphasizing scalability and feeless transactions. In the world of 6G IoT, DAGs have the potential to revolutionize communication between IoT devices. They offer a lightweight and efficient solution for high-throughput data transfer

and microtransactions, eliminating the need for intermediaries. The asynchronous nature of Directed Acyclic Graphs (DAGs) also fits perfectly with the sporadic data transmission patterns often seen in Internet of Things (IoT) deployments.

3. *Hashgraph*: Hashgraph is a consensus algorithm (Bessant, Y. A. et al., 2023) and data structure that achieves distributed consensus through the exchange of information and virtual voting. It provides excellent performance, minimal delay, equal treatment, and robustness against malicious attacks. Hashgraph offers rapid and effective transaction validation, facilitating instantaneous data exchange and secure communication among IoT devices in the realm of IoT and 6G.

4. *Holochain*: Holochain is a cutting-edge technology that empowers distributed applications (dApps) to operate without relying on centralization. Utilizing a distributed hash table (DHT) in conjunction with cryptographic signatures, every node within the network is empowered to uphold its own unalterable ledger. Within the realm of 6G IoT, Holochain has the ability to enhance edge computing and facilitate peer-to-peer communication among IoT devices, promoting the development of decentralized and resilient networks.

5. *MultiChain*: MultiChain is a powerful platform that empowers organizations to create and deploy their own private blockchains, giving them full control over who can access their distributed ledgers. In the realm of 6G IoT, MultiChain enables secure and permissioned data sharing among trusted entities, guaranteeing confidentiality and adherence to regulatory standards. It offers the ability to create personalized consensus mechanisms and data access controls that are perfect for individual IoT applications.

Figure 1. DLT types for IoT in 6G

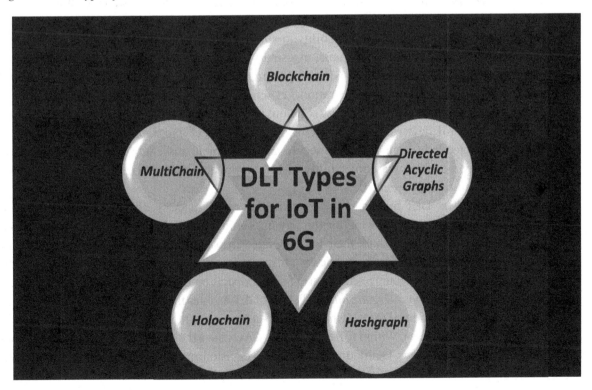

Every kind of DLT has its own set of benefits and compromises when it comes to scalability, security, decentralization, and transaction speed. Through the use of the right DLT solution, 6G networks have the potential to seamlessly integrate IoT devices, allowing for the development of innovative applications and services that offer improved connectivity, efficiency, and reliability.

DLT Benefits in IoT

Distributed Ledger Technology (DLT) has a wide range of applications and advantages in Internet of Things (IoT) ecosystems, especially in the context of 6G networks. Here are a few important applications and advantages of DLT in IoT ecosystems are shown in the figure 2.

- *Secure Data Exchange*: DLT offers a reliable and unalterable platform for facilitating data exchange among IoT devices. Through the use of cryptographic techniques and decentralized consensus mechanisms, DLT guarantees the security and reliability of data exchanged within the network. This is essential for sensitive applications like healthcare, smart cities, and industrial automation.
- *Transparent Supply Chains*: DLT facilitates the creation of transparent and traceable supply chains by documenting the complete journey of products, starting from their production to their final delivery. IoT devices with sensors can effortlessly capture real-time data on the movement and condition of goods, which is then securely recorded on the distributed ledger. This level of transparency contributes to increased accountability, decreased instances of fraud, and enhanced efficiency in supply chain management.
- *Smart Contracts*: DLT platforms, such as blockchain, enable the use of smart contracts. These contracts are designed to be self-executing agreements with predetermined conditions. In IoT ecosystems, smart contracts have the ability to automate and enforce business rules, eliminating the need for intermediaries. For example, in a smart grid scenario, IoT devices can independently carry out energy trading agreements according to predetermined conditions, resulting in improved resource allocation and decreased operational expenses.
- *Access Management*: DLT-based identity solutions provide strong authentication and access control mechanisms for IoT devices. Every device is given a distinct cryptographic identity that is stored on the distributed ledger, guaranteeing secure and unalterable authentication. By implementing robust security measures, the vulnerability of IoT networks to unauthorized access and identity spoofing is significantly reduced.
- *Data Monetization*: DLT facilitates the monetization of IoT data by offering a platform that is transparent and auditable, allowing for the buying, selling, and licensing of data. Organizations have the ability to securely share their IoT-generated data with third parties, while maintaining control over access rights and usage policies. This creates opportunities for data owners to generate additional income and encourages the sharing and cooperation of data among different sectors.
- *Decentralized Autonomous Organizations* (DAOs): DLT enables the development of decentralized autonomous organizations (DAOs), which are self-governing entities governed by smart contracts and distributed consensus. Within IoT ecosystems, DAOs have the ability to coordinate the actions of independent IoT devices, facilitating decentralized decision-making and allocation of resources. This enhances scalability, resilience, and agility in IoT deployments.

- *Data Integrity and Auditability*: Ensuring the integrity and auditability of data: DLT guarantees the integrity and auditability of data by creating an unchangeable and visible record of every transaction and data transfer. This feature is especially important in industries (Malhotra, P. et al., 2021) that have strict regulations, such as healthcare and finance. Ensuring data integrity and meeting audit requirements is of utmost importance in these sectors.

By distributing data across network nodes, Distributed Ledger Technology (DLT) transforms asset recording and transactions. DLT uses a decentralized architecture to store and handle data, unlike traditional databases. DLT technology and protocols allow distributed ledgers to access, validate, and update records simultaneously. These ledgers ensure transparency and resilience by using a network of entities, locations, or nodes. Each node in a distributed ledger system processes and verifies transactions independently, creating a shared record. This technique promotes participant consensus on data accuracy and validity. Distribution ledgers can store static data like registries and dynamic data like financial transactions. Blockchain's sequential chaining of transaction data blocks makes it a popular DLT implementation. DLT goes beyond blockchain to include other technologies and methods that improve transactional security, efficiency, and transparency.

Figure 2. Advantages of in IoT

IoT THREAT LANDSCAPE

Managing security in IoT (Internet of Things) environments can be quite challenging due to the numerous vulnerabilities and risks present in the threat landscape. There is a growing issue with the increasing number of connected devices in various sectors, including smart homes and critical infrastructure. Given the vast number of interconnected IoT devices, the potential for malicious actors to exploit multiple entry points has significantly increased. Managing the complexity of IoT security can be quite challenging due to the wide range of devices, each with its own set of characteristics and potential vulnerabilities. These devices frequently lack strong security measures, leaving them vulnerable to a range of cyber threats including malware, ransomware, and distributed denial-of-service (DDoS) attacks. Numerous IoT (Vinodhini, V. et al., 2022) devices are deployed with default or weak passwords, rendering them vulnerable to exploitation by malicious actors (Bikos, A. N., & Sklavos, N., 2012). The nature of IoT ecosystems, where devices communicate and share data over networks, brings about extra security challenges. Any vulnerabilities in a single device or component have the potential to compromise the entire network, resulting in a series of security breaches. In addition, the absence of standardized security protocols and the reliance on outdated systems worsen the security risks linked to IoT environments. Data privacy and confidentiality in IoT deployments are a major concern that needs to be addressed (Pandey, B. K. et al., 2021). Similar to overseeing an IT project, it is crucial to recognize that numerous IoT devices gather valuable data, encompassing personal and proprietary information. In the event of a breach, the potential repercussions for individuals and organizations can be quite severe. Ensuring the security of IoT data is crucial to safeguard against privacy breaches, identity theft, and financial harm. It is imperative to implement strong data protection measures to prevent unauthorized access. Implementing robust authentication mechanisms, encryption protocols, and access controls is crucial for ensuring the security of IoT devices and networks. In order to effectively detect and mitigate security breaches in real-time, it is crucial to have ongoing monitoring, threat intelligence, and incident response capabilities (Mohapatra, S. K. et al., 2015).

Mitigating Threats in IoT Environments

Ensuring the security of the Internet of Things (IoT) poses distinct difficulties because of the extensive network of interconnected devices, each exposed to possible cyber risks. In order to effectively tackle these difficulties, it is imperative to adopt a comprehensive security strategy that encompasses several components, including Distributed Ledger Technology (DLT), encryption, authentication, and resilient monitoring and response methods. Distributed Ledger Technology (DLT), sometimes linked with blockchain, provides a decentralized and immutable method for documenting transactions and data throughout a network. Within the realm of IoT security, Distributed Ledger Technology (DLT) can have a substantial impact on guaranteeing the authenticity and reliability of device data and transactions. An important benefit of DLT is its capacity to establish a clear and unchangeable record of IoT device operations. The distributed ledger technology (DLT) guarantees the prompt detection of any illegal alteration or manipulation of data by systematically documenting each transaction across many nodes inside the network. The maintenance of integrity in IoT systems, particularly in sectors such as healthcare, supply chain management, and critical infrastructure, relies heavily on the essential aspects of openness and auditability. It can bolster the security of Internet of Things (IoT) systems by facilitating secure identity management and access control. DLT has the capability to authenticate devices, grant access to resources, and enforce predetermined rules and permissions by utilizing cryptographic algorithms and smart contracts. This measure serves to mitigate illegal access and guarantees that solely authenticated devices are able to engage with the Internet of Things (IoT) network. Encryption is an essential component in ensuring the preservation of data security and privacy during the transmission process between Internet of Things (IoT) devices and backend systems. IoT communications maintain security by employing robust cryptographic techniques to encrypt data, ensuring protection even in the event of interception by hostile individuals. The implementation of end-to-end encryption holds significant importance in Internet of Things (IoT) settings, as it facilitates the secure sharing of sensitive data, including personal health information and financial transactions. Ensuring the confidentiality of data throughout its lifecycle, from transmission to storage, is achieved by using encryption techniques at both the device and network levels. Authentication systems play a crucial role in validating the identification of both devices and people inside an Internet of Things (IoT) environment (Menon, V. et al., 2022). In order to mitigate the risk of unauthorized access and safeguard against identity spoofing attacks, it is imperative to implement robust authentication methods, including mutual authentication and multi-factor authentication. The continuous monitoring of Internet of Things (IoT) devices (Iyyanar, P. et al., 2023) and networks is necessary in order to promptly identify and address security concerns. By implementing comprehensive monitoring solutions, organizations may acquire a deeper understanding of the behavior and performance of Internet of Things (IoT) (Pramanik, S. et al., 2023) devices, while also identifying any irregularities or potentially malicious behaviors. Potential security breaches or unauthorized access attempts can be identified by the utilization of real-time threat detection technologies, such as intrusion detection systems (IDS) and anomaly detection algorithms. In order to detect potential malicious activities, these systems employ the analysis of network traffic, device behavior, and system records to find trends. The act of monitoring, it is imperative for enterprises to build comprehensive incident response procedures in order to immediately resolve security incidents. This entails delineating specific duties and obligations, creating effective lines of communication, and executing incident response guidelines customized for dangers specific to the Internet of Things (IoT) (Panwar et al., 2016).

Proactive Threat Measures

Ensuring the security of IoT protocols is crucial in safeguarding connected devices and networks against potential cyber threats. These protocols establish guidelines and processes for ensuring secure communication, encrypting data, and controlling access within IoT ecosystems. Some IoT security protocols that are commonly used are Transport Layer Security (TLS), Datagram Transport Layer Security (DTLS), Message Queuing Telemetry Transport (MQTT), and Lightweight M2M (LwM2M). Through the implementation of strong security measures, organizations can effectively reduce the likelihood of unauthorized access, data breaches, and other security incidents in IoT environments. Using advanced algorithms and machine learning techniques (Anand, R. et al., 2024), predictive analytics can analyze historical data to identify patterns that may indicate potential security threats. When it comes to IoT security, predictive analytics can come in handy for identifying anomalies, abnormal behavior, and suspicious activities across connected devices and networks. By staying ahead of potential security threats, organizations can strengthen their incident response capabilities and avoid expensive security breaches.

Ensuring a thorough risk assessment is an essential part of developing a strong IoT security strategy (Pandey, J. K. et al., 2022). One of the tasks is to identify and assess possible security risks and vulnerabilities related to IoT deployments. This includes examining device misconfigurations, software vulnerabilities, insecure communication channels, and physical security threats. Through a comprehensive risk assessment, organizations can effectively prioritize security measures, allocate resources efficiently, and implement controls to mitigate identified risks. Machine learning algorithms have the potential to significantly improve security in IoT ecosystems. By analyzing large volumes of data, these algorithms can detect and respond to emerging threats, ensuring that security measures are constantly updated in real-time. Machine learning techniques such as anomaly detection, pattern recognition, and behavioral analysis are effective in identifying and addressing complex cyber threats, such as malware, ransomware, and insider attacks. Through the use of machine learning in security, organizations have the opportunity to enhance their ability to detect threats, minimize false positives, and strengthen the overall security of IoT environments (Akpakwu, G. A. et al., 2017). By integrating cutting-edge technologies and innovative methodologies, organizations can effectively address the ever-changing landscape of cyber threats. This approach enhances the security of IoT systems, safeguarding valuable data and ensuring the reliability and accessibility of connected devices and networks. By implementing a proactive and comprehensive strategy for IoT security, organizations can effectively protect their assets, ensure regulatory compliance, and establish trust with customers and stakeholders.

CONCLUSION

IoT and 6G technology open up a huge range of new ways to improve and innovate in many fields. IoT devices are becoming more common, and 6G networks offer speed, reliability, and connectivity that have never been seen before. It is important to deal with the problems that come with these developments in a smart way. Protecting their IoT environments from threats is possible by putting in place strong security measures such as DLT, encryption, authentication, and proactive threat management. IoT settings safer and more resilient by using predictive analytics, thorough risk assessments, and machine learning algorithms. Groups can find and stop threats before they get worse by following the security-by-design principles and using cutting-edge technologies like DLT and machine learning. It is also important to stay ahead of changing risks and weaknesses, as shown in the proactive steps described in this paper. People and businesses can protect their IoT systems from hacking and keep them safe by being proactive about security.

REFERENCES

Akpakwu, G. A., Silva, B. J., Hancke, G. P., & Abu-Mahfouz, A. M. (2017). A survey on 5G networks for the Internet of Things: Communication technologies and challenges. *IEEE Access : Practical Innovations, Open Solutions*, 6, 3619–3647. 10.1109/ACCESS.2017.2779844

Anand, R., Khan, B., Nassa, V. K., Pandey, D., Dhabliya, D., Pandey, B. K., & Dadheech, P. (2023). Hybrid convolutional neural network (CNN) for kennedy space center hyperspectral image. *Aerospace Systems*, 6(1), 71–78. 10.1007/s42401-022-00168-4

Anand, R., Lakshmi, S. V., Pandey, D., & Pandey, B. K. (2024). An enhanced ResNet-50 deep learning model for arrhythmia detection using electrocardiogram biomedical indicators. *Evolving Systems*, 15(1), 83–97. 10.1007/s12530-023-09559-0

Bessant, Y. A., Jency, J. G., Sagayam, K. M., Jone, A. A. A., Pandey, D., & Pandey, B. K. (2023). Improved parallel matrix multiplication using Strassen and Urdhvatiryagbhyam method. *CCF Transactions on High Performance Computing*, 5(2), 102–115. 10.1007/s42514-023-00149-9

Bikos, A. N., & Sklavos, N. (2012). LTE/SAE security issues on 4G wireless networks. *IEEE Security and Privacy*, 11(2), 55–62. 10.1109/MSP.2012.136

David, S., Duraipandian, K., Chandrasekaran, D., Pandey, D., Sindhwani, N., & Pandey, B. K. (2023). Impact of blockchain in healthcare system. In *Unleashing the Potentials of blockchain technology for healthcare industries* (pp. 37–57). Academic Press. 10.1016/B978-0-323-99481-1.00004-3

Du John, H. V., Ajay, T., Reddy, G. M. K., Ganesh, M. N. S., Hembram, A., Pandey, B. K., & Pandey, D. (2023b). Design and simulation of SRR-based tungsten metamaterial absorber for biomedical sensing applications. *Plasmonics*, 18(5), 1903–1912. 10.1007/s11468-023-01910-0

Du John, H. V., Ajay, T., Reddy, G. M. K., Ganesh, M. N. S., Hembram, A., Pandey, B. K., & Pandey, D. (2023c). Design and simulation of SRR-based tungsten metamaterial absorber for biomedical sensing applications. *Plasmonics*, 18(5), 1903–1912. 10.1007/s11468-023-01910-0

Du John, H. V., Jose, T., Sagayam, K. M., Pandey, B. K., & Pandey, D. (2024). Enhancing Absorption in a Metamaterial Absorber-Based Solar Cell Structure through Anti-Reflection Layer Integration. *Silicon*, 1-11.

Du John, H. V., Moni, D. J., Ponraj, D. N., Sagayam, K. M., Pandey, D., & Pandey, B. K. (2021). Design of Si based nano strip resonator with polarization-insensitive metamaterial (MTM) absorber on a glass substrate. *Silicon*, 1-10.

Du John, H. V., Sagayam, K. M., Jose, T., Pandey, D., Pandey, B. K., Kotti, J., & Kaur, P. (2023a). Design simulation and parametric investigation of a metamaterial light absorber with tungsten resonator for solar cell applications using silicon as dielectric layer. *Silicon*, 15(9), 4065–4079. 10.1007/s12633-023-02321-w

Ekong, M. O., George, W. K., Pandey, B. K., & Pandey, D. (2023). Enhancing the Fundamentals of Industrial Safety Management in TVET for Metaverse Realities. In *Applications of Neuromarketing in the Metaverse* (pp. 19-41). IGI Global. 10.4018/978-1-6684-8150-9.ch002

George, W. K., Ekong, M. O., Pandey, D., & Pandey, B. K. (2023). Pedagogy for Implementation of TVET Curriculum for the Digital World. In *Applications of Neuromarketing in the Metaverse* (pp. 117-136). IGI Global. 10.4018/978-1-6684-8150-9.ch009

Gupta, A. K., Sharma, R., Pandey, D., Nassa, V. K., Pandey, B. K., George, A. S., & Dadheech, P. (2023). Performance analysis of eight-channel WDM optical network with different optical amplifiers for industry 4.0. In *Innovation and Competitiveness in Industry 4.0 Based on Intelligent Systems* (pp. 197–212). Springer International Publishing. 10.1007/978-3-031-29775-5_9

Huang, T., Yang, W., Wu, J., Ma, J., Zhang, X., & Zhang, D. (2019). A survey on green 6G network: Architecture and technologies. *IEEE Access : Practical Innovations, Open Solutions*, 7, 175758–175768. 10.1109/ACCESS.2019.2957648

Iyyanar, P., Anand, R., Shanthi, T., Nassa, V. K., Pandey, B. K., George, A. S., & Pandey, D. (2023). A real-time smart sewage cleaning UAV assistance system using IoT. In *Handbook of Research on Data-Driven Mathematical Modeling in Smart Cities* (pp. 24–39). IGI Global.

Jefferies, N. (1995, February). Security in third-generation mobile systems. In *IEE Colloquium on Security in Networks (Digest No. 1995/024)* (pp. 8-1). IET. 10.1049/ic:19950136

Karjaluoto, H. (2006). An investigation of third generation (3G) mobile technologies and services. *Contemporary Management Research*, 2(2), 91–91. 10.7903/cmr.653

Kato, N., Mao, B., Tang, F., Kawamoto, Y., & Liu, J. (2020). Ten challenges in advancing machine learning technologies toward 6G. *IEEE Wireless Communications*, 27(3), 96–103. 10.1109/MWC.001.1900476

Kirubasri, G., Sankar, S., Pandey, D., Pandey, B. K., Nassa, V. K., & Dadheech, P. (2022). Software-defined networking-based Ad hoc networks routing protocols. In *Software defined networking for Ad Hoc networks* (pp. 95–123). Springer International Publishing. 10.1007/978-3-030-91149-2_5

Kirubasri, G., Sankar, S., Pandey, D., Pandey, B. K., Singh, H., & Anand, R. (2021, September). A recent survey on 6G vehicular technology, applications and challenges. In *2021 9th International Conference on Reliability, Infocom Technologies and Optimization (Trends and Future Directions)(ICRITO)* (pp. 1-5). IEEE.

Kumar Pandey, B., Pandey, D., Nassa, V. K., Ahmad, T., Singh, C., George, A. S., & Wakchaure, M. A. (2021). Encryption and steganography-based text extraction in IoT using the EWCTS optimizer. *Imaging Science Journal*, 69(1-4), 38–56. 10.1080/13682199.2022.2146885

KVM, S., Pandey, B. K., & Pandey, D. (2024). Design of Surface Plasmon Resonance (SPR) Sensors for Highly Sensitive Biomolecular Detection in Cancer Diagnostics. *Plasmonics*, 1–13.

Letaief, K. B., Chen, W., Shi, Y., Zhang, J., & Zhang, Y. J. A. (2019). The roadmap to 6G: AI empowered wireless networks. *IEEE Communications Magazine*, 57(8), 84–90. 10.1109/MCOM.2019.1900271

Malhotra, P., Pandey, D., Pandey, B. K., & Patra, P. M. (2021). Managing agricultural supply chains in COVID-19 lockdown. *International Journal of Quality and Innovation*, 5(2), 109–118. 10.1504/IJQI.2021.117181

Menon, V., Pandey, D., Khosla, D., Kaur, M., Vashishtha, H. K., George, A. S., & Pandey, B. K. (2022). A Study on COVID–19, Its Origin, Phenomenon, Variants, and IoT-Based Framework to Detect the Presence of Coronavirus. In *IoT Based Smart Applications* (pp. 1–13). Springer International Publishing.

Mohapatra, S. K., Swain, B. R., & Das, P. (2015). Comprehensive survey of possible security issues on 4G networks. *International Journal of Network Security & its Applications*, 7(2), 61–69. 10.5121/ijnsa.2015.7205

Pandey, B. K., Mane, D., Nassa, V. K. K., Pandey, D., Dutta, S., Ventayen, R. J. M., & Rastogi, R. (2021). Secure text extraction from complex degraded images by applying steganography and deep learning. In *Multidisciplinary approach to modern digital steganography* (pp. 146–163). IGI Global. 10.4018/978-1-7998-7160-6.ch007

Pandey, B. K., & Pandey, D. (2023). Parametric optimization and prediction of enhanced thermoelectric performance in co-doped CaMnO3 using response surface methodology and neural network. *Journal of Materials Science Materials in Electronics*, 34(21), 1589. 10.1007/s10854-023-10954-1

Pandey, B. K., Pandey, D., Gupta, A., Nassa, V. K., Dadheech, P., & George, A. S. (2023). Secret data transmission using advanced morphological component analysis and steganography. In *Role of data-intensive distributed computing systems in designing data solutions* (pp. 21–44). Springer International Publishing. 10.1007/978-3-031-15542-0_2

Pandey, B. K., Pandey, D., & Sahani, S. K. (2024). Autopilot control unmanned aerial vehicle system for sewage defect detection using deep learning. *Engineering Reports*, 12852. 10.1002/eng2.12852

Pandey, D., Nassa, V. K., Jhamb, A., Mahto, D., Pandey, B. K., George, A. H., & Bandyopadhyay, S. K. (2021a). An integration of keyless encryption, steganography, and artificial intelligence for the secure transmission of stego images. In *Multidisciplinary approach to modern digital steganography* (pp. 211–234). IGI Global. 10.4018/978-1-7998-7160-6.ch010

Pandey, D., & Pandey, B. K. (2022). An efficient deep neural network with adaptive galactic swarm optimization for complex image text extraction. In *Process mining techniques for pattern recognition* (pp. 121–137). CRC Press. 10.1201/9781003169550-10

Pandey, D., Pandey, B. K., & Wairya, S. (2021b). Hybrid deep neural network with adaptive galactic swarm optimization for text extraction from scene images. *Soft Computing*, 25(2), 1563–1580. 10.1007/s00500-020-05245-4

Pandey, J. K., Jain, R., Dilip, R., Kumbhkar, M., Jaiswal, S., Pandey, B. K., & Pandey, D. (2022). Investigating role of iot in the development of smart application for security enhancement. In *IoT Based Smart Applications* (pp. 219–243). Springer International Publishing.

Panwar, N., Sharma, S., & Singh, A. K. (2016). A survey on 5G: The next generation of mobile communication. *Physical Communication*, 18, 64–84. 10.1016/j.phycom.2015.10.006

Pramanik, S., Pandey, D., Joardar, S., Niranjanamurthy, M., Pandey, B. K., & Kaur, J. (2023, October). An overview of IoT privacy and security in smart cities. In *AIP Conference Proceedings* (*Vol. 2495*, No. 1). AIP Publishing 10.1063/5.0123511

Raja, D., Kumar, D. R., Santhiyakumari, N., Kumarganesh, S., Sagayam, K. M., Thiyaneswaran, B., Pandey, B. K., & Pandey, D. (2024). A compact dual-feed wide-band slotted antenna for future wireless applications. *Analog Integrated Circuits and Signal Processing*, 118(2), 1–15. 10.1007/s10470-023-02233-0

Rupprecht, D., Dabrowski, A., Holz, T., Weippl, E., & Pöpper, C. (2018). On security research towards future mobile network generations. *IEEE Communications Surveys and Tutorials*, 20(3), 2518–2542. 10.1109/COMST.2018.2820728

Sahani, K., Khadka, S. S., Sahani, S. K., Pandey, B. K., & Pandey, D. (2023). A possible underground roadway for transportation facilities in Kathmandu Valley: A racking deformation of underground rectangular struct es. *Engineering Reports*, 12821. 10.1002/eng2.12821

Saxena, N., & Chaudhari, N. S. (2014). Secure-aka: An efficient aka protocol for umts networks. *Wireless Personal Communications*, 78(2), 1345–1373. 10.1007/s11277-014-1821-0

. Sengupta, R., Sengupta, D., Pandey, D., Pandey, B. K., Nassa, V. K., & Dadeech, P. (2021). A Systematic review of 5G opportunities, architecture and challenges. *Future Trends in 5G and 6G*, 247-269.

Sharma, S., Pandey, B. K., Pandey, D., Anand, R., Sharma, A., & Saini, S. (2023, March). Character Recognition Technique Implementation for Complicated Deteriorated Scene. In *2023 6th International Conference on Information Systems and Computer Networks (ISCON)* (pp. 1-4). IEEE. 10.1109/ISCON57294.2023.10112185

Sheth, K., Patel, K., Shah, H., Tanwar, S., Gupta, R., & Kumar, N. (2020). A taxonomy of AI techniques for 6G communication networks. *Computer Communications*, 161, 279–303. 10.1016/j.comcom.2020.07.035

Singh, S., Madaan, G., Kaur, J., Swapna, H. R., Pandey, D., Singh, A., & Pandey, B. K. (2023). Bibliometric Review on Healthcare Sustainability. *Handbook of Research on Safe Disposal Methods of Municipal Solid Wastes for a Sustainable Environment*, 142-161. 10.4018/978-1-6684-8117-2.ch011

Swapna, H. R., Bigirimana, E., Madaan, G., Hasan, A., Pandey, B. K., & Pandey, D. (2023). Impact of neuromarketing on consumer psychology in digitally connected networks. In *Applications of Neuromarketing in the Metaverse* (pp. 193–205). IGI Global. 10.4018/978-1-6684-8150-9.ch015

Tareke, S. A., Lelisho, M. E., Hassen, S. S., Seid, A. A., Jemal, S. S., Teshale, B. M., & Pandey, B. K. (2022). The prevalence and predictors of depressive, anxiety, and stress symptoms among Tepi town residents during the COVID-19 pandemic lockdown in Ethiopia. *Journal of Racial and Ethnic Health Disparities*, 1–13.35028903

Tripathi, R. P., Sharma, M., Gupta, A. K., Pandey, D., Pandey, B. K., Shahul, A., & George, A. H. (2023). Timely prediction of diabetes by means of machine learning practices. *Augmented Human Research*, 8(1), 1. 10.1007/s41133-023-00062-4

Vinodhini, V., Kumar, M. S., Sankar, S., Pandey, D., Pandey, B. K., & Nassa, V. K. (2022). IoT-based early forest fire detection using MLP and AROC method. *International Journal of Global Warming*, 27(1), 55–70. 10.1504/IJGW.2022.122794

Yang, H., Alphones, A., Xiong, Z., Niyato, D., Zhao, J., & Wu, K. (2020). Artificial-intelligence-enabled intelligent 6G networks. *IEEE Network*, 34(6), 272–280. 10.1109/MNET.011.2000195

Chapter 11
Defending the Digital Twin Machine Learning Strategies for 6G Protection

Freddy Ochoa-Tataje
Universidad César Vallejo, Lima, Peru

Joel Alanya-Beltran
Universidad San Ignacio de Loyola, Peru

Jenny Ruiz-Salazar
Universidad Privada del Norte, Lima, Peru

Juan Paucar-Elera
Universidad Nacional Federico Villarreal, Lima, Peru

Michel Mendez-Escobar
Universidad Autónoma del Perú, Lima, Peru

Frank Alvarez-Huertas
Universidad Nacional Mayor de San Marcos, Lima, Peru

ABSTRACT

The idea of the digital twin has become a game-changer in today's quickly developing technological environment, revolutionizing everything from manufacturing to urban planning. A digital twin replicates the behavior and qualities of real-world physical objects, processes, services, or environments in a virtual space. This essay explores the nuances of the digital twin idea, its uses, and its importance in influencing technology in the future. A digital twin is a dynamic simulation that uses real-world data to predict behavior and performance, going beyond a static representation.

INTRODUCTION

The idea of the "Digital Twin" has become a game-changer in today's quickly developing technological environment, revolutionizing everything from manufacturing to urban planning. A digital twin replicates the behavior and qualities of real-world physical objects, processes, services, or environments in a virtual space. This essay explores the nuances of the Digital Twin idea, its uses, and its importance in influencing technology in the future. A digital twin is a dynamic simulation that uses real-world data to predict behavior and performance, going beyond a static representation. A versatile platform for modeling and analysis is provided by digital twin technology, which can be used to replicate wind farms,

DOI: 10.4018/979-8-3693-2931-3.ch011

jet engines, buildings, or entire cities. It goes even farther than that, though, by enabling the simulation of procedures in order to obtain predictive insights. A digital twin is essentially an advanced computer program powered by real-world data (Bessant, Y. A. et al., 2023). These programs simulate situations and forecast results by utilizing a combination of past data and present asset conditions. By combining cutting-edge technologies like software (Kirubasri, G. et al., 2022) analytics, artificial intelligence (AI), and the Internet of Things (IoT) (Pramanik, S. et al., 2023), this predictive capability is further improved. IoT sensors (Dhanasekar, S. et al., 2023) are essential for gathering data in real time, which makes it possible to create precise virtual representations. Digital twins are now at the forefront of contemporary engineering practices thanks to the convergence of machine learning (Tripathi, R. P. et al., 2023) and big data. These virtual models are extremely useful for promoting innovation and enhancing output in a variety of sectors. Organizations can reduce risks, increase efficiency, and speed up decision-making by utilizing sophisticated analytical, monitoring, and predictive capabilities. The capacity of digital twins to anticipate and resolve possible malfunctions in physical objects is one of its main benefits. Through the use of simulations and performance metrics analysis, engineers are able to detect weaknesses and take corrective action before they become expensive problems. In addition to protecting assets, this proactive approach reduces maintenance costs and downtime. With the use of digital twins, businesses can reduce the risks involved in real-world experimentation by testing services and processes in a controlled setting. Digital twins provide an innovation and iterative improvement sandbox for applications such as supply chain logistics optimization (Pandey, D. et al., 2021a) and manufacturing workflow refinement. The emergence of digital twin technology signals the beginning of a new phase in risk reduction and strategic innovation. Organizations can gain access to previously unattainable insights and expansion opportunities by bridging the gap between the physical and digital domains. This transformative potential does, however, come with a unique set of considerations and challenges. Organizations must address issues with data security, privacy (Pandey, B. K. et al., 2023a). and interoperability as they adopt digital twins. Digital twins can only be fully utilized if disparate systems and data sources are seamlessly integrated. Standardized frameworks and protocols are also required in order to promote industry-wide cooperation and knowledge exchange (Wu, Y., et al., 2021).

Emergence of 6G Technology

6G technology represents the beginning of the next stage in wireless communication (Devasenapathy, D. et al., 2024), offering unparalleled speed, capacity, and connectivity. Expanding on the groundwork established by previous generations, 6G signifies a significant advancement in technological progress, capable of transforming various sectors and restructuring the digital environment. 6G technology fundamentally seeks to meet the escalating requirements for fast, dependable, and minimal delay connectivity in an ever more interconnected global environment. The 5G technology has already revolutionized communication and interaction, and the 6G technology aims to expand these advancements by enabling immersive experiences, intelligent automation, and ubiquitous connectivity. The speed of 6G technology is considered to be one of its defining characteristics. With anticipated data rates exceeding terabits per second, 6G (Kirubasri, G. et al., 2021) networks hold the potential to provide uninterrupted streaming, instantaneous downloads, and instantaneous communication on an unparalleled magnitude. The exponential growth in bandwidth presents a multitude of opportunities, encompassing various domains such as ultra-high-definition video streaming and immersive virtual and augmented reality encounters. Through the utilization of sophisticated network architectures and intelligent resource management techniques

(Khan, B. et al., 2021), the implementation of 6G networks will facilitate the widespread adoption of intelligent sensors, autonomous vehicles, and interconnected devices (Nguyen, H. X. et al., 2021). This will mark the advent of an era characterized by unparalleled connectivity and digital transformation (Pandey, B. K. et al., 2023b). The implementation of 6G technology holds the potential to significantly reduce latency to an unprecedented extent, thereby enabling the development of real-time applications like remote surgery, autonomous vehicles, and tactile internet experiences. The implementation of 6G networks will facilitate the development of highly responsive applications that were previously envisioned in the realm of science fiction, primarily through the reduction of latency between data transmission and reception (Pandey, B. K. et al., 2022). The anticipated consequences of the advent of 6G technology on society, economy, and culture are extensive, owing to its advanced technical capabilities. 6G has the capacity to close the gap in digital access and foster innovation and prosperity by facilitating remote education, telemedicine, empowering marginalized communities, and stimulating economic development. Nonetheless, possessing significant authority entails a substantial obligation. With the ongoing advancement of 6G technology, it is crucial to tackle issues pertaining to privacy, security, and ethical considerations (Pandey, D. et al., 2021b). 6G networks, being the foundation of the digital economy, will have a crucial impact on shaping the future of society. Therefore, it is crucial to ensure that these technologies are deployed in a responsible and ethical manner (Zhou, C. et al., 2021).

Protecting Digital Twins in 6G

Due to the pivotal role that digital twins play in a variety of fields, protecting them is of the utmost importance in the context of 6G technology. Real-time monitoring, analysis, and optimization (Pandey, D., & Pandey, B. K., 2022) are all made possible by digital twins, which are virtual representations of physical objects or processes. It is an essential component of 6G networks. Digital Twins will improve connectivity with 6G by providing real-time asset status and performance data. Virtual replicas facilitate data exchange and communication between. Predictive maintenance strategies use digital twins to predict and prevent physical asset failures. Digital Twins use historical data and real-time sensor readings to detect equipment degradation, abnormal patterns, and maintenance needs before they become costly. Proactive maintenance reduces downtime, costs, and asset lifespan, improving operational efficiency and reliability. Digital Twins also optimise maintenance schedules, resource allocation, and task prioritisation, maximizing operational uptime and minimizing disruptions by continuously monitoring asset performance and analysing data trends. Digital Twins optimise 6G network resource management. Simulating scenarios and analyzing data in real time helps organizations allocate resources, consume energy, and use assets. Digital Twins can model energy production and consumption, predict demand fluctuations, and optimize energy distribution in smart grid systems to reduce waste and increase efficiency. Digital Twins can also simulate traffic flows, optimize route planning, and manage fleet operations in transportation systems (Sahani, K. et al., 2023) to reduce congestion, fuel consumption, and efficiency. Organizations can save money, use resources better, and promote sustainability by using Digital Twins in 6G networks (Grieves, M., 2016). 6G networks must protect Digital Twins to reduce cybersecurity risks and protect sensitive data. Cyberattacks and malicious activities target Digital Twins as they become more interconnected and essential to critical infrastructure. To prevent unauthorized access, data breaches, and other cybersecurity threats, encryption (Kumar Pandey, B. et al., 2021), authentication, and access controls are necessary. To stay resilient to evolving cyber threats, organizations must continuously monitor and assess Digital Twin security, identify vulnerabilities, and remediate them. Organizations can protect sensitive data, critical infrastructure, and digital ecosystem integrity by prioritizing cybersecurity in 6G Digital Twin design and implementation. Through virtual experimentation, collaboration, and knowledge sharing in 6G networks, Digital Twins promote innovation and collaboration. Virtual replicas let researchers, engineers, and developers test new ideas, simulate complex scenarios, and iterate quickly in a controlled environment. Organizations can accelerate innovation, reduce time-to-market for new products and services, and improve across industries by using Digital Twins. Digital Twins enable cross-disciplinary teams to collaborate, share insights, and co-create solutions to complex problems. Digital Twins accelerate innovation and collaboration in 6G-enabled ecosystems, encouraging creativity, experimentation, and continuous learning in smart city design, supply chain optimization, and advanced healthcare system development (Tuegel, E. J. et al., 2011).

UNDERSTANDING DIGITAL TWINS

Components of Digital Twins

Digital twins are computerized models that closely resemble real-world counterparts of tangible systems, environments, processes, or objects. In the digital sphere, they act as virtual copies of their physical counterparts, mimicking their traits, interactions, and behavior. Since then, the aerospace sector has embraced the idea of digital twins, which are now used in manufacturing, healthcare (Deepa, R. et al., 2022). Comprehensive performance), transportation, and urban planning, among other industries Components of Digital Twins are as show in the figure 1.

1. *Physical System or Object*: Every Digital Twin starts with a physical object or system. It includes sensors, machines, manufacturing plants, transportation networks, and even cities. Every physical object or system has unique behaviors, characteristics, and interactions that its Digital Twin mirrors. Digital Twin fidelity depends on the detail and complexity needed to accurately represent the physical counterpart. Industrial robot Digital Twins may include geometrical models, kinematic simulations, and operational parameters to simulate their movements and performance in real time.
2. *Sensors and IoT Devices*: To feed the Digital Twin real-time data from the physical world, sensors and IoT devices (Sharma, M. et al., 2022) are essential. These devices record temperature (Sharma, M. et al., 2022), pressure, humidity, vibration, motion, and more. In smart factories, sensors in machinery and equipment monitor temperature, pressure, and energy consumption. IoT devices (Du John, H. V. et al., 2022) throughout the facility track inventory, production workflows, and equipment performance. This abundance of real-time data enables accurate and responsive Digital Twins that simulate and optimize physical processes.
3. *Data Processing and Integration*: Digital Twins aggregate, clean, and analyze sensor and IoT data. Sensor data is turned into actionable insights for decision-making and optimization. Data integration unites sensor, database, and external system data to create a single physical environment view. Filtering (Bruntha, P. M. et al., 2023), normalization, and feature extraction prepare the data for analysis. Data is analyzed using machine learning (Vinodhini, V. et al., 2022) and statistical modeling to identify patterns, trends, and anomalies. This analytical process lets Digital Twins detect performance issues, predict failures, and optimize processes in real time.

4. *Models and Simulations*: Digital Twins use modelling and simulation to create virtual representations that match physical objects and systems. Mathematical equations, algorithms, and simulation techniques capture the dynamics, interactions, and responses of the real-world counterpart in these virtual models. Wind turbine Digital Twins may use aerodynamic models, mechanical simulations, and environmental factors to predict performance under different operating conditions. The Digital Twin is kept accurate by updating and refining these models using real-time data from sensors and IoT devices. Stakeholders can test "what-if" scenarios, strategies, and performance without disrupting the physical environment with simulation.

5. *Analytics/Visualization*: Digital Twins require analytics and visualization to gain insights and present information clearly. Statistical analysis, machine learning, and predictive modeling are used to find patterns, trends, and correlations in sensor and IoT (Pandey, J. K. et al., 2022) data. Analytics algorithms may identify equipment behavior anomalies, predict maintenance needs, or optimize energy consumption using historical data. Stakeholders receive these insights in an intuitive and actionable manner using dashboards, charts, and 3D models. Visualisations help stakeholders analyse complex data, track KPIs, and make informed decisions to improve efficiency and effectiveness.

6. *Feedback Loop*: Digital Twins monitor, analyze, and optimize physical processes using the feedback loop. When sensors and IoT devices send real-time data, the Digital Twin updates its model and simulates physical environment changes. The Digital Twin simulates the effects of a sensor detecting an abnormality or deviation from expected behavior in a manufacturing process and recommends actions to mitigate risks or optimize performance. Due to this closed-loop feedback mechanism, Digital Twins can adapt to changing conditions and make proactive decisions to reduce downtime, improve quality, and boost performance.

7. *Interconnectivity*: Connectivity allows Digital Twins to communicate and share data with other systems, databases, and platforms. To enable cross-functional data sharing and collaboration, digital twins are often integrated with enterprise systems like MES, ERP, and asset management systems. To improve energy efficiency, occupant comfort, and safety, a smart building's Digital Twin may communicate with energy management, security, and building automation systems. To improve decision-making, Digital Twins can be connected to external data sources like weather forecasts, market trends, and supply chain data. Interoperability allows Digital Twins to use data from multiple sources to create holistic insights that drive value creation and innovation across domains.

Figure 1. Components of digital twins

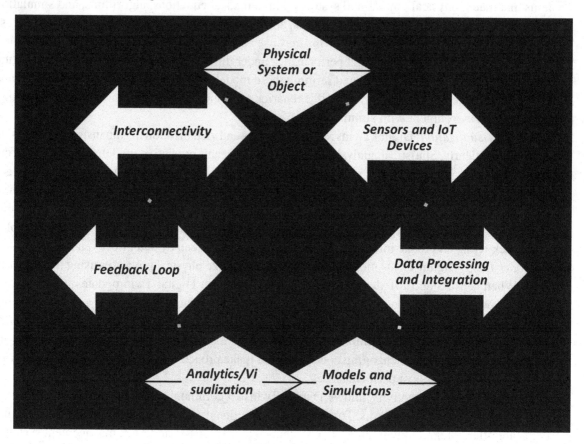

Diverse Industries Harnessing Digital Twins

Various industries are now using Digital Twins to enhance operations, improve processes, and foster innovation. These digital representations of tangible items, systems, or surroundings allow companies to acquire valuable knowledge, enhance their decision-making processes, and boost performance in different areas. In manufacturing, Digital Twins are transforming traditional processes by allowing for predictive maintenance, production optimization, and supply chain management. Manufacturers have the ability to monitor equipment health in real-time (Sasidevi, S. et al., 2024), anticipate potential failures, and proactively schedule maintenance. This helps minimize downtime and reduce costs. Additionally, Digital Twins simulate manufacturing processes, allowing for the optimization (Jayapoorani, S. et al., 2023) of production lines, resource utilization, and workflow efficiency, leading to improved productivity and quality. In addition, Digital Twins simulate supply chain operations, offering valuable information on inventory levels, demand forecasting, and logistics optimization, guaranteeing seamless and effective supply chain operations. The healthcare industry (Pandey, D. et al., 2023) is also reaping the rewards of Digital Twins, enabling personalized medicine, simulating medical devices, and optimizing healthcare facilities (Tareke, S. A. et al., 2022). Utilizing advanced technology, digital twins of patients combine various data sources such as medical records, genomic data, and physiological parameters (Rathore, M. M. et al., 2021). This integration allows for the creation of personalized health profiles, which in turn enable the development of tailored treatment plans and proactive healthcare management. In addition, Digital Twins can simulate the behavior of medical devices in different scenarios (Pandey, B. K. et al., 2011), enabling virtual testing, optimization, and customization to cater to the unique requirements of each patient. In addition, Digital Twins of healthcare facilities simulate patient flows, resource utilization, and facility layouts to optimize operations, enhance patient experience, and improve staff efficiency. Smart cities are utilizing Digital Twins to assist with urban planning, public safety, and smart infrastructure management. Using digital twins, cities can create models that simulate infrastructure, transportation networks, and environmental factors. These models provide valuable insights for urban planning decisions, optimizing traffic flow, designing sustainable buildings, and mitigating environmental risks. In addition, Digital Twins combine data from surveillance systems, sensors, and emergency response services to monitor public safety and security. This enables quick incident detection, coordination of response efforts (Abdulkarim, Y. I. et al., 2024) and preparation for disasters. In addition, Digital Twins of infrastructure assets allow for predictive maintenance, asset monitoring, and resilience planning, which helps ensure the effective operation and long-term durability of vital infrastructure. Within the energy and utilities sector, Digital Twins play a crucial role in enhancing asset performance management, facilitating the integration of renewable energy, and optimizing water management initiatives. Using digital twins of energy assets allows for the monitoring of performance metrics, prediction of equipment failures, and optimization of energy production. This helps to maximize uptime, improve efficiency, and reduce operational costs. In addition, Digital Twins are used to model renewable energy sources in order to optimize generation, storage, and distribution. This helps with the integration of renewable energy into the grid and improves energy sustainability. In addition, Digital Twins of water systems can simulate water distribution networks, monitor water quality, and forecast demand patterns. This allows for more effective management of water resources, early detection of leaks, and conservation efforts. In transportation industries, Digital Twins are transforming fleet management, smart mobility solutions, and the development of autonomous vehicles. Organizations have the ability to monitor vehicle health, optimize routes, and manage logistics operations, which can lead to improved fleet efficiency, reduced fuel consumption, and enhanced driver safety. In addition, Digital Twins are capable of modeling transportation networks to enhance traffic flow, alleviate congestion, and enhance the overall commuting experience. They play a crucial role in the advancement of smart city mobility solutions. In addition, Digital Twins can simulate the behavior of autonomous vehicles, allowing for virtual testing, training, and validation of autonomous driving systems. This helps speed up the development and deployment of autonomous vehicle technologies (Ai, Y. et al., 2018).

Importance in Various Domains

Incidents that compromise security can have far-reaching consequences not only for businesses but also for society as a whole. The financial losses that affected organizations have incurred are one of the effects that are the most immediate and tangible they have experienced. Remediation efforts, legal fees, and regulatory fines are examples of direct costs that can quickly add up, which can significantly impact profitability and drain resources. Indirect financial consequences may also be experienced by businesses, including a loss of revenue as a result of downtime, a decrease in customer trust that results in a decrease in sales, and an increase in insurance premiums paid. In particular, smaller businesses that may lack the resources to recover from such setbacks are particularly vulnerable to the financial implications that can threaten the financial stability and viability of businesses. A significant amount of reputational damage is inflicted upon businesses as a result of security breaches, in addition to the immediate financial impact. Customers' trust in the affected organization is damaged whenever sensitive data is compromised or when personal information is exposed. It is possible for the name of the organization to be tarnished by the perception of negligence or incompetence, which can result in a loss of confidence in the products or services that the organization offers. It can be a time-consuming and difficult process to rebuild trust and restore reputation (Lu, Y. et al., 2021a). This process frequently requires extensive communication efforts, transparency, and proactive measures to reassure stakeholders. Data privacy and personal security are endangered when there is a breach in security, which has repercussions not only for individuals but also for society as a whole. A number of different types of cybercrime can be committed as a result of unauthorized access to personal information or theft of that information. A decrease in trust in digital systems and an increase in concerns regarding privacy and security are being experienced by society as a whole, in addition to the victims who are directly affected by such crimes.

SECURITY CHALLENGES IN 6G

Digital Twin Vulnerabilities

Although Digital Twins have many advantages, they also come with certain risks and vulnerabilities that can have serious consequences if exploited by malicious individuals. One of the main concerns with Digital Twins is the risk of data breaches. These virtual representations depend on gathering and analyzing large quantities of data from sensors, IoT devices, and other sources (Iyyanar, P. et al., 2023). If the security measures guarding this data are compromised, it could result in data breaches, which would expose sensitive information to unauthorized parties. This may involve sensitive business data, personal identifiable information (PII), or operational details of critical infrastructure. Another important concern is the vulnerability to cyberattacks. Digital Twins can be susceptible to a range of cyber threats, such as malware, ransomware, and denial-of-service (DoS) attacks. Malicious actors can take advantage of weaknesses in the digital twin's software or network infrastructure to gain unauthorized access, disrupt operations, or pilfer valuable data. Such attacks can have significant impacts, including financial losses, operational downtime, and damage to reputation. In addition, the interconnected nature of Digital Twins within larger networks amplifies the vulnerability to cyberattacks and their potential consequences. Control system manipulation is a significant concern when it comes to vulnerabilities in Digital Twins, especially in industrial environments. Digital Twins are commonly employed for the purpose of monitoring and controlling physical processes and machinery. If unauthorized individuals gain access to the Digital Twin, they have the potential to manipulate control systems, resulting in physical damage, safety hazards, or disruptions to operations. As an electrical engineer, you understand the potential risks associated with tampering with Digital Twins in critical infrastructure like power plants (Govindaraj, V. et al., 2023) or transportation networks. The consequences could be catastrophic, posing a threat to public safety and national security. Interactions with external systems, databases, and platforms can introduce vulnerabilities at different touchpoints for Digital Twins. Potential vulnerabilities in these integrations or reliance on external providers could potentially compromise the security of Digital Twins and lead to unauthorized access or data breaches. In addition, the intricate nature of Digital Twin ecosystems poses a challenge in effectively identifying and addressing vulnerabilities, thereby increasing the risk of exploitation by malicious individuals.

Security Breaches on Businesses and Society

Security breaches can result in far-reaching consequences for society and businesses, not limited to urgent financial setbacks. Organizational-wide security intrusions frequently lead to substantial financial setbacks. Direct costs consist of expenditures associated with remediation activities, legal charges, and regulatory sanctions. Indirect financial consequences, such as increased insurance premiums, revenue loss due to downtime, and diminished consumer confidence resulting in decreased sales, can be equally detrimental. Particularly for lesser enterprises, these monetary setbacks may present existential perils, putting their financial sustainability and enduring existence at risk. Businesses suffer significant reputational harm as a result of security breaches. The exposure of sensitive data or personal information results in a loss of consumer confidence in the organization that is compromised. Deterioration of customer confidence in the organization's offerings may ensue as a consequence of this erosion of trust; customer loyalty and retention may subsequently decline. In addition, adverse publicity pertaining to a security lapse (David, S. et al., 2023) may elicit unwelcome scrutiny from regulatory entities and the media, thereby inflicting additional harm upon the organization's standing. Restoring reputation and re-establishing trust frequently necessitate substantial investments of time and resources, with enduring consequences for the brand value and market position of the organization. The immediate business repercussions, security breaches have far-reaching societal consequences. One of the foremost concerns pertains to the breach of privacy regarding personal data. Personal identifiers, financial records, medical histories, and other sensitive data become susceptible to identity theft, fraud, and various other forms of exploitation when compromised by cybercriminals. This undermines confidence in online transactions and erodes trust in digital systems (Pandey, D. et al., 2021c), in addition to causing financial damage to individuals. Cyberattacks that are directed at utility providers, government agencies, or healthcare systems have the potential to cause extensive damage and disruption, impacting millions of individuals and communities.

Security vulnerabilities can potentially result in significant financial and reputational ramifications for organizations. Direct costs associated with a security violation may consist of legal fees, regulatory fines, and remediation expenses. Revenue loss attributable to downtime, diminished consumer confidence, and increased insurance premiums are all potential indirect costs. The financial repercussions can be particularly detrimental to small enterprises, potentially jeopardizing their sustainability (Dong, R. et al., 2019). The consequences of a security breach transcend mere monetary detriment. Sensitive data compromise has the potential to cause substantial harm to the reputation of a business. Loss of customer confidence in the organization may result in diminished customer loyalty and sales. Adverse publicity pertaining to the breach has the potential to further damage the company's reputation, drawing unwelcome scrutiny from regulatory agencies and the media. The process of restoring trust and mending reputation can be an extended one, requiring significant resources, and may have enduring consequences for the brand value and market position of the organization. Businesses must prioritize cybersecurity measures in order to protect against security vulnerabilities in light of these threats. This entails the establishment of resilient security protocols, the consistent updating of software and systems, and the provision of cybersecurity best practices training for personnel. It is imperative to foster collaboration with industry partners, government agencies, and cybersecurity professionals in order to safeguard digital infrastructure against security breaches and minimize their far-reaching societal consequences (Groshev, M. et al., 2021).

ML APPROACHES FOR PROTECTION

Advanced methods for safeguarding Digital Twins from security risks are provided by machine learning techniques are shown in the figure 2. Organizations can identify vulnerabilities, foresee hazards, and put adaptive defense mechanisms in place to protect Digital Twins in dynamic environments, guaranteeing resilience and dependability, by utilizing anomaly detection, predictive analytics, and reinforcement learning, and self-learning systems (Saxena, A. et al., 2021).

Figure 2. ML techniques for security

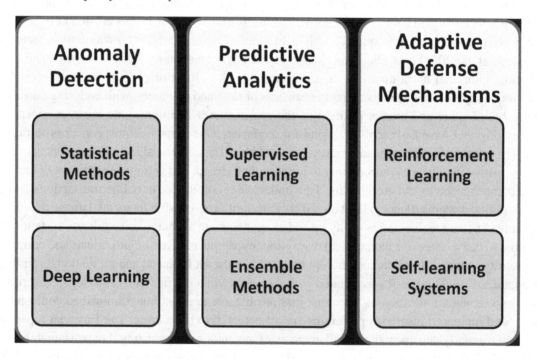

Anomaly Detection

Statistical Methods

Effective tools for data analysis, pattern recognition, and evidence-based decision-making, statistical methods are applied in many domains. These methods cover a broad spectrum of approaches to meaningfully gather, summarize, analyze, and present data. Statistical techniques are vital to research, industry, healthcare, the social sciences, and many other fields. They range from basic descriptive statistics to sophisticated inferential procedures. Statistical analysis, at its core, is the process of analyzing data and making inferences about populations from sample data by applying mathematical concepts and probability theory. The features of a dataset are summarized and described using descriptive statistics, which include measures of central tendency (such as mean, median, and mode) and measures of dispersion (such as variance and standard deviation). Gain important insights into the distribution, variability, and form of the data with these summary statistics. To generate hypotheses, test predictions, and draw conclusions about populations from sample data, inferential statistical techniques are employed in addition to descriptive statistics. Regression analysis, analysis of variance (ANOVA), confidence intervals, and hypothesis testing are some of these techniques. Researchers can assess whether observed variations between groups are merely the result of chance or statistically significant by using hypothesis testing, for instance. By analyzing the relationship between variables, regression analysis, on the other hand, can be used to find factors that influence an outcome or to make predictions.

In organizations and industries, statistical approaches are also frequently employed in quality control and process improvement activities. Statistical tools are used by methodologies like Six Sigma and statistical process control (SPC) to track operations, find causes of variance, and enhance productivity and quality. Organizations can spot trends, spot performance that deviates from expectations, and take remedial action to minimize errors and maximize efficiency by evaluating data gathered at different phases of a process. Factorial designs and randomized controlled trials are two experimental design strategies that assist researchers in identifying causal links between variables and accounting for confounding variables. Researchers can test hypotheses and assess the importance of observed effects by using data analysis techniques like chi-square tests and analysis of variance (ANOVA). Clinical trials in the medical field employ statistical approaches to compare interventions, measure patient outcomes, and determine whether novel therapies are effective. Utilizing statistical methods, epidemiological studies look into the distribution and incidence of diseases within communities and pinpoint risk factors linked to the development of disease. Furthermore, patient survival rates are estimated, treatment choices are guided, and disease progression is predicted through the use of predictive modeling techniques like survival analysis and logistic regression (Zhao, L. et al., 2020).

Deep Learning

Deep learning models (Anand, R. et al., 2023) are machine learning algorithms inspired by brain anatomy and function. These models can learn complicated patterns and representations from enormous volumes of data, making them ideal for image, natural language, and audio recognition. Deep learning models have multilayered architectures with interconnected neurons or units. Many deep learning methods use the artificial neural network (ANN) (Pandey, B. K., & Pandey, D., 2023), which has layers of interconnected neurons. Each neuron processes incoming signals, computes, and outputs a signal that is transferred to the next layer of neurons. Deep neural networks (DNNs) learn complicated input data characteristics and representations using numerous layers of neurons. Image and spatial data challenges are ideal for convolutional neural networks (CNNs). CNNs use convolutional layers to extract features from input images and pooling layers to downsample feature maps and simplify computation. These models excel at picture classification, object recognition, and segmentation. Recurrent neural networks (RNNs) are another deep learning model (Revathi, T. K. et al., 2022) used for sequential data like time series or natural language sequences. RNNs can remember prior inputs because they build directed cycles, unlike feedforward neural networks. This lets RNNs understand temporal connections and detect sequential data patterns. Lengthy short-term memory (LSTM) networks and gated recurrent units (GRUs) are RNN versions that address the vanishing gradient problem while training deep networks on lengthy data sequences (Pandey, B. K. et al., 2021). Gated mechanisms control information flow through the network, allowing these models to learn long-range dependencies and capture context over extended periods. GANs are deep learning models (Saxena, A. et al., 2024). Examination of the Criticality of Customer) that train a generator and a discriminator neural network simultaneously in a competitive way. Generator network generates synthetic data samples, while discriminator network learns to differentiate actual and synthetic data. Adversarial training can teach GANs to generate realistic images, sounds, and text. Transformer models, such as BERT and GPT, are another deep learning architecture that excels in natural language processing. Transformers use self-attention processes to record long-range relationships in sequential data and process inputs in parallel to learn word or token contextual representations.

Predictive Analytics

Supervised Learning

Supervised learning algorithms play a crucial role in machine learning, working with labeled datasets that consist of input-output pairs. These algorithms strive to understand patterns and relationships in the given data in order to make predictions or decisions about unseen data. Linear regression is a straightforward supervised learning algorithm that is often utilized for regression tasks. It involves fitting a linear equation to the data in order to model the relationship between independent and dependent variables. It is especially valuable in situations where a consistent result is anticipated from one or more input features. Logistic regression is a commonly used algorithm for classification tasks. Although it may seem counterintuitive, this particular model is designed to accurately predict probabilities ranging from 0 to 1 for binary outcomes. As a result, it is highly effective for solving binary classification problems. On the other hand, decision trees provide a flexible approach that can handle both classification and regression tasks. They divide the feature space by using feature thresholds and assign labels or values to different regions, making them easy to understand and visualize. Random forests are a widely used ensemble method that significantly improves the performance of decision trees. This is achieved by combining multiple trees and aggregating their predictions. Through the process of training each tree on a random subset of the data and then combining their outputs through voting or averaging, random forests effectively address the issue of overfitting and enhance the ability to make accurate predictions on unseen data. SVMs are highly effective algorithms for classification tasks, especially when faced with intricate decision boundaries. They excel at identifying the best hyperplane to separate various classes, maximizing the margin between them and proving their effectiveness in high-dimensional spaces. K-nearest neighbors (KNN) is a straightforward yet powerful algorithm for both classification and regression tasks. It predicts by identifying the k nearest neighbors to a given input and then averaging or voting their labels or values. Naive Bayes is a probabilistic classifier that calculates the probability of each class based on the input features, using Bayes' theorem. Although Naive Bayes makes the "naive" assumption of feature independence, it is highly efficient and frequently achieves good performance, particularly in text classification tasks. Gradient boosting machines (GBMs) are a powerful ensemble method that constructs a robust predictive model by iteratively combining weak learners, like decision trees, and fine-tuning them to minimize the loss function. This iterative process gradually enhances the model's performance, ensuring its resilience and exceptional accuracy. Neural networks, with their structure and function inspired by the human brain, have the remarkable ability to learn intricate patterns from data. Neural networks possess interconnected layers of neurons, enabling them to tackle a wide range of tasks such as classification, regression, image recognition, and natural language processing. This versatility makes them an incredibly potent tool in the realm of supervised learning algorithms (Spirent Communications, 2019).

Ensemble Methods

Ensemble machine learning approaches increase predicted performance and resilience by combining many models. These strategies use the diversity of models to provide more accurate and stable predictions than any one model. Ensemble approaches have excelled in competitions, real-world applications, and research. Bootstrap aggregating (bagging) is a prominent ensemble approach. Bagging involves training multiple models on distinct subsets of the training data using bootstrapping to produce random samples with replacement. Each model's forecasts are then averaged or voted on to make the final projection. Random forests, which use bagging to train numerous decision trees, are a common example. Boosting sequentially trains weak learners and emphasizes instances that prior models misclassified or poorly forecasted. AdaBoost (Adaptive Boosting) and Gradient Boosting Machines (GBMs) iteratively train models to focus on earlier model faults, boosting predicted accuracy. An ensemble method called stacking generalization uses a meta-learner to merge many models. Stacking trains a meta-learner, usually a linear regression or logistic regression model, to combine base model predictions instead of average or voting them. This lets stacking take use of model strengths and improve forecast performance. Ensembles have advantages over individual models. Ensemble models prevent overfitting and increase generalization, especially when the models are different. Ensemble approaches are more resilient to data noise and outliers because other models generally counterbalance model faults. Ensemble approaches can also detect complicated data patterns and linkages that individual models miss, improving prediction accuracy. Ensemble approaches combine numerous models to make a forecast, making them harder to understand. Ensemble approaches may not increase predictive accuracy if the models are strongly coupled or the data is unsuitable for ensemble learning.

Adaptive Defense Mechanisms

Reinforcement Learning

Cybersecurity, emergency response, military operations, and critical infrastructure protection can benefit from reinforcement learning (RL) for dynamic threat response. An agent in RL learns to maximize cumulative rewards by adopting activities that lead to desirable outcomes. This paradigm works well in dynamic environments where real-time judgments are needed to mitigate hazards. may use RL to create adaptive defenses that adapt to new threats. RL agents identify and respond to intrusions quickly by monitoring network traffic and system logs. Based on threat landscapes and reaction effects, these agents can alter firewall rules, isolate compromised systems, or launch countermeasures. In uncertain emergency scenarios like natural disasters or terrorist attacks, RL can optimize resource allocation and decision-making. RL agents improve response efforts by dynamically assigning emergency personnel, resources, and equipment based on changing conditions and priorities. In disaster scenarios, RL agents may prioritize places with the most casualties or secondary dangers RL can help military operations with adaptive mission planning, tactical decision-making, and autonomous weapon systems. RL agents improve military operations by anticipating enemy moves, adapting to changing battlefield conditions, and optimizing methods to fulfill mission objectives while avoiding friendly force risks. RL can create autonomous drones that can navigate complex surroundings, identify targets, and launch accurate strikes without human intervention. Critical infrastructure like power plants, transportation networks, and communication systems can be protected from physical and cyber threats with RL. Infrastructure assets can be monitored by RL agents to detect anomalies or security breaches and prevent or minimize disturbances. In reaction to equipment failures or cyberattacks, RL agents in power grids could optimize load balancing, reroute power flows, or activate backup systems.

Self-Learning Systems

Self-learning systems use superior artificial intelligence (AI) (Anand, R. et al., 2024) to steadily increase performance without human intervention, revolutionizing continual adaptation in different fields. These systems learn from experience, adapt, and optimise performance in response to changing conditions and new information. Self-learning systems have great potential in robotics, manufacturing, healthcare, and cybersecurity. Self-learning robots and automation systems can transform industrial processes by adapting to changing environments and tasks. To improve efficiency, productivity, and dependability, these systems can automatically optimise settings, develop control tactics, and learn from prior experiences. In manufacturing, self-learning robots can adjust their movements, grasping, and assembling methods to match product designs and production situations, boosting flexibility and adaptability. It can transform healthcare by offering tailored treatment regimens, predictive diagnoses, and autonomous decision-making. These algorithms can detect patterns, trends, and correlations in massive patient data, medical records, and research findings that clinicians may miss. Self-learning systems improve patient outcomes and healthcare delivery by learning from fresh data and feedback to improve diagnostic accuracy, treatment suggestions, and prognostic predictions. These systems may learn to recognize abnormal behavior, detect new threats, and adjust defensive measures in real time to reduce risks and preserve important assets. Self-learning systems can increase cybersecurity and resilience by learning from past assaults, security breaches, and security updates. These systems can automatically evaluate financial data, predict market trends, and optimize financial investment plans. Self-learning systems can increase autonomous vehicle navigation, traffic adaptability, and passenger safety. Self-learning systems can improve crop management, soil health, and yields while conserving resources. To promote sustainable environmental management, self-learning algorithms evaluate sensor data, identify pollution trends, and predict environmental concerns.

CONCLUSION

Digital Twins and 6G technologies bring huge prospects and challenges. Digital Twins enable unprecedented virtual simulation, predictive analytics, and real-time monitoring across sectors. In the 6G era, Digital Twins are more necessary to key infrastructure and essential services, making their security and resilience crucial. Digital Twins, 6G technology, and the importance of safeguarding them in 6G networks. The security risks of 6G networks and Digital Twins, highlighting the necessity for strong security frameworks and proactive defenses. Organizations may secure Digital Twins and reduce risks using machine learning methods including anomaly detection, predictive analytics, reinforcement learning, and self-learning systems. Stakeholders, legislators, and security professionals must work together to solve 6G dangers. In the dynamic and interconnected world of 6G, protecting Digital Twins and vital infrastructure requires extensive security solutions, machine learning, and cybersecurity awareness. Through innovative technologies, advanced analytics, and cybersecurity, we can use Digital Twins and 6G technology to drive sustainable growth, improve operational efficiency, and improve quality of life for people and communities worldwide. A safe and robust 6G future starts with preventive actions now.

REFERENCES

Abdulkarim, Y. I., Awl, H. N., Muhammadsharif, F. F., Saeed, S. R., Sidiq, K. R., Khasraw, S. S., Dong, J., Pandey, B. K., & Pandey, D. (2024). Metamaterial-based sensors loaded corona-shaped resonator for COVID-19 detection by using microwave techniques. *Plasmonics*, 19(2), 595–610. 10.1007/s11468-023-02007-4

Ai, Y., Peng, M., & Zhang, K. (2018). Edge computing technologies for Internet of Things: A primer. *Digital Communications and Networks*, 4(2), 77–86. 10.1016/j.dcan.2017.07.001

Anand, R., Khan, B., Nassa, V. K., Pandey, D., Dhabliya, D., Pandey, B. K., & Dadheech, P. (2023). Hybrid convolutional neural network (CNN) for kennedy space center hyperspectral image. *Aerospace Systems*, 6(1), 71–78. 10.1007/s42401-022-00168-4

Anand, R., Lakshmi, S. V., Pandey, D., & Pandey, B. K. (2024). An enhanced ResNet-50 deep learning model for arrhythmia detection using electrocardiogram biomedical indicators. *Evolving Systems*, 15(1), 83–97. 10.1007/s12530-023-09559-0

Bessant, Y. A., Jency, J. G., Sagayam, K. M., Jone, A. A. A., Pandey, D., & Pandey, B. K. (2023). Improved parallel matrix multiplication using Strassen and Urdhvatiryagbhyam method. *CCF Transactions on High Performance Computing*, 5(2), 102–115. 10.1007/s42514-023-00149-9

Bruntha, P. M., Dhanasekar, S., Hepsiba, D., Sagayam, K. M., Neebha, T. M., Pandey, D., & Pandey, B. K. (2023). Application of switching median filter with L 2 norm-based auto-tuning function for removing random valued impulse noise. *Aerospace systems, 6*(1), 53-59.

David, S., Duraipandian, K., Chandrasekaran, D., Pandey, D., Sindhwani, N., & Pandey, B. K. (2023). Impact of blockchain in healthcare system. In *Unleashing the Potentials of blockchain technology for healthcare industries* (pp. 37–57). Academic Press. 10.1016/B978-0-323-99481-1.00004-3

Deepa, R., Anand, R., Pandey, D., Pandey, B. K., & Karki, B. (2022). Comprehensive performance analysis of classifiers in diagnosis of epilepsy. *Mathematical Problems in Engineering*, 2022, 2022. 10.1155/2022/1559312

Devasenapathy, D., Madhumathy, P., Umamaheshwari, R., Pandey, B. K., & Pandey, D. (2024). Transmission-efficient grid-based synchronized model for routing in wireless sensor networks using Bayesian compressive sensing. *SN Computer Science*, 5(1), 1–11.

Dhanasekar, S., Martin Sagayam, K., Pandey, B. K., & Pandey, D. (2023). Refractive Index Sensing Using Metamaterial Absorbing Augmentation in Elliptical Graphene Arrays. *Plasmonics*, 1–11. 10.1007/s11468-023-02152-w

Dong, R., She, C., Hardjawana, W., Li, Y., & Vucetic, B. (2019). Deep learning for hybrid 5G services in mobile edge computing systems: Learn from a digital twin. *IEEE Transactions on Wireless Communications*, 18(10), 4692–4707. 10.1109/TWC.2019.2927312

Du John, H. V., Jose, T., Jone, A. A. A., Sagayam, K. M., Pandey, B. K., & Pandey, D. (2022). Polarization insensitive circular ring resonator based perfect metamaterial absorber design and simulation on a silicon substrate. *Silicon*, 14(14), 9009–9020. 10.1007/s12633-021-01645-9

. Govindaraj, V., Dhanasekar, S., Martinsagayam, K., Pandey, D., Pandey, B. K., & Nassa, V. K. (2023). Low-power test pattern generator using modified LFSR. *Aerospace Systems*, 1-8.

Grieves, M., & Vickers, J. (2016). *Origins of the digital twin concept. Florida Institute of Technology, 8*, 3-20.

Groshev, M., Guimarães, C., Martín-Pérez, J., & de la Oliva, A. (2021). Toward intelligent cyber-physical systems: Digital twin meets artificial intelligence. *IEEE Communications Magazine*, 59(8), 14–20. 10.1109/MCOM.001.2001237

Iyyanar, P., Anand, R., Shanthi, T., Nassa, V. K., Pandey, B. K., George, A. S., & Pandey, D. (2023). A real-time smart sewage cleaning UAV assistance system using IoT. In *Handbook of Research on Data-Driven Mathematical Modeling in Smart Cities* (pp. 24–39). IGI Global.

Jayapoorani, S., Pandey, D., Sasirekha, N. S., Anand, R., & Pandey, B. K. (2023). Systolic optimized adaptive filter architecture designs for ECG noise cancellation by Vertex-5. *Aerospace Systems*, 6(1), 163–173. 10.1007/s42401-022-00177-3

Khan, B., Hasan, A., Pandey, D., Ventayen, R. J. M., Pandey, B. K., & Gowwrii, G. (2021). Fusion of datamining and artificial intelligence in prediction of hazardous road accidents. In *Machine learning and iot for intelligent systems and smart applications* (pp. 201–223). CRC Press. 10.1201/9781003194415-12

Kirubasri, G., Sankar, S., Pandey, D., Pandey, B. K., Nassa, V. K., & Dadheech, P. (2022). Software-defined networking-based Ad hoc networks routing protocols. In *Software defined networking for Ad Hoc networks* (pp. 95–123). Springer International Publishing. 10.1007/978-3-030-91149-2_5

Kirubasri, G., Sankar, S., Pandey, D., Pandey, B. K., Singh, H., & Anand, R. (2021, September). A recent survey on 6G vehicular technology, applications and challenges. In *2021 9th International Conference on Reliability, Infocom Technologies and Optimization (Trends and Future Directions)(ICRITO)* (pp. 1-5). IEEE.

Kumar Pandey, B., Pandey, D., Nassa, V. K., Ahmad, T., Singh, C., George, A. S., & Wakchaure, M. A. (2021). Encryption and steganography-based text extraction in IoT using the EWCTS optimizer. *Imaging Science Journal*, 69(1-4), 38–56. 10.1080/13682199.2022.2146885

Lu, Y., Maharjan, S., & Zhang, Y. (2021). Adaptive edge association for wireless digital twin networks in 6G. *IEEE Internet of Things Journal*, 8(22), 16219–16230. 10.1109/JIOT.2021.3098508

Nguyen, H. X., Trestian, R., To, D., & Tatipamula, M. (2021). Digital twin for 5G and beyond. *IEEE Communications Magazine*, 59(2), 10–15. 10.1109/MCOM.001.2000343

Pandey, B. K., Mane, D., Nassa, V. K. K., Pandey, D., Dutta, S., Ventayen, R. J. M., & Rastogi, R. (2021). Secure text extraction from complex degraded images by applying steganography and deep learning. In *Multidisciplinary approach to modern digital steganography* (pp. 146–163). IGI Global. 10.4018/978-1-7998-7160-6.ch007

Pandey, B. K., & Pandey, D. (2023). Parametric optimization and prediction of enhanced thermoelectric performance in co-doped CaMnO3 using response surface methodology and neural network. *Journal of Materials Science Materials in Electronics*, 34(21), 1589. 10.1007/s10854-023-10954-1

Pandey, B. K., Pandey, D., & Agarwal, A. (2022). Encrypted information transmission by enhanced steganography and image transformation. [IJDAI]. *International Journal of Distributed Artificial Intelligence*, 14(1), 1–14. 10.4018/IJDAI.297110

Pandey, B. K., Pandey, D., Alkhafaji, M. A., Güneşer, M. T., & Şeker, C. (2023a). A reliable transmission and extraction of textual information using keyless encryption, steganography, and deep algorithm with cuckoo optimization. In *Micro-Electronics and Telecommunication Engineering: Proceedings of 6th ICMETE 2022* (pp. 629–636). Springer Nature Singapore. 10.1007/978-981-19-9512-5_57

Pandey, B. K., Pandey, D., Gupta, A., Nassa, V. K., Dadheech, P., & George, A. S. (2023b). Secret data transmission using advanced morphological component analysis and steganography. In *Role of data-intensive distributed computing systems in designing data solutions* (pp. 21–44). Springer International Publishing. 10.1007/978-3-031-15542-0_2

Pandey, B. K., Pandey, S. K., & Pandey, D. (2011). A survey of bioinformatics applications on parallel architectures. *International Journal of Computer Applications*, 23(4), 21–25. 10.5120/2877-3744

Pandey, D., Hasan, A., Pandey, B. K., Lelisho, M. E., George, A. H., & Shahul, A. (2023). COVID-19 epidemic anxiety, mental stress, and sleep disorders in developing country university students. *CSI Transactions on ICT*, 11(2), 119–127. 10.1007/s40012-023-00383-0

Pandey, D., Nassa, V. K., Jhamb, A., Mahto, D., Pandey, B. K., George, A. H., & Bandyopadhyay, S. K. (2021a). An integration of keyless encryption, steganography, and artificial intelligence for the secure transmission of stego images. In *Multidisciplinary approach to modern digital steganography* (pp. 211–234). IGI Global. 10.4018/978-1-7998-7160-6.ch010

. Pandey, D., Ogunmola, G. A., Enbeyle, W., Abdullahi, M., Pandey, B. K., & Pramanik, S. (2021b). COVID-19: A framework for effective delivering of online classes during lockdown. *Human Arenas*, 1-15.

Pandey, D., & Pandey, B. K. (2022). An efficient deep neural network with adaptive galactic swarm optimization for complex image text extraction. In *Process mining techniques for pattern recognition* (pp. 121–137). CRC Press. 10.1201/9781003169550-10

Pandey, D., Pandey, B. K., & Wairya, S. (2021c). Hybrid deep neural network with adaptive galactic swarm optimization for text extraction from scene images. *Soft Computing*, 25(2), 1563–1580. 10.1007/s00500-020-05245-4

Pandey, J. K., Jain, R., Dilip, R., Kumbhkar, M., Jaiswal, S., Pandey, B. K., & Pandey, D. (2022). Investigating role of iot in the development of smart application for security enhancement. In *IoT Based Smart Applications* (pp. 219–243). Springer International Publishing.

Pramanik, S., Pandey, D., Joardar, S., Niranjanamurthy, M., Pandey, B. K., & Kaur, J. (2023, October). An overview of IoT privacy and security in smart cities. In *AIP Conference Proceedings (Vol. 2495,* No. 1). AIP Publishing. 10.1063/5.0123511

Rathore, M. M., Shah, S. A., Shukla, D., Bentafat, E., & Bakiras, S. (2021). The role of ai, machine learning, and big data in digital twinning: A systematic literature review, challenges, and opportunities. *IEEE Access: Practical Innovations, Open Solutions,* 9, 32030–32052. 10.1109/ACCESS.2021.3060863

Revathi, T. K., Sathiyabhama, B., Sankar, S., Pandey, D., Pandey, B. K., & Dadeech, P. (2022). An intelligent model for coronary heart disease diagnosis. In *Networking Technologies in Smart Healthcare* (pp. 309–327). CRC Press. 10.1201/9781003239888-15

Sahani, K., Khadka, S. S., Sahani, S. K., Pandey, B. K., & Pandey, D. (2023). A possible underground roadway for transportation facilities in Kathmandu Valley: A racking deformation of underground rectangular structures. *Engineering Reports*, 12821. 10.1002/eng2.12821

Sasidevi, S., Kumarganesh, S., Saranya, S., Thiyaneswaran, B., & Shree, K. V. M. (2024, May 15). Martin, Sagayam. K., Pandey, Binay, Kumar., Pandey, Digvijay, (2024). Design of Surface Plasmon Resonance (SPR) Sensors for Highly Sensitive Biomolecular Detection in Cancer Diagnostics. *Plasmonics*. Advance online publication. 10.1007/s11468-024-02343-z

Saxena, A., Agarwal, A., Pandey, B. K., & Pandey, D. (2024). Examination of the Criticality of Customer Segmentation Using Unsupervised Learning Methods. *Circular Economy and Sustainability*, 1–14. 10.1007/s43615-023-00336-4

Saxena, A., Sharma, N. K., Pandey, D., & Pandey, B. K. (2021). Influence of tourists satisfaction on future behavioral intentions with special reference to desert triangle of Rajasthan. *Augmented Human Research*, 6(1), 13. 10.1007/s41133-021-00052-4

Sharma, M., Pandey, D., Khosla, D., Goyal, S., Pandey, B. K., & Gupta, A. K. (2022). Design of a GaN-based Flip Chip Light Emitting Diode (FC-LED) with au Bumps & Thermal Analysis with different sizes and adhesive materials for performance considerations. *Silicon*, 14(12), 7109–7120. 10.1007/s12633-021-01457-x

Sharma, M., Pandey, D., Palta, P., & Pandey, B. K. (2022). Design and power dissipation consideration of PFAL CMOS V/S conventional CMOS based 2: 1 multiplexer and full adder. *Silicon*, 14(8), 4401–4410. 10.1007/s12633-021-01221-1

Tareke, S. A., Lelisho, M. E., Hassen, S. S., Seid, A. A., Jemal, S. S., Teshale, B. M., & Pandey, B. K. (2022). The prevalence and predictors of depressive, anxiety, and stress symptoms among Tepi town residents during the COVID-19 pandemic lockdown in Ethiopia. *Journal of Racial and Ethnic Health Disparities*, 1–13.35028903

Tripathi, R. P., Sharma, M., Gupta, A. K., Pandey, D., Pandey, B. K., Shahul, A., & George, A. H. (2023). Timely prediction of diabetes by means of machine learning practices. *Augmented Human Research*, 8(1), 1. 10.1007/s41133-023-00062-4

Tuegel, E. J., Ingraffea, A. R., Eason, T. G., & Spottswood, S. M. (2011). Reengineering aircraft structural life prediction using a digital twin. *International Journal of Aerospace Engineering*, 2011, 2011. 10.1155/2011/154798

Vinodhini, V., Kumar, M. S., Sankar, S., Pandey, D., Pandey, B. K., & Nassa, V. K. (2022). IoT-based early forest fire detection using MLP and AROC method. *International Journal of Global Warming*, 27(1), 55–70. 10.1504/IJGW.2022.122794

Wu, Y., Zhang, K., & Zhang, Y. (2021). Digital twin networks: A survey. *IEEE Internet of Things Journal*, 8(18), 13789–13804. 10.1109/JIOT.2021.3079510

Zhao, L., Han, G., Li, Z., & Shu, L. (2020). Intelligent digital twin-based software-defined vehicular networks. *IEEE Network*, 34(5), 178–184. 10.1109/MNET.011.1900587

Zhou, C., Yang, H., Duan, X., Lopez, D., Pastor, A., Wu, Q., & Jacquenet, C. (2021). *Digital twin network: Concepts and reference architecture*. Internet Engineering Task Force.

Chapter 12
Future Directions of Digital Twin Architectures for 6G Communication Networks

Binay Kumar Pandey
https://orcid.org/0000-0002-4041-1213
College of Technology, Govind Ballabh Pant University of Agriculture and Technology, India

Mukundan Appadurai Paramashivan
https://orcid.org/0009-0009-5608-4788
Aligarh Muslim University, India & Champions Group, India

Digvijay Pandey
https://orcid.org/0000-0003-0353-174X
Department of Technical Education, Government of Uttar Pradesh, India

A. Shaji George
https://orcid.org/0000-0002-8677-3682
Almarai Company, Riyadh, Saudi Arabia

Ashi Agarwal
Department of Computer Science, ABES Engineering College, Ghaziabad, India

Darshan A. Mahajan
https://orcid.org/0000-0002-1239-6343
NICMAR University, Pune, India

Pankaj Dadheech Dadheech
https://orcid.org/0000-0001-5783-1989
Swami Keshvanand Institute of Technology, Management, and Gramothan, India

Sabyasachi Pramanik
https://orcid.org/0000-0002-9431-8751
Department of Computer Science and Engineering, Haldia Institute of Technology, India

ABSTRACT

Initiating the study into digital twin technology, the planning and implementation of the 6G network necessitates real-time interaction and alignment between physical systems and their virtual representation. From simple parts to intricate systems, the digital twin's flexibility and agility improve design and operational procedure efficiency in a predictable manner. It can validate policies, give a virtual representation of a physical entity, or evaluate how a system or entity behaves in a real-time setting. It evaluates the effectiveness and suitability of QoS regulations in 6G communication, in addition to the creation and management of novel services. Physical system maintenance costs and security threats can also be reduced, but doing so requires standardization efforts that open the door to previously unheard-of difficulties with fault tolerance, efficiency, accuracy, and security. The fundamental needs of a digital twin that are focused on 6G communication are covered in this chapter. These include decoupling, scalable

DOI: 10.4018/979-8-3693-2931-3.ch012

intelligent analytics, data management using blockchain.

INTRODUCTION: BACKGROUND AND DRIVING FORCES

Initiating the study into Digital Twin Technology, the planning and implementation of the 6G (Kirubasri, G. et al., 2021) network necessitates real-time interaction and alignment between physical systems and their virtual representation. From simple parts to intricate systems, the Digital Twin's (**Tao, Y. et al., 2023**) flexibility and agility improve design and operational procedure efficiency in a predictable manner. It can validate policies, give a virtual representation of a physical entity, or evaluate how a system or entity behaves in a real-time setting. It evaluates the effectiveness and suitability of QoS regulations in 6G communication (Gupta, A. K. et al., 2023), in addition to the creation and management of novel services. Physical system maintenance costs and security threats (Pandey, B. K. et al., 2023) can also be reduced, but doing so requires standardization efforts that open the door to previously unheard-of difficulties with fault tolerance, efficiency, accuracy, and security (Pandey, B. K. et al., 2021a). The fundamental needs of a digital twin that are focused on 6G communication are covered in this work. These include decoupling, scalable intelligent analytics (Revathi, T. K. et al., 2022), data management using blockchain (David, S. et al., 2023), scalability and dependability, delays, responsiveness, connectivity, protection, etc. Subsequently, the literature presents basic and enhanced Digital (George, W. K. et al., 2023) Twin designs, together with comparisons across several dimensions. This focuses on the methods used to store, retrieve, and manage the data (Vinodhini, V. et al., 2022) as well as how often it is collected. We also talk about the specifics of the models utilized to create digital twins, the many kinds of interfaces, and their unique characteristics that make them ideal for 6G communications. A methodical process is followed in order to investigate the operational steps. The chapter ends with a discussion of the future directions, with particular emphasis on mobility management for Edge-based Twins, twin-based forensics, integrated virtual twins, and isolation between twins-based services for the World Wide Web of Anything.

DIGITAL TWIN

A virtual depiction of the components and dynamics of a real-world system is called a digital twin. It models the dynamics of a particular system using machine learning (Pandey, B. K., & Pandey, D., 2023), data analytics, and multiphysics simulation.

The digital twins are categorized into

(i) Digital twin monitoring, which keeps an eye on a physical system's condition.

(ii) By utilizing simulation techniques based on machine learning (Tripathi, R. P. et al., 2023), simulation digital twins forecast future states.

(iii) An operational digital twin allows operators to communicate with a cyber-physical system and conducts system analysis and design.

Figure 1. A reference 6G digital twin network architecture

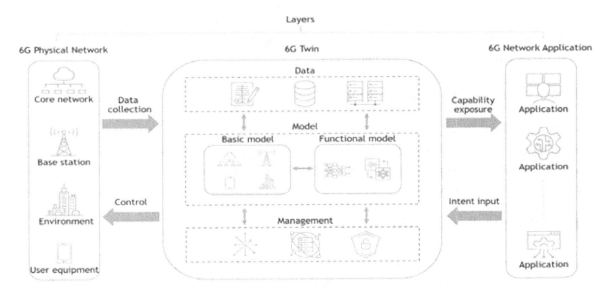

Figure 1 shows an image of the 6G model. The physical network layer, twin layer, and application layer are its three layers. The term "6G physical network layer" refers to a real-world 6G network made up of physical network components and their operational environments, such as a core network, transport network, radio cell, and radio access network (Sengupta, R. et al., 2021). These networks exchange control messages and data (JayaLakshmi, G. et al., 2024) with the 6G dual layer.

A 6G DTN consists of three domains: data (Muniandi, B. et al., 2024), model, and management domains, which make up the 6G twin layer. Data domain is focused on data administration, service, storage spaces and retrieval within the data repository subsystem. A service-mapping subsystem known as the "model domain" is made up of models that, using the data gathered, represent actual objects (Pandey, D. et al., 2022) in the 6G physical network. Basic and functional models are among the categories that fall under the model domain. The target physical network's topology and networking constituents are described using basic models. Analytical models for deriving insights from the DTN, such as network arranging, traffic analysis, defect detection, network emulation, and prediction (Kumar Pandey, B. et al., 2021) are referred to as functional models. Management domain is responsible for the management function of the 6G twin layer used for model creation, configuration, update, and monitoring, etc and security management i.e., authentication, authorization, encryption (Pandey, B. K. et al., 2022), and integrity protection.

The 6G network application layer includes various network applications that exploit the capabilities exposed by the 6G twin layer that can also send control messages to the 6G physical network layer i.e., network operations, administration, and maintenance (OAM), network optimization, and network visualization.

DTN USE CASES

DTNs can serve a wide range of use cases for 6G networks and network operations, such as what-if analysis, AI training and interference, data generation through simulation, network modeling and planning, and network operation and administration.

Network Simulation and Planning

Large-scale, physically accurate digital replicas of city blocks are used by DTNs to improve network modeling and planning. This enables network construction teams to evaluate and enhance base station installation and placement and configurations first in the virtual world, saving money and resources when compared to the real world.

Network Operation and Management

With software-defined front-haul networks, the softwarization of 6G modifies networks. In order to ensure system resilience and robustness, a virtual drive-test is created using the DTN while any configuration changes are applied to the actual network in accordance with the needs of the client.

Data Generation by Simulation

Real-world data access that takes into account a variety of system operating variables, such as the operating setting, the weather, and network operational parameters, is crucial for AI-native 6G networks. In order to train AI/ML models for data, artificial information (Anand, R. et al., 2023) is generated in a cyber-sister world of the real world for optimization and network control.

AI training and Inference

Models used in AI/ML for image and video processing are offline-trained using datasets like ImageNet, Urban environments, etc. For example, a deep reinforcement learning (DRL) (Guo, Q. et al., 2022) based closed-loop management uses inference engines to produce a priori model parameters. DTN trains and infers pipelines using hardware accelerators that can be reprogrammed.

What-If-Analysis

Due to the dynamic nature of 6G networks, network operators must make changes to the design and crucial network parameters in a matter of hours or minutes as opposed to months or years. DTN is a virtual sandbox that helps anticipate future performance by locating security flaws, connection bottlenecks, and incorrect network setups.

DTN Requirements

6G DTNs need to meet a number of specifications, such as security, interpretability, flexibility, flexibility, generalization, and reliability as well as latency.

Reliability and Latency

A DTN's operational infrastructure must be robust in order for it to be considered trustworthy and reliable. This includes the ability to respond quickly in the event of a malicious attack, human error, equipment failure, or other latency-critical adverse situation, as well as the capacity to ensure robust data collection, storage, modeling, and information exchange (Sharma, S. et al., 2023) between the DTN and other network entities. Additionally, a high level of availability, disaster recovery capabilities with backup power (Govindaraj, V. et al., 2023), and the capacity to restore important historical states or data points are all necessary.

Scalability

Depending on the dimensions and complexity of its physical counterpart, DTN can be scaled over an extensive range.

Agility

With versatility in its functions for cross-domain communication (Raja, D. et al., 2024), exchange of data, and service coordination amongst many DT entities, the DTN provides on-demand service to the various network applications.

Generalizability

For data collecting, storage, modeling, and backward compatibility between updated and previous versions, DTNs need to be somewhat compatible. Additionally, their interfaces must be able to allow multi-vendor, multi-standard interoperability.

Security

To maintain integrity, a strong DTN needs sufficient defense against possible attacks on the data, models, interactive interfaces, and network infrastructure as a whole.

Interpretability

Easy-to-use asset administration services provided by DTNs provide the user-friendly control of the twin entities' whole lifecycle, including network equipment, linkages, traffic flows, etc.

DIGITAL TWIN KEY DESIGN REQUIREMENTS

Figure 2. Digital twin enabled 6G key design requirements

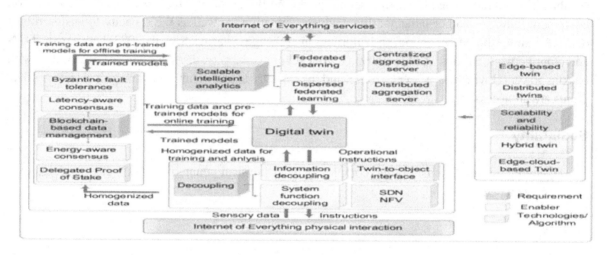

Figure 3. Benefits of digital twins based architecture

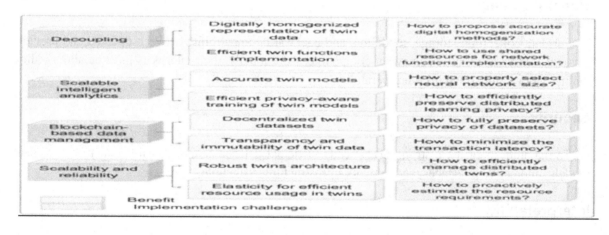

As seen in Figures 2 and 3, decoupling, scalable intelligent analytics, and blockchain-based data management with scalable and dependable architecture and algorithms are the essential design criteria for DTs for wireless 6G systems. In 6G, the digital twin will essentially depict the physical wireless system (Kirubasri, G. et al., 2022), such as intelligent reflecting surfaces with backhaul links, a typical physical system for a 6G application, such as an Industry 4.0 (George, W. K. et al., 2024) plant, a healthcare system (Singh, S. et al., 2023), or an autonomous vehicle, as well as specific modules, such as edge caching and computational offloading at the edge.

Decoupling

The process of turning a physical system into a digital (Swapna, H. R. et al., 2023) duplicate is called decoupling. There are two varieties of it: decoupling of system functions and decoupling of information. Information decoupling makes it possible to convert data from physical systems or systems with dynamically changing states into a homogenized digital representation that will facilitate the use of digital twins and provide universality. Base station (BSs), intelligent reflecting surfaces, intelligent devices/sensors (Du John, H. V. et al., 2022), edge/cloud servers, and haptics sensors make up a 6G physical interaction space. The data these components collect includes industrial process control, holographic images, haptics sensors, 6G spectrum usage, resource management data from edge servers, and more. From hardware to software, many system operations like as resource allocation, mobility management, edge caching, and so forth, must be handled. SDN divides the control and data planes to enable the digital twin, whereas NFV provides system deployment that is less expensive. With the help of real-time control via network slicing and stored data on a blockchain, the DT conducts proactive or offline analytics, such as data analytics and pre-training of twin models.

Scalable Intelligent Analytics

Effective machine learning algorithms (Saxena, A. et al., 2024) are utilized for huge datasets to support heterogeneous system needs, network configurations, and hardware architectures. Mobility management, resource allocation, edge caching, and other highly dynamic situations suffer from performance degradation caused by DTs because of their enormous size, complicated machine learning model (Pandey, B. K. et al., 2023), and high processing power. Distributed deep learning-based twin models that vary in training time and space are utilized to get around issues. The model's (Pandey, B. K. et al., 2024) processing time reduces and its communication time grows with the number of distributed computers. Federated learning with sparsification is employed in order to maintain scalability.

Blockchain-Based Data Management

Blockchain is used by DTs to handle distributed datasets in a freely accessible and unchangeable way since it allows data to be stored in an unaltered state without network majority collusion and allows for the retroactive modification of all subsequent blocks with no interruption in service and no loss of data fidelity. Miners on edge servers run a blockchain consensus algorithm. Scalability, high latency, high energy consumption, and privacy concerns are a few of the difficulties.

Scalability and Reliability

Large-scale ultra-reliable low-latency communication (mURLLC) services are implemented using distributed twin architecture, which lowers latency. Distributed DTs continue to have more difficult management. Hybrid centralized and distributed DTs are used to balance latency and processing power.

ARCHITECTURE OF DT ENABLED 6G

This section discusses the components of the DT-based 6G architecture and its potential for deployment.

Twin Objects

One or more Twin objects use transient-based virtual machines (TVM) to build and terminate 6G services dynamically. They also do optimization, control, and artificial intelligence (Anand, R. et al., 2024) model training. TVM-based twin objects provide proactive resource customization made possible by post-use cleanup and proactive intelligent analytics based on ML (Deepa, R. et al., 2022) methods.

Figure 4. Twin objects deployment trends

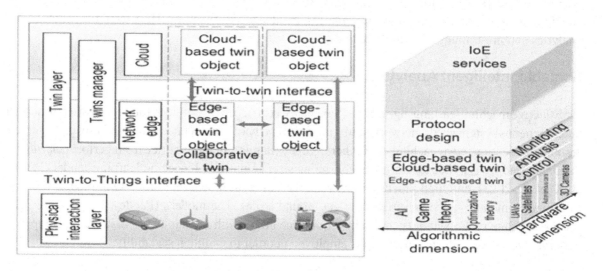

Twin Object Deployment Trends

As seen in Figure 4, twin objects can be implemented according to architectures such as edge-based digital twin, cloud-based digital twin, and edgecloud-based collaborative digital twin. While cloud-based twin objects are delay tolerant and employed in high computing power applications, edge-based twin objects, such as large URLLC, are better suited for 6G applications with stringent latency requirements. The edge-cloud-based twins provide a trade-off between processing power and latency by utilizing both edge and cloud resources. It is necessary to suggest a variety of interfaces, including twins-to-things, twins-to-twins, and twins-to-service interfaces, for smooth, isolation-based, complicated interaction in a scalable and dependable way. The IoE devices (Dhanasekar, S. et al., 2023) can be effectively detached from the twin layer of the twin-to-object interfaces. Different twins at different degrees of communication will employ a twin-to-twin interface.

Digital Twin Operational Steps

There are two stages to the twin surgery: training and operation. Distributed machine learning (Pandey, B. K. et al., 2024) is employed in the training phase, sending the local learning models to the twin layer so the blockchain miner can aggregate them. After the global models are computed (Bessant, Y. A. et al., 2023), IoE devices update their local learning models. Either synchronous or asynchronous learning is possible. In a synchronous fashion, devices must transmit their local learning models to the miner for global aggregation within a set amount of time; in an asynchronous fashion, devices will only send their local learning models when they establish a connection with miners. First, authentication is performed in response to a service request from BS. Following validation, the request is translated using semantic reasoning techniques. To facilitate reliable data sharing, the twin objects at the BS based on TVM are instantiated to associate with blockchain miners.

6G DIGITAL TWIN FUTURE DIRECTIONS

As an enabler and a use-case, the 6G DTs have some research directions in the configuration and transformation of 6G networks. Here are some of them in more depth.

DT Ownership Issues

Due to the possible ownership mismatch between the actual company and the DT platform, DT ownership gives rise to technical, financial, and legal issues. For instance, even though the generated data is owned by and kept on the application provider's cloud, a fitness tracker gadget can be owned by an individual. This interest in an IoT (Iyyanar, P. et al., 2023) situation is protected by the General Data Protection Regulation (GDPR). The owner of an appliance has a practical choice: they can rent cloud, fog, or edge services to install and maintain the DT, or they can buy a home gateway with storage and security capabilities (Pandey, J. K. et al., 2022).

Ultra-Low-Latency and Reliable Communication Between DT and PT

Critical systems, such as remote surgery systems, require continuous, dependable, ultra-low latency data interchange between the DT and the PT in real time.

Federated DT in the Cloud/Edge

A system like the DT, which combines connectivity, data analysis, and AI-based processing, requires intensive resource management due to limited resources like electricity, storage, and high-speed memory. Performance bottlenecks may occur during DT distribution or even replication. If server or network link faults could prevent a PT and its DT from communicating seamlessly, multiple copies of the DT can be shared. Then, federated DT can be utilized to build automated and intelligent operations for synchronized and collaborative AI algorithms across the network's nodes by exchanging data and/or training AI models.

DT of an Entire Network

Using AI-assisted Network Function Virtualization (NFV) and Software Defined Networking (SDN), DT technology creates autonomous and automated communications networks. Software that is native to the cloud can be used to construct the physical infrastructure, such as transceivers, antennas, optical fibers, filters (Bruntha, P. M. et al., 2023), etc. With its DT, the many settings can be improved, managed, and troubleshooted. Research topics include ownership concerns using smart contracts hosted in a blockchain and network monitoring and troubleshooting using AI-based analytics (Pramanik, S. et al., 2023).

Experimental Investigation of DTs

Commercial software components like AmarisoftTM LTE 100 eNodeB, UE from software radio systems (srsUETM), and a generic RF front end can be used to create the entire LTE network. Python and Linux-based code can be used to control the network on and off. In order to visualize the functions of every component, a DT must plot different performance curves using an integrated Graphical User Interface (GUI) and gather data from the video in real-time for analysis.

Isolation Between Twins-Based Services

Effective use of network resources is required for a variety of twin-based Internet of Everything applications in order to increase performance by explicit resource allocation (i.e., compute and communication resources). It is possible to create new optimization strategies for twin objects that effectively leverage shared resources across numerous twin-based services while meeting the isolation requirements.

Mobility Management for Edge-Based Twins

While employing a backhaul link might mitigate interrupted service caused by restricted access point/ BS coverage with the twin, greater latency and intermittent service are still present. Therefore, new, efficient prediction techniques or services can be designed that migrate.

Digital Twin Forensics

Players in the 6G system that are enabled by digital twins, such as end devices, TVM-based twin-objects, communication interfaces, etc., are susceptible to a number of security risks, including those involving the identification, gathering, and presenting of evidence. It is possible to look into efficient forensic methods to oppose them. The difficulties would come from attacks.

CONCLUSION

This section describes digital twins as a major enabler for networks designed with the 6G architecture and their upcoming research. The edge-based digital twins guarantee scalability and dependability through dispersed deployment. DTNs are an interesting and powerful technology for 6G wireless network design (Devasenapathy, D. et al., 2024), analysis, diagnostics, simulation, and control because of their unique characteristics. When DT and AI are combined, 6G network design, deployment, and operation can be facilitated, leading to high network resilience and increased speed and reliability. In addition, the need for DT across industries like as aerospace, Industry 4.0 (Saxena, A. et al., 2021), and healthcare is a key factor in the development of 6G. Future cutting-edge platforms and technologies are made possible by the stricter security and performance standards (Sharma, M. et al., 2024).

REFERENCES

Anand, R., Khan, B., Nassa, V. K., Pandey, D., Dhabliya, D., Pandey, B. K., & Dadheech, P. (2023). Hybrid convolutional neural network (CNN) for kennedy space center hyperspectral image. *Aerospace Systems*, 6(1), 71–78. 10.1007/s42401-022-00168-4

Anand, R., Lakshmi, S. V., Pandey, D., & Pandey, B. K. (2024). An enhanced ResNet-50 deep learning model for arrhythmia detection using electrocardiogram biomedical indicators. *Evolving Systems*, 15(1), 83–97. 10.1007/s12530-023-09559-0

Bessant, Y. A., Jency, J. G., Sagayam, K. M., Jone, A. A. A., Pandey, D., & Pandey, B. K. (2023). Improved parallel matrix multiplication using Strassen and Urdhvatiryagbhyam method. *CCF Transactions on High Performance Computing*, 5(2), 102–115. 10.1007/s42514-023-00149-9

. Bruntha, P. M., Dhanasekar, S., Hepsiba, D., Sagayam, K. M., Neebha, T. M., Pandey, D., & Pandey, B. K. (2023). Application of switching median filter with L 2 norm-based auto-tuning function for removing random valued impulse noise. *Aerospace systems, 6*(1), 53-59.

David, S., Duraipandian, K., Chandrasekaran, D., Pandey, D., Sindhwani, N., & Pandey, B. K. (2023). Impact of blockchain in healthcare system. In *Unleashing the Potentials of blockchain technology for healthcare industries* (pp. 37–57). Academic Press. 10.1016/B978-0-323-99481-1.00004-3

Deepa, R., Anand, R., Pandey, D., Pandey, B. K., & Karki, B. (2022). Comprehensive performance analysis of classifiers in diagnosis of epilepsy. *Mathematical Problems in Engineering*, 2022, 2022. 10.1155/2022/1559312

Devasenapathy, D., Madhumathy, P., Umamaheshwari, R., Pandey, B. K., & Pandey, D. (2024). Transmission-efficient grid-based synchronized model for routing in wireless sensor networks using Bayesian compressive sensing. *SN Computer Science*, 5(1), 1–11.

Dhanasekar, S., Martin Sagayam, K., Pandey, B. K., & Pandey, D. (2023). Refractive Index Sensing Using Metamaterial Absorbing Augmentation in Elliptical Graphene Arrays. *Plasmonics*, 1–11. 10.1007/s11468-023-02152-w

Du John, H. V., Jose, T., Jone, A. A. A., Sagayam, K. M., Pandey, B. K., & Pandey, D. (2022). Polarization insensitive circular ring resonator based perfect metamaterial absorber design and simulation on a silicon substrate. *Silicon*, 14(14), 9009–9020. 10.1007/s12633-021-01645-9

George, W. K., Ekong, M. O., Pandey, D., & Pandey, B. K. (2023). Pedagogy for Implementation of TVET Curriculum for the Digital World. In *Applications of Neuromarketing in the Metaverse* (pp. 117–136). IGI Global. 10.4018/978-1-6684-8150-9.ch009

George, W. K., Silas, E. I., Pandey, D., & Pandey, B. K. (2024). Utilization of Industry 4.0 Technologies in Nigerian Technical and Vocational Education: A Conundrum for Educators. In *Examining the Rapid Advance of Digital Technology in Africa* (pp. 270–293). IGI Global. 10.4018/978-1-6684-9962-7.ch014

. Govindaraj, V., Dhanasekar, S., Martinsagayam, K., Pandey, D., Pandey, B. K., & Nassa, V. K. (2023). Low-power test pattern generator using modified LFSR. *Aerospace Systems*, 1-8.

Guo, Q., Tang, F., & Kato, N. (2022). Federated reinforcement learning-based resource allocation for D2D-aided digital twin edge networks in 6G industrial IoT. *IEEE Transactions on Industrial Informatics*.

Gupta, A. K., Sharma, R., Pandey, D., Nassa, V. K., Pandey, B. K., George, A. S., & Dadheech, P. (2023). Performance analysis of eight-channel WDM optical network with different optical amplifiers for industry 4.0. In *Innovation and Competitiveness in Industry 4.0 Based on Intelligent Systems* (pp. 197–212). Springer International Publishing. 10.1007/978-3-031-29775-5_9

Iyyanar, P., Anand, R., Shanthi, T., Nassa, V. K., Pandey, B. K., George, A. S., & Pandey, D. (2023). A real-time smart sewage cleaning UAV assistance system using IoT. In *Handbook of Research on Data-Driven Mathematical Modeling in Smart Cities* (pp. 24–39). IGI Global.

JayaLakshmi. G., Pandey, D., Pandey, B. K., Kaur, P., Mahajan, D. A., & Dari, S. S. (2024). Smart Big Data Collection for Intelligent Supply Chain Improvement. In *AI and Machine Learning Impacts in Intelligent Supply Chain* (pp. 180-195). IGI Global.

Kirubasri, G., Sankar, S., Pandey, D., Pandey, B. K., Nassa, V. K., & Dadheech, P. (2022). Software-defined networking-based Ad hoc networks routing protocols. In *Software defined networking for Ad Hoc networks* (pp. 95–123). Springer International Publishing. 10.1007/978-3-030-91149-2_5

Kirubasri, G., Sankar, S., Pandey, D., Pandey, B. K., Singh, H., & Anand, R. (2021, September). A recent survey on 6G vehicular technology, applications and challenges. In *2021 9th International Conference on Reliability, Infocom Technologies and Optimization (Trends and Future Directions)(ICRITO)* (pp. 1-5). IEEE.

Kumar Pandey, B., Pandey, D., Nassa, V. K., Ahmad, T., Singh, C., George, A. S., & Wakchaure, M. A. (2021). Encryption and steganography-based text extraction in IoT using the EWCTS optimizer. *Imaging Science Journal*, 69(1-4), 38–56. 10.1080/13682199.2022.2146885

Muniandi, B., Nassa, V. K., Pandey, D., Pandey, B. K., Dadheech, P., & George, A. S. (2024). Pattern Analysis for Feature Extraction in Complex Images. In *Using Machine Learning to Detect Emotions and Predict Human Psychology* (pp. 145–167). IGI Global. 10.4018/979-8-3693-1910-9.ch007

Pandey, B. K., & Pandey, D. (2023). Parametric optimization and prediction of enhanced thermoelectric performance in co-doped CaMnO3 using response surface methodology and neural network. *Journal of Materials Science Materials in Electronics*, 34(21), 1589. 10.1007/s10854-023-10954-1

Pandey, B. K., Pandey, D., & Agarwal, A. (2022). Encrypted information transmission by enhanced steganography and image transformation. [IJDAI]. *International Journal of Distributed Artificial Intelligence*, 14(1), 1–14. 10.4018/IJDAI.297110

Pandey, B. K., Pandey, D., Alkhafaji, M. A., Güneşer, M. T., & Şeker, C. (2023). A reliable transmission and extraction of textual information using keyless encryption, steganography, and deep algorithm with cuckoo optimization. In *Micro-Electronics and Telecommunication Engineering: Proceedings of 6th ICMETE 2022* (pp. 629–636). Springer Nature Singapore. 10.1007/978-981-19-9512-5_57

Pandey, B. K., Pandey, D., Gupta, A., Nassa, V. K., Dadheech, P., & George, A. S. (2023). Secret data transmission using advanced morphological component analysis and steganography. In *Role of data-intensive distributed computing systems in designing data solutions* (pp. 21–44). Springer International Publishing. 10.1007/978-3-031-15542-0_2

Pandey, B. K., Pandey, D., & Sahani, S. K. (2024). Autopilot control unmanned aerial vehicle system for sewage defect detection using deep learning. *Engineering Reports*, 12852. 10.1002/eng2.12852

Pandey, B. K., Pandey, D., Wairya, S., & Agarwal, G. (2021). An advanced morphological component analysis, steganography, and deep learning-based system to transmit secure textual data. [IJDAI]. *International Journal of Distributed Artificial Intelligence*, 13(2), 40–62. 10.4018/IJDAI.2021070104

Pandey, B. K., Paramashivan, M. A., Kanike, U. K., Mahajan, D. A., Mahajan, R., George, A. S., & Hameed, A. S. H. (2024). Impacts of Artificial Intelligence and Machine Learning on Intelligent Supply Chains. In *AI and Machine Learning Impacts in Intelligent Supply Chain* (pp. 57–73). IGI Global. 10.4018/979-8-3693-1347-3.ch005

Pandey, D., Wairya, S., Sharma, M., Gupta, A. K., Kakkar, R., & Pandey, B. K. (2022). An approach for object tracking, categorization, and autopilot guidance for passive homing missiles. *Aerospace Systems*, 5(4), 553–566. 10.1007/s42401-022-00150-0

Pandey, J. K., Jain, R., Dilip, R., Kumbhkar, M., Jaiswal, S., Pandey, B. K., & Pandey, D. (2022). Investigating role of iot in the development of smart application for security enhancement. In *IoT Based Smart Applications* (pp. 219–243). Springer International Publishing.

Pramanik, S., Pandey, D., Joardar, S., Niranjanamurthy, M., Pandey, B. K., & Kaur, J. (2023, October). An overview of IoT privacy and security in smart cities. In *AIP Conference Proceedings* (*Vol. 2495*, No. 1). AIP Publishing. 10.1063/5.0123511

Raja, D., Kumar, D. R., Santhiyakumari, N., Kumarganesh, S., Sagayam, K. M., Thiyaneswaran, B., Pandey, B. K., & Pandey, D. (2024). A compact dual-feed wide-band slotted antenna for future wireless applications. *Analog Integrated Circuits and Signal Processing*, 118(2), 1–15. 10.1007/s10470-023-02233-0

Revathi, T. K., Sathiyabhama, B., Sankar, S., Pandey, D., Pandey, B. K., & Dadeech, P. (2022). An intelligent model for coronary heart disease diagnosis. In *Networking Technologies in Smart Healthcare* (pp. 309–327). CRC Press. 10.1201/9781003239888-15

Saxena, A., Agarwal, A., Pandey, B. K., & Pandey, D. (2024). Examination of the Criticality of Customer Segmentation Using Unsupervised Learning Methods. *Circular Economy and Sustainability*, 1–14. 10.1007/s43615-023-00336-4

Saxena, A., Sharma, N. K., Pandey, D., & Pandey, B. K. (2021). Influence of tourists satisfaction on future behavioral intentions with special reference to desert triangle of Rajasthan. *Augmented Human Research*, 6(1), 13. 10.1007/s41133-021-00052-4

. Sengupta, R., Sengupta, D., Pandey, D., Pandey, B. K., Nassa, V. K., & Dadeech, P. (2021). A Systematic review of 5G opportunities, architecture and challenges. *Future Trends in 5G and 6G*, 247-269.

Sharma, M., Talwar, R., Pandey, D., Nassa, V. K., Pandey, B. K., & Dadheech, P. (2024). A Review of Dielectric Resonator Antennas (DRA)-Based RFID Technology for Industry 4.0. *Robotics and Automation in Industry*, 4(0), 303–324.

Sharma, S., Pandey, B. K., Pandey, D., Anand, R., Sharma, A., & Saini, S. (2023, March). Character Recognition Technique Implementation for Complicated Deteriorated Scene. In *2023 6th International Conference on Information Systems and Computer Networks (ISCON)* (pp. 1-4). IEEE. 10.1109/IS-CON57294.2023.10112185

Singh, S., Madaan, G., Kaur, J., Swapna, H. R., Pandey, D., Singh, A., & Pandey, B. K. (2023). Bibliometric Review on Healthcare Sustainability. *Handbook of Research on Safe Disposal Methods of Municipal Solid Wastes for a Sustainable Environment*, 142-161.

Swapna, H. R., Bigirimana, E., Madaan, G., Hasan, A., Pandey, B. K., & Pandey, D. (2023). Impact of neuromarketing on consumer psychology in digitally connected networks. In *Applications of Neuromarketing in the Metaverse* (pp. 193–205). IGI Global. 10.4018/978-1-6684-8150-9.ch015

Tao, Y., Wu, J., Lin, X., & Yang, W. (2023). *DRL-Driven Digital Twin Function Virtualization for Adaptive Service Response in 6G Networks*. IEEE Networking Letters.

Tripathi, R. P., Sharma, M., Gupta, A. K., Pandey, D., Pandey, B. K., Shahul, A., & George, A. H. (2023). Timely prediction of diabetes by means of machine learning practices. *Augmented Human Research*, 8(1), 1. 10.1007/s41133-023-00062-4

Vinodhini, V., Kumar, M. S., Sankar, S., Pandey, D., Pandey, B. K., & Nassa, V. K. (2022). IoT-based early forest fire detection using MLP and AROC method. *International Journal of Global Warming*, 27(1), 55–70. 10.1504/IJGW.2022.122794

Chapter 13
Beyond Data Breaches:
Enhancing Security in 6G Communications

Binay Kumar Pandey
https://orcid.org/0000-0002-4041-1213

College of Technology, Govind Ballabh Pant University of Agriculture and Technology, India

Digvijay Pandey
https://orcid.org/0000-0003-0353-174X

Department of Technical Education, Government of Uttar Pradesh, India

Ashi Agarwal

Department of Computer Science, ABES Engineering College, Ghaziabad, India

Darshan A. Mahajan
https://orcid.org/0000-0002-1239-6343

NICMAR University, Pune, India

Pankaj Dadheech Dadheech
https://orcid.org/0000-0001-5783-1989

Swami Keshvanand Institute of Technology, Management, and Gramothan, India

A. Shaji George
https://orcid.org/0000-0002-8677-3682

Almarai Company, Riyadh, Saudi Arabia

Pankaj Kumar Rai
Loxoft, USA

ABSTRACT

This chapter describes the different newly adopted 6G technologies, along with any security risks and potential fixes. The primary 6G technologies that will open up a whole new universe of possibilities are AI/ML, DLT, quantum computing, VLC, and THz communication. The emergence of new generation information and communication technologies, including blockchain technology, virtual reality/augmented reality/extended reality, internet of things, and artificial intelligence, gave rise to the 6G communication network. The intelligence process of communication development, which includes holographic, pervasive, deep, and intelligent connectivity, is significantly impacted by the development of 6G.

INTRODUCTION

The emergence of new generation information and communication technologies, including blockchain technology, virtual reality/augmented reality/extended reality, Internet of Things, and artificial intelligence, gave rise to the 6G communication network(Kirubasri, G.,et al.,2021). The intelligence process of

DOI: 10.4018/979-8-3693-2931-3.ch013

communication development, which includes holographic, pervasive, deep, and intelligent connectivity, is significantly impacted by the development of 6G. A few security concerns are depicted in Figure 1.

Figure 1. 6G security issues

Expected 6G requirements, security key performance indicators (KPIs), unique network design, new applications, and enabling technologies might all provide security and privacy challenges (Kirubasri, G., et al., 2022). The problems stem from the mobility of UAVs, holographic telepresence, digital twins, connected autonomous vehicles, extended reality, smart grid 2.0, industry 5.0, intelligent healthcare, etc. (Govindaraj, V., et al.,2023). Distributed ledger technology (DLT), physical layer security, quantum communication, and distributed AI/ML security solutions are offered. The dangers associated with Distributed Ledger Technology encompass a variety of tactics such as the majority attack, also known as the 51% attack, double spending, re-entrency, Sybil, privacy leaks, and other attacks(Sharma, M.,et al.,2022). Appropriate access control and authentication methods, choosing the right kind of blockchain or DLT based on the 6G application and services, and so on are some of the potential answers. Line-of-sight transmission is necessary for Physical Layer Security, but this can be achieved through multipath transmission (Pandey, D., et al.,2021) in Terahertz technology; the broadcast nature of visible light communications can be overcome by improving secrecy performance through the use of multiple input multiple output technology; quantum cloning and collision attacks in Quantum Computing can be prevented by using lattice-based, code-based, hash-based, and multivariate-based cryptography; and attacks such as poisoning, evasion, and API-based attacks can affect Distributed and Scalable AI/ML. These can be prevented by using Adversarial training injects and defensive distillation, among other techniques. Lastly, a road map is presented for bringing 6G security ideas to life (Pandey, B. K., & Pandey, D. (2023)).

6G SECURITY

The goal of the 6G communication network is to enable humankind to access a multitude of services that demand high security. Figure 9.2, which lists the numerous industries that require security at varying levels, illustrates the essential criteria. The information is condensed based on the applications (Pandey, D., et al., (2021)).

Figure 2. 6G key security requirements

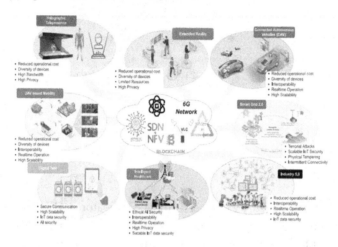

UAV Based Mobility

Due to their limited processing and power management capabilities, unmanned aerial vehicles (UAVs) equipped with artificial intelligence (AI) for services like logistics, military operations, and medicine delivery over 6G are vulnerable to possible threats. Security measures that are lightweight are employed to justify low latency requirements (Saxena, A.,et al., 2023). Collision avoidance, path planning, and route optimization are all supported by 6G. In particular, safeguarded control data integrity is essential for optimal functioning. Due to their unmanned nature, physical attacks and coordinated drone operations ranging from cyberattacks to physical terrorist attacks are the security challenges that occur (Kumar Pandey, B., et al.,2021).

Holographic Telepresence

Holographic telepresence, which demands extremely high bandwidth, is used to realistically project the physical presence of faraway persons during 3D video conferences or news broadcasts (Tareke, S. A., et al., 2022). If planned at a remote site where there is no control over the prevailing environmental conditions, the low operational cost of various devices networked for privacy gives rise to the security requirement.

Extended Reality

Augmented reality (AR), virtual reality (VR), mixed reality (MR), and other technologies are combined in extended reality (XR) and are utilized in virtual travel, online gaming, entertainment, online education, healthcare, and robot control, among other applications. The use of a variety of devices with minimal overhead and high scalability requirements in XR presents significant risks, including personal data breaches and phony or falsified data (Anand, R., et al.,2024).

Connected Autonomous Vehicles (CAV)

The new era expects safe and dependable autonomous vehicles that are commercially viable for use in public transportation. The stages of data collection, the CAV supply chain, and the vehicle level are where security problems arise. Vehicle-level attacks can involve V2X communications, sensor hijacking, and physical control theft. These attacks necessitate integrated auto stop options and continuous monitoring. There are privacy concerns for both drivers and passengers (Pandey, J. K., et al.,2022).

Smart Grid 2.0

Intelligent dynamic pricing, intelligent line loss analysis, automated distribution grid management, self-healing electric power delivery, automated meter data analysis, etc. are some of the aspects of smart grid 2.0. The security flaws include network-based assaults, software-related threats, threats aimed at control elements, physical attacks, and attacks involving AI and ML.

Industry 5.0

Industry 5.0 uses automated robots and smart machines to increase efficiency; in these situations, security is essential to preserving integrity, availability, audit aspects, authentication, etc. When designing security measures, factors like low operating costs, device diversity, and high scalability must be taken into account (Iyyanar, P., et al., (2023)).

Intelligent Healthcare

Body Area Networks (BANs) for individualized health monitoring and management are used in AI-driven intelligent healthcare with Intelligent Wearable Devices (IWD), Intelligent Internet of Medical Things (IIoMT), Hospital-to-Home (H2H) services, etc. to improve quality of life. Protecting privacy and upholding the morality of user data or electronic health records would be the main security problems. The regulatory authorities must implement strict privacy laws and guidelines (Pandey, D., & Pandey, B. K. (2022)).

Digital Twin

By gathering real-time data from IoT devices connected to the physical system and storing it in locally decentralized servers or centralized cloud servers for real-time analysis, the digital twin bridges the virtual and physical worlds. The ability of an attacker to intercept, alter, and replay any communication between the physical and digital domains is the largest security threat. Therefore, blockchain technology and extremely scalable, secure communication routes provided by 6G are essential.

6G NETWORK BASED CHALLENGES

Three layers—the physical layer, the connection/network layer, and the service/application layer—are where the 6G enabling technologies are integrated. There are security concerns at every tier.

At Physical Layer

Physical layer security uses wireless channel fading, noise, and other phenomena to improve confidentiality and carry out lightweight authentication. The following is a description of the various recent advances along with some possible threats.

Security in 6G mmWave Communications

Massive MIMO and mmWave can be employed in non-standalone 6G networks that are vulnerable to three primary types of assaults: pilot contamination attacks (PCA), jamming, and eavesdropping. Beamforming technology can prevent eavesdropping, which is accomplished by inferring and wiretapping (sniffing) open (unsecured) wireless communications (Sahani, K.,et al.,2023). A denial of service is caused by the jamming assault, which injects radio signals to occupy a shared wireless channel. In PCA, the attacker deliberately sends out duplicate pilot signals, also known as spoofing uplink signals, to taint the transmitter's user detection and channel estimation phases, leading to signal leakage. Increasing the secrecy rate, adding extra randomness to the modulation, producing multiple random frequency shifts, pilot sequences, or frequency hopping, using friendly jamming or covert communication with artificial noise to confuse the eavesdropper, using AI/ML techniques like channel hopping, etc. are some examples of security solutions.

Security in 6G Large Intelligent Surface

To improve communication performance, Large Intelligent Surface (LIS) dynamically tunes the transmission signal phase shift using a planar array of inexpensive reflecting elements. Even if multi-path propagation channels are utilized to give authorized users access to several secure communication lines, eavesdropping remains a security risk.

Security in NOMA for 6G Massive Connectivity

Non-Orthogonal Multiple Access (NOMA), which enables both multicast transmission (to cluster users) and unicast transmission (to single user), equitably distributes channel resources for the receivers by adding power for "weak signal" users and deducting power for strong signal users. Eavesdropping, jamming assaults, and other security risks with NOMA can be prevented by optimizing transmission secrecy rate and using the successive interference cancellation (SIC) procedure.

6G Holographic Radio Technology with Large Intelligent Surface

For 6G indoor and outdoor communications, holographic radio, also known as beamforming and MIMO, is a disruptive radio technology based on software-defined antenna or photonics-defined antenna arrays; that is, it does not require phase shifters or active amplification during the beam-steering process. Through the cancellation of reflections to eavesdroppers, these gain from the optimization of secrecy rate of base station transmissions. However, hardware design and electromagnetic wave scattering pose a concern as well.

Security in 6G Terahertz Communications

Ultra-high data rates (up to terabits per second) are made possible by Terahertz (THz) for 6G applications such as XR/AR services, haptic Internet, etc. Low penetration power, high absorption resonance, limited coverage area, and other factors provide obstacles for THz. comprises dense networks of THz-capable devices for efficient line-of-sight (LOS) transmission communications. THz is protected from frequency hoping over numerous sub-channels, jamming, and eavesdropping assaults. To defend against these attacks, employ beam encryption, spatial modeling, and lightweight encryption. Scattered signal gathering during re-verification during device hand-off is one of the other enemies.

Security in 6G VLC Communications

Visible light communication (VLC) enables high-speed communication using visible light with a frequency range of 400–800 THz. It is a cost-effective substitute for 6G in undersea applications, interior facilities (such as hospitals and private rooms), and electromagnetically sensitive places (such as nuclear plants). The super-high bandwidth is an advantage, but the inability to pass through walls is a drawback. While eavesdropping poses a risk, it can be mitigated by employing techniques like beamforming signals, spatial modulation with the zeroforcingprecoding scheme, and maximizing secrecy rate.

Security in 6G Molecular Communications

For communication between nano/cellscale entities, such as drug delivery in blood vessels in healthcare, water/fuel distribution monitoring in industry, etc., molecular communication uses chemical signals or molecules as an information channel. They are helpful for extremely stealthy channels in hostile settings, such as extremely absorbent channels or networks with a high risk of jamming. Healthcare data leaks pose a risk, and biomachines could be remotely assaulted as a result of IoT device vulnerabilities. In order to get around this, a stringent access control system or a powerful firewall at the gateway portal can be used in conjunction with biochemical cryptography, which protects information integrity by encoding information using biological molecules such as DNA/RNA or protein structure.

Other Prospective Technologies for 6G Physical Layer Security

The three main physical-layer-based techniques, which include spoofing messages and manipulating physical data bits that happen in the network and application levels, are employed to counter unique assaults in 6G applications.

1) Spoofing and impersonation attempts are countered by physical layer authentication.
2) Physical key creation uses the entropy of randomness in transmit-receive channels like CSI and Received-Signal-Strength (RSS) to generate secrecy keys for communications, preventing eavesdropping on the confidentiality of communications between UEs and stations.
3) Physical coding to safeguard PHY data integrity from manipulation and increase the rate at which valid users can transmit data. For instance, quasi-cyclic multi-user LDPC, polar coding, space-time coding, and Low-Density Parity-check Code (LDPC) coding.

At Connection or Network Layer

The network and transport layers are combined to form the connection layer. This layer of security deals with DoS assaults and spoofing attacks. Key security issues and its authentication are as follows.

For Mutual Authentication Between the Subscriber and the Network, 6G Authentication and Key Management (6G-AKA) is Used

Network access control to meet the needs of holographic telepresence use and highly customized services 6G AKA aims to address problems with subscriber authentication, a dual authentication model for large heterogeneous networks, a new subscriber identification privacy paradigm, and other related challenges.

Quantum-Safe Algorithms and Quantum Communication Networks for 6G Secure Communication

use cipher techniques to keep unauthorized interceptors from accessing messages. However, quantum is capable of breaking any cryptosystem based on discrete logarithms and intricate integer factoring. These can be avoided by employing post-quantum algorithms in the long run, or by improving current ciphersuites and related protocols in the medium term. Alternatively, one can use quantum key distribution (QKD), public-key quantum-resistant algorithms like lattice-based encryption, etc.

Enhanced Security Edge Protection Proxy (SEPP) for Securing Interconnect Between 6G Networks

TLS-based protocols provide enhanced cryptographic techniques to enable gigabit networks' high-performance TCP/IP communications, hence enabling roaming security in 6G. Attacks that downgrade protocols are among the dangers.

Blockchain and Distributed Ledger Technologies for a Vision of 6G Trust Networks

Maintaining the value of information exchange while blocking fraudulent or dishonest sources, ensuring a low probability of unwanted events, and eliminating single points of failure are all part of trust networks and 6G services. Use distributed ledger technologies (DLT), blockchain, digital signatures and certificates, quantum-safe encryption, and other techniques to get around the difficulties in securing all of these.

SD-WAN Security

Control over 6G network management Software-Defined Local Area Network (SDLAN) and Software-Defined Wide Area Network (SD-WAN) are two examples of SDN. DoS/DDoS assaults, insider adversaries, and other things are among the threats. Use Moving Target Defense (MTD), anomalous Intrusion Detection System (IDS), ML/DL to improve detection engines, and supporting systems like FlowVisor, FlowChecker, and FlowGuard, among others, to safeguard them.

Deep Slicing and Open RAN for 6G Network Security Isolation

RAN/Core Network Slicing

Because network slicing avoids cross-talk between slices, it improves security. The difficulty lies in creating a security-isolation solution despite the absence of guidelines for creating a framework. DoS attacks with a high volume are one of the threats.

Virtualized RAN, Cloud-RAN, and Open RAN

Open RAN and virtualized radio access networks (vRAN) are utilized to increase modularity, decrease interdependencies, and provide more precise security attestation. If the source code is used again as a library to create new codes, they become susceptible.

Next-Generation Firewalls/Intrusion Detection for 6G Network Endpoint and Multi-Access Edge Security

In order to prevent illegal traffic, endpoints include perimeter routers, IP core network gateways, web application firewalls (WAF), IDS, antivirus software, VPN, and service-oriented architecture API protection. Use deep learning models, activate stronger protection, etc. to ward off threats.

Security in the Service Layer

The edge, fog, and cloud technologies that make up the service layer serve as middleware for the provision of value-added services to third parties. Authentication, data encryption, firewalls, hardware security, application security protocols, service identity access management, reinforcement of operation/kernel systems, data-center network protection, etc. are among the activities that are necessary. The following authentications are employed:

6G Application Authentication: Distributed PKI and Blockchain-Based PKI

The public-key infrastructure (PKI) uses decentralized, quantum-safe cryptography to facilitate user and application authentication. A public blockchain based PKI based on Ethereum, a complex model such pseudonym certificate production, Certificate Revocation List, etc., can be used to combat the threat of a Certificate Authority breach.

Using Service Access Authentication (6G AKA) for Application Authentication

In heterogeneous networks, credentials such as session keys will be preserved at both the application server and the user endpoint to ensure integrity and confidentiality. Use OAuth or SSO schemes to guard against AKMA maintenance.

6G Biometric Authentication for 6G-Enabled IoT and Implantable Devices

In order to get unique biological traits for more accuracy and flexibility, biometric authentication systems using wearable and implantable equipment are employed to avoid having users type complex codes or memorize usernames and passwords, especially for individuals with the disabilities (Pandey, B. K., et al.,(2023)).

In the future, scams might be prevented by brain/heart signal-based verification. A potential threat is biometric spoofing attacks, which use potentially exploited fake or synthetic face/iris samples. Use cancellable biometric models, secure enclave equipment, or pseudo-biometric identities to get around this (Sharma, M., Pandey, et al.,2022).

OAuth 3.0: New Authorization Protocol for 6G Applications and Network Function Services

With additional features like multi-user delegation, multi-device processing, key proofing procedures, etc., OAuth 3.0 is employed in 6G end-to-end service-based architecture. Threats to networking, security, and spectrum share allocation can all be swiftly implemented as services with on-demand authorization.

Enhanced HTTP/3 Over QUIC for Secure Data Exchange in 6G Low-Latency Applications

Web applications and mobile platforms use Hypertext Transfer Protocol Version 3 (HTTP/3) to enable secure communication over the interexchange/roaming links between the home network and the serving network. In order to manage congestion, it is based on Quick UDP Internet Connections (QUIC). The hazards posed by monitoring attacks are neutralized by quantum-safe cryptography methods.

Quantum Homomorphic Encryption for Secure Computation

Encryption techniques such as identity-based encryption, attributed-based encryption, homomorphic encryption, etc. are used in service computing nodes to enable quantum-safe standards and enable secure computation to safeguard confidentiality and integrity in computing nodes like edge servers. Secure range query, secure multi-keyword semantic search, searchable symmetric encryption, and other difficult issues are included.

Liquid Software Security: A Step to 6G Platform-Agnostic Security

Liquid software allows data and applications to flow from one node to the others, use enhancing containerization architecture like Kubernetes and cloudization of edge/fog nodes to accommodate multiple applications and enable interactions through API calls. The attacks include platform-agnostic security attacks for multiple devices and in combination with hardware solutions like Trusted Platform Module (TPM) and Hardware Security Module (HSM).

AI-Empowered Security-as-a-Service Transition for 6G "Service Everywhere" Architecture

Every 6G communication hop makes use of edge computing, autonomous driving, holographic telepresence, and enormous IoT applications. Threats to distributed computing nodes include data leaks, virus, and intrusion behaviors. Use super-intelligent AI models and the security-as-a-service (SECaaS) model on the computer nodes to get around these.

POTENTIAL SOLUTIONS FOR 6G SECURITY THREATS

This article discusses the threat landscape and potential security fixes for a few 6G technologies.

Distributed Ledger Technology (DLT)

Among the DLTs that provide reliable and secure 6G networks is the Blockchain technology, which offers benefits like disintermediation, immutability, non-repudiation, provenance evidence, integrity, pseudonymity, etc. The deployment of AI/ML in 6G has made networks susceptible to a number of novel assaults, such as poisoning and evasion attempts, which target the training and testing phases of the network. To meet 6G standards, the AI-driven solutions in a multi-tenant/multi-domain environment provide services like automated Security SLA management, secure slice brokering, safe roaming and offloading handling, scalable IoT PKI management, secure VNF management, and user privacy protection.

The threats that can occur include

(i) majority attack / 51% attack i.e., which happens when a group of malicious users capture $\geq 51\%$ nodes can take control of the blockchain which uses majority voting consensus.

(ii) Double spending attacks happens when a user spends a singke token multiple times due to lack of physical notes.

(iii) Re-entrency Attack can occur when a smart contract invokes another iteratively and the secondary smart contract invoked can be malicious.

(iv) Sybil attacks happen when a attacker hijacks the blockchain peer network by conceiving fake identities explicitly.

(v) Broken authentication and access control foreseen in authentication and access control mechanisms.

(vi) Security misconfiguration due to insecure or outdated configurations usage.

(vii) Privacy leakages of user with transaction data, smart contract logic privacy etc.

(viii) Other vulnerabilities include destroyable contracts, exception disorder, call stack vulnerability, bad randomness, underflow/Overflow errors, unbounded computational power intensive operations, etc.

Strong remedies include making sure the smart contract is accurate, appropriately validating its right functionality by spotting semantic errors, utilizing security check tools, carrying out formal verification, and so on. Because blockchain and DLT offer a variety of architecture types, including public, private, consortium, hybrid, and more, choosing the right blockchain or DLT with roaming, spectrum management, and other features can reduce the impact of some attacks (Pramanik, S., et al.,2023).

Quantum Computing

A novel strategy to combat the risks posed by the discrete logarithmic problem for asymmetric cryptography—which might be solved in polynomial time—is quantum secure cryptography. In order to enhance the transmission quality of supervised and unsupervised learning for classification and clustering tasks, quantum machine learning algorithms also provide complete randomization and security. As a communication protocol, quantum key distribution (QKD) is used to create a secret key between two authorized parties.

The risks that are anticipated include quantum cloning assaults, OT, and quantum-based attacks on Internet of Things devices. A few post-quantum cryptographic primitives, such as hash-, multivariate-, lattice-, and code-based encryption with verification security in the quantum-accessible random oracle model, are among the powerful options (Bessant,Y. A., et al.,2023).

Distributed and Scalable AI/ML

Self-monitoring, self-configuration, self-optimization, and self-healing are anticipated with autonomous networks. Intent-based interfaces, closed-loop functionality, and AI/ML approaches are all part of the ZSM architecture, which enables complete automation of network management tasks, including security. The benefits of AI/ML-driven cybersecurity include increased accuracy, predictability, and autonomy in security analytics.

The risks include physical attacks that target infrastructure and meddle with communication, as well as attacks connected to AI/ML, such as evasion and poisoning, as well as traditional attack vectors that target software, firmware, and hardware elements. Defensive distillation based on shifting target defense, input validation, using blockchain for enhancing security, and knowledge transfer across neurons via soft labels are some of the effective methods.

Physical Layer Security

In cases when resources are limited, these approaches enhance confidentiality and execute lightweight authentication and key exchange with flexibility and adaptability by depending on random and noisy wireless networks. Among the many technologies employed in physical layer security are

Terahertz Technology

It makes use of higher carrier frequencies in the terahertz region (1 GHz to 10 THz) to increase future wireless networks' capacity and spectral efficiency and to deliver high-speed Internet access to all. The risks include disclosure of data transmission, listening in on conversations, attacks on access control, etc. (David, S., et al.,2023). Effective methods include the use of multipath nature to improve information theoretic security, distance dependent-path loss-based authentication, and the use of intercept signals, which involve putting an object in the transmission path to get dispersed radiations and backscatter of the channel.

Visible Light Communication Technology

High data rates, a vast spectrum of usable frequencies, resistance against interference, and minimal implementation costs are all provided by VLC. Potential dangers include eavesdropping assaults, for which effective countermeasures include employing multiple-input multiple-output (MIMO) VLC systems with linear precoding to improve secrecy performance. Watermarking approach with spread spectrum, etc.

Reconfigurable Intelligent Surface

A planar array of numerous passive, inexpensive reflecting elements that can dynamically adjust their reflective coefficients to control the amplitude and/or phase shift of reflected signals and improve wireless propagation performance makes up the software-controlled metasurface known as RIS. Traditional PLS measures, including the installation of friendly jammers or active relays that use artificial noise (AN) for security provisioning, are part of the threat environment and may result in higher hardware and energy costs. Effective methods include using RIS-assisted PLS for secure and affordable 6G networks and intelligent phase shift control (Anand, R., et al., (2023)).

Molecular Communication

In MC, bionanomachines interact with one another in an aquatic environment by chemical signals or molecules, particularly in applications related to healthcare (Saxena, A., et al.,2021). Threats include a number of issues with communication, authentication, and encryption that raise security and privacy concerns. These issues can be resolved by applying biochemical cryptography (Revathi, T. K., et al.,2022).

Privacy

The growing trend of end devices sharing local data with centralized entities poses challenges to digital privacy. Processing and storing this massive amount of data while implementing additional privacy protection measures will be challenging. The types of privacy, privacy violations, privacy protection, and related technologies are described in depth in Figure 3.

Figure 3. 6G privacy types, issues, and solutions

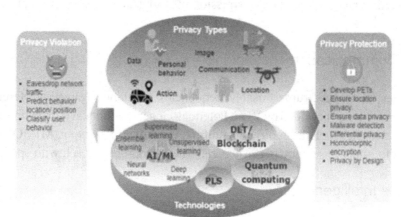

The three main obstacles to 6G privacy protection are: enforcing regulations and rights regarding extremely large data exchanges that result in a large number of small data accumulations; integrating lightweight privacy-protecting mechanisms in devices with limited resources; and regulating data access, ownership, and supervision through differential privacy (Bruntha, P. M.,et al.,2023). It offers defense against a variety of privacy-related threats, including deception, linking, and differencing (Pandey, B. K., et al 2022). Risk signatures, zero-knowledge augments and coin mixing, using AI to mimic human brain function with cooperative/collaborative robots (cobots), adding quantum noise to protect quantum data will drive the security concept of DP towards quantum differential privacy, and other techniques can be used to mitigate the privacy risks (Pandey, D., et al.,2021).

SECURITY STANDARDIZATION

Active standardization is necessary for digital services and networks of the future. Concerning 6G security, the different Standards Developing Organizations (SDOs) that are relevant include.

ETSI

Several Industry Specification Groups (ISG) have started it to study 5G component technologies like network automation (ETSI ISG Zero touch and service management - ZSM), AI (ETSI ISG Securing Artificial Intelligence - SAI, ETSI ISG Experiential Network Intelligence - ENI), and NFV (ETSI NFV for group specifications and reports).

ITU-T

ITU-T Focus Group on ML for Future Networks (FG-ML5G) is developing technical specifications, such as interfaces, network architectures, protocols, algorithms, and data formats, for machine learning for future networks on a worldwide scale.

3GPP

The Network Data Analytics Function, which 3GPP introduces to address the use of AI/ML in the 5G Core Service Based Architecture (SBA), provides notifications and analytics to other network functions about user behavior and network status for solutions to slice networks and perform Network Data Analytics.

NIST

For digital signatures, public-key encryption, and cryptographic key establishment, the National Institute of Standards and Technology (NIST) standardizes post-quantum cryptographic algorithms to identify candidates and then specifies quantum-resistant algorithms for each.

IETF

An architecture based on entities, or components, that communicate by exchanging information via policy repositories, vulnerability definition data repositories, and security information repositories is defined by the IETF Security Automation and Continuous Monitoring (SACM) Architecture RFC. This architecture facilitates a cooperative SACM ecosystem.

5G PPP

The 5G PPP Security Work Group addresses 5G security threats and issues, offering guidance on how to handle 5G security. It directly affects security KPIs, intelligent network security, new risks, threats, and countermeasures for Beyond 5G networks.

NGMN

The requirements for network entities and functions for the capabilities of an end-to-end framework, which also includes security for the end-to-end protection of the various network features and enabling capabilities, are outlined in the NGMN 5G End-to-End Architecture Framework v4.3 (2020).

IEEE

With an eye on end users, network operators, service/content providers, etc., IEEE P1915.1 Standard for Software Defined Networking and Network Function Virtualization (SDN/NFV) Security aims to offer a framework for creating and managing secure SDN/NFV systems. IEEE P1917.1 Standard for Network Function Virtualization and Software Defined Networking dependability creates a framework for dependable SDN/NFV service delivery architecture and focuses on dependability requirements (Tripathi, R. P.,et al.,2023). The SDQC protocol, which enables configuration of quantum endpoints in a communication network to allow dynamic creation, modification, or removal of quantum protocols or applications in a software-defined setting, is defined by IEEE P1913.1 (Draft) Standard for Software-Defined Quantum Communication (SDQC).

CONCLUSION OF 6G SECURITY

We need to find solutions to deployment problems by the year 2030 for 6G to be a technology that enables the future. This chapter provides an overview of the various 6G technologies that have recently been implemented, as well as any potential security problems and potential solutions. AI/ML, distributed ledger technology (DLT), quantum computing, visual light communication (VLC), and THz communication are the major 6G technologies that will open up an entirely new universe of possibilities.

REFERENCES

Anand, R., Khan, B., Nassa, V. K., Pandey, D., Dhabliya, D., Pandey, B. K., & Dadheech, P. (2023). Hybrid convolutional neural network (CNN) for kennedy space center hyperspectralimage. *Aerospace Systems*, 6(1), 71–78. 10.1007/s42401-022-00168-4

Anand, R., Lakshmi, S. V., Pandey, D., & Pandey, B. K. (2024). An enhanced ResNet-50 deep learningmodel for arrhythmia detection using electrocardiogram biomedical indicators. *Evolving Systems*, 15(1), 83–97. 10.1007/s12530-023-09559-0

Bessant, Y. A., Jency, J. G., Sagayam, K. M., Jone, A. A. A., Pandey, D., & Pandey, B. K. (2023). Improved parallel matrix multiplication using Strassen and Urdhvatiryagbhyam method. *CCF Transactions on High Performance Computing*, 5(2), 102–115. 10.1007/s42514-023-00149-9

. Bruntha, P. M., Dhanasekar, S., Hepsiba, D., Sagayam, K. M., Neebha, T. M., Pandey, D., & Pandey, B.K. (2023). Application of switching median filter with L 2 norm-based auto-tuning function for removing random valued impulse noise. *Aerospace systems, 6*(1), 53-59.

. David, S., Duraipandian, K., Chandrasekaran, D., Pandey, D., Sindhwani, N., & Pandey, B. K. (2023).Impact of blockchain in healthcare system. In *Unleashing the Potentials of blockchain technology forhealthcare industries* (pp. 37-57). Academic Press..

. Govindaraj, V., Dhanasekar, S., Martinsagayam, K., Pandey, D., Pandey, B. K., & Nassa, V. K. (2023).Low-power test pattern generator using modified LFSR. *Aerospace Systems*, 1-8.

. Iyyanar, P., Anand, R., Shanthi, T., Nassa, V. K., Pandey, B. K., George, A. S., & Pandey, D. (2023). Areal-time smart sewage cleaning UAV assistance system using IoT. In *Handbook of Research on Data-Driven Mathematical Modeling in Smart Cities* (pp. 24-39). IGI Global.

. Kirubasri, G., Sankar, S., Pandey, D., Pandey, B. K., Nassa, V. K., & Dadheech, P. (2022). Software-defined networking-based Ad hoc networks routing protocols. In *Software defined networking for AdHoc networks* (pp. 95-123). Cham: Springer International Publishing.

. Kirubasri, G., Sankar, S., Pandey, D., Pandey, B. K., Singh, H., and Anand, R. (2021, September). A recentsurvey on 6G vehicular technology, applications and challenges. In *2021 9th InternationalConference on Reliability, Infocom Technologies and Optimization (Trends and FutureDirections)(ICRITO)* (pp. 1-5). IEEE.

Kumar Pandey, B., Pandey, D., Nassa, V. K., Ahmad, T., Singh, C., George, A. S., & Wakchaure, M. A. (2021). Encryption and steganography-based text extraction in IoT using the EWCTS optimizer. *Imaging Science Journal*, 69(1-4), 38–56. 10.1080/13682199.2022.2146885

Pandey, B. K., & Pandey, D. (2023). Parametric optimization and prediction of enhancedthermoelectric performance in co-doped CaMnO3 using response surface methodology and neuralnetwork. *Journal of Materials Science Materials in Electronics*, 34(21), 1589. 10.1007/s10854-023-10954-1

Pandey, B. K., Pandey, D., & Agarwal, A. (2022). Encrypted information transmission by enhanced steganography and image transformation. [IJDAI]. *International Journal of Distributed Artificial Intelligence*, 14(1), 1–14. 10.4018/IJDAI.297110

Pandey, B. K., Pandey, D., Gupta, A., Nassa, V. K., Dadheech, P., & George, A. S. (2023). Secret data transmission using advanced morphological component analysis and steganography. In *Role of data-intensive distributed computing systems in designing data solutions* (pp. 21-44). Cham: Springer International Publishing.

Pandey, D. (2022). An efficient deep neural network with adaptive galactic swarm optimization for complex image text extraction. In *Process mining techniques for pattern recognition* (pp. 121-137). CRC Press.

. Pandey, D., Nassa, V. K., Jhamb, A., Mahto, D., Pandey, B. K., George, A. H., & Bandyopadhyay, S. K.(2021). An integration of keyless encryption, steganography, and artificial intelligence for the securetransmission of stego images. In *Multidisciplinary approach to modern digital steganography* (pp.211-234). IGI Global.

. Pandey, D., Ogunmola, G. A., Enbeyle, W., Abdullahi, M., Pandey, B. K., & Pramanik, S. (2021). COVID-19: A framework for effective delivering of online classes during lockdown. *Human Arenas*, 1-15.

Pandey, D., Pandey, B. K., & Wairya, S. (2021). Hybrid deep neural network with adaptive galacticswarm optimization for text extraction from scene images. *Soft Computing*, 25(2), 1563–1580. 10.1007/s00500-020-05245-4

. Pandey, J. K., Jain, R., Dilip, R., Kumbhkar, M., Jaiswal, S., Pandey, B. K., & Pandey, D. (2022). Investigating role of iot in the development of smart application for security enhancement. In *IoT Based Smart Applications* (pp. 219-243). Cham: Springer International Publishing.

. Pramanik, S., Pandey, D., Joardar, S., Niranjanamurthy, M., Pandey, B. K., & Kaur, J. (2023, October). An overview of IoT privacy and security in smart cities. In *AIP Conference Proceedings* (*Vol. 2495*, No.1). AIP Publishing .

. Revathi, T. K., Sathiyabhama, B., Sankar, S., Pandey, D., Pandey, B. K., & Dadeech, P. (2022). An intelligent model for coronary heart disease diagnosis. In *Networking Technologies in Smart Healthcare* (pp. 309-327). CRC Press.

Sahani, K., Khadka, S. S., Sahani, S. K., Pandey, B. K., & Pandey, D. (2023). A possible underground roadway for transportation facilities in Kathmandu Valley: A racking deformation of undergroundrectangular struct es. *Engineering Reports*, 12821. 10.1002/eng2.12821

Saxena, A., Agarwal, A., Pandey, B. K., & Pandey, D. (2024). Examination of the Criticality of Customer Segmentation Using Unsupervised Learning Methods. *Circular Economy and Sustainability*, 1–14. 10.1007/s43615-023-00336-4

Saxena, A., Sharma, N. K., Pandey, D., & Pandey, B. K. (2021). Influence of tourists satisfaction onfuture behavioral intentions with special reference to desert triangle of Rajasthan. *AugmentedHuman Research*, 6(1), 13. 10.1007/s41133-021-00052-4

Sharma, M., Pandey, D., Khosla, D., Goyal, S., Pandey, B. K., & Gupta, A. K. (2022). Design of a GaN- based Flip Chip Light Emitting Diode (FC-LED) with au Bumps & Thermal Analysis with different sizes and adhesive materials for performance considerations. *Silicon*, 14(12), 7109–7120. 10.1007/s12633-021-01457-x

Sharma, M., Pandey, D., Palta, P., & Pandey, B. K. (2022). Design and power dissipation consideration of PFAL CMOS V/S conventional CMOS based 2: 1 multiplexer and full adder. *Silicon*, 14(8), 4401–4410. 10.1007/s12633-021-01221-1

Tareke, S. A., Lelisho, M. E., Hassen, S. S., Seid, A. A., Jemal, S. S., & Teshale, B. M., & Pandey, B. K. (2022). The prevalence and predictors of depressive, anxiety, and stress symptoms among Tepi town-residents during the COVID-19 pandemic lockdown in Ethiopia. *Journal of Racial and Ethnic Health Disparities*, 1–13.35028903

Tripathi, R. P., Sharma, M., Gupta, A. K., Pandey, D., Pandey, B. K., Shahul, A., & George, A. H. (2023). *Timely prediction of diabetes by means of machine learning practices*. Augmented Human Research.

Chapter 14
Exploring Machine Learning Solutions for Anomaly Detection in 6G Communication Systems

Brajesh Kumar Khare
Harcourt Butler Technical University, India

Deshraj Sahu
Dr. A.P.J. Abdul Kalam Technical University, India

Digvijay Pandey
https://orcid.org/0000-0003-0353-174X
Dr. A.P.J. Abdul Kalam Technical University, India

Mamta Tiwari
Chhatrapati Shahu Ji Maharaj University, India

Hemant Kumar
https://orcid.org/0000-0003-0603-4394
Chhatrapati Shahu Ji Maharaj University, India

Nigar Siddiqui
Dr. Virendra Swarup Memorial Trust Group of Institutions, India

ABSTRACT

The chapter explores the use of machine learning (ML) in detecting and addressing anomalies in advanced 6G communication systems. It emphasizes the drawbacks of conventional approaches and delves into ML algorithms that are appropriate for identifying anomalies, such as clustering, classification, and deep learning. The study focuses on the difficulties of choosing important features from various data sources in 6G networks, including network traffic and device behavior. It also explores possible attacks on ML models and suggests ways to improve their resilience. Exploring integration with network slicing and highlighting the adaptability of ML to dynamic virtualized networks. The chapter highlights the importance of ML-based anomaly detection in strengthening 6G network security and suggests areas for future research.

INTRODUCTION

The advancement of 6G (Kirubasri, G. et al., 2021) communications is expected to enhance the ability of humans to engage with virtual worlds through the internet in the near future (Viswanathan et al., 2020). The digital transformation, which is now being led by the implementation of the 5G network, is expected

DOI: 10.4018/979-8-3693-2931-3.ch014

to continue and develop in the following decade (M. M. Saeed et al., 2021, Banafaa et al., 2023). In the future, it is anticipated that networks would require advanced technologies to provide interconnected intelligence in digital virtual environments, tackling the intricacies associated with networking and communication (Saad et al., 2020). According to recent research, 6G systems are expected to introduce new application areas, such as distributed robotic systems, self-governing systems, and various extended reality (XR) applications. However, traditional applications like multimedia streaming will continue to exist (Alwis et al.,2021). Furthermore, other 6G applications necessitate fast data transfer speeds, minimal latency, and exceptional network stability. These applications include holographic communication, telemedicine, and biomedical (Pandey, B. K. et al., 2011) sensor networks. Anomaly detection emerges as a critical technique for fortifying the resilience of 6G networks against various threats, including cyberattacks, equipment failures, and performance degradation. By leveraging advanced machine learning algorithms (Tripathi, R. P. et al., 2023), statistical models, and data-driven approaches (Khan, B. et al., 2021), anomaly detection mechanisms enable the identification of abnormal behaviors or deviations from expected patterns (Govindaraj, V. et al., 2023) within network traffic, system parameters, and communication protocols. This survey paper aims to provide a comprehensive overview of the state-of-the-art in anomaly detection techniques tailored specifically for 6G communication systems. Through an in-depth analysis of existing literature, we seek to elucidate the fundamental principles, methodologies, challenges, and future directions in this burgeoning field. By synthesizing insights from diverse sources, we aim to offer valuable guidance for researchers, practitioners, and stakeholders in navigating the complexities of anomaly detection within the context of 6G communications. The advent of 6G communication systems heralds a transformative era in digital connectivity, promising unprecedented data rates, ultra-low latency, and massive semiconductor device (Sharma, M. et al., 2022) connectivity that can empower new applications from autonomous vehicles to immersive augmented realities. As these networks become more complex and pervasive, ensuring their reliability and security becomes crucial. Anomaly detection, in this context, emerges as a critical field of study, aimed at identifying deviations from normal behavior in network traffic, which could signify potential faults, cyberattacks, or other security threats (Pandey, B. K. et al., 2023).

Identifying zero-day attacks presents a significant obstacle in this ever-changing environment. Every day, many dubious activities arise, potentially containing intricate invasions, which, if not managed, could result in substantial dangers (Giordani et al., 2021, Ziegler et al., 2020). Intrusion Detection Systems (IDSs) play a crucial role in protecting information systems by delivering alerts when they identify abnormal behavior or recognized dangers, ensuring the systems' integrity (Liang et al., 2021). Intrusion Detection Systems (IDSs) are designed to observe computer systems for unusual behaviors, detecting indications of possibly malicious operations, cyber disturbances, and unauthorized entry into network packets. In order to enhance accuracy and minimize false alarm rates (FAR), intrusion detection systems (IDSs) utilize two main approaches to detect unauthorized access: anomaly-based intrusion detection systems (AIDS) and signature-based intrusion detection systems (SIDS) (Rajagopal et al.,2020). A new Intrusion Detection System (IDS) model has been suggested to overcome the constraints related to AIDS. This model combines both Intrusion Detection Systems (SIDS) and AIDS to improve the ability to detect AIDS (Mokhtari et al., 2021).

BACKGROUND

The study by Alatabani et al., (2022) stands out for its innovative approach. They employed KNN reliability-enhanced voting classifiers (Deepa, R. et al., 2022), linear decision boundary analysis, and machine learning-based binary classification models (Anand, R., 2024) to address magnitudes. The researchers also used the SMOTE approach to mitigate data imbalance and trained the model using recently generated training samples. With 16 carefully chosen characteristics and experimental assessments on the NSL-KDD dataset, they achieved an accuracy rate of 83.24%. They also reported a false alarm rate of 4.83%, a true positive rate of 82%, and a false positive rate of 5.43%.

The study by Pajouh et al., (2017) has significant practical implications. They used GAR-Forest to employ feature subsets obtained from the Association, Knowledge gain, Balanced uncertainty, and NSL-KDD datasets. Multi-class classification accuracy was 78.9035% when using ten features based on information gain. In contrast, binary class classification demonstrated an accuracy rate of 85.0559% while employing 32 characteristics. These findings suggest the potential for effective intrusion detection in real-world scenarios. Kanakarajan and Muniasamy (2016) recommended utilizing the NSL-KDD dataset to detect assaults on wireless sensor networks. The authors reported precision, recall, and F1-Score values of 94.00%, 98.00%, and 96.00%, respectively, resulting in an accuracy rate of 95%. Gawali and Ranjan (2023) introduced a customized feature extraction and detection solution for 5G networks. This system incorporates deep belief networks (DBN) and BiLSTM as components of a hybrid classifier (Anand, R. et al., 2023). They enhanced the detecting effectiveness and accuracy by optimizing weights utilizing a novel deer hunting updated sunflower optimization (DHSFO) model.

In their study, Koursioumpas et al. (2022) introduced an innovative architectural design to identify traffic anomalies in 5G networks or forecast potential anomalies using a BiLSTM autoencoder. The evaluation findings demonstrated a prediction accuracy of up to 90.02%, highlighting the approach's effectiveness and practicality. Lazar et al., (2021) have proposed a novel approach to anomaly identification that holds promise for future applications. Their method, inspired by the autoencoder principle and LSTM, demonstrated a notable accuracy rate of 87% in detecting anomalies across different periods. This suggests its potential for effectively identifying a wide range of anomaly kinds, paving the way for more robust intrusion detection systems in the future. Oleiwi et al. (2022) introduced an innovative methodology for identifying abnormalities in communication networks. This approach employed an ensemble learning algorithm-based technique (Bessant, Y. A. et al., 2023), demonstrating notable accuracy across various datasets such as CIC_IDS2017, NSL_KDD, and UNSW_NB2015.

The study conducted by Ozpoyraz et al. (2022) sought to showcase the progress made in deep learning (Pandey, D. et al., 2021) based physical layer techniques for novel applications in the context of 6G. Johnson et al. (2015) introduced an intrusion detection system that uses decision trees and rule learners in an ensemble-based approach. The system demonstrated high accuracy, with 80% accuracy, an 81% detection rate (DR), and a 15.1% false alarm rate (FAR). Mhawi et al. (2021) and R. A. Saeed et al. (2022) have investigated distributed machine-learning approaches (Saxena, A. et al., 2024), which have shown notable decreases in data complexity. The authors demonstrated the advantages of utilizing machine learning models in the pre-processing phase of intrusion detection systems (IDSs) to manage substantial volumes of data. Additionally, they highlighted the potential of deep learning (DL) algorithms in revealing hidden characteristics during anomaly classification, thereby facilitating the detection of previously unnoticed attacks.

Data mining techniques are crucial in intrusion detection (ID) as they employ statistical analytical methods to identify, extract, and differentiate between benign and malicious attributes. According to Safaldin et al. (2021), feature selection (FS) techniques play a critical role in the intrusion detection system (IDS) process by effectively detecting important characteristics and eliminating duplicate ones in order to enhance performance. According to Loey et al. (2021), the Correlation-based Feature Selection (CFS) approach entails prioritizing features based on their significance and examining additional attributes based on their shared associations with various aspects.

Table 1. Different approaches proposed by researchers

Reference	Methodology/Technique	Dataset/Scenario	Results/Findings
Zhang et al. (2023)	Deep Learning	Synthetic	Achieved a detection accuracy of 95% using LSTM-based approach
Wang and Liu (2022)	Statistical Analysis	Field Trials	Identified anomalous patterns in real-world 6G field trial data, leading to the development of robust statistical anomaly detection
Chen and Kim (2023)	Graph-Based Approaches	Network Traffic	Achieving a detection rate of 87% on a simulated dataset.
Zhang et al. (2023)	Machine Learning	Synthetic	Achieving an accuracy of 93% on synthetic data.
Smith et al. (2022)	Time-Series Analysis	Laboratory Experiments	achieving a detection accuracy of 88% on laboratory datasets.
Kanakarajan and Muniasamy,2016	Wireless sensor network	NSL-KDD	Achieved an F1 score 96%
Koursioumpas et al.,2022	BiLSTM Autoencoder	NSL-KDD	Achieved an accuracy of up to 90.02%
Johnson et al.,2015	Decision trees and Rule learners	Synthetic	Achieved 80% accuracy for intrusion detection system.
Oleiwi et al.,2022	ensemble learning algorithm-based technique	CIC_IDS2017, NSL_KDD, and UNSW_NB2015.	99.6% accuracy with a 0.004 false-alarm rate for NSL_KDD 99.1% accuracy with a 0.008 false-alarm rate for UNSW_NB2015 99.4% accuracy with a 0.0012 false-alarm rate for CIC_IDS2017.
Alatabani et al.,2022	KNN reliability-enhanced voting classifiers	NSL-KDD dataset	Achieved an accuracy rate of 83.24%

6G COMMUNICATION

The 6G communication system, projected for deployment around 2030, represents the next evolutionary step beyond 5G, promising transformative changes across global telecommunications. 6G is expected to achieve breakthroughs in speed and capacity, offering data rates up to 1 terabyte per second and significantly lower latency, potentially below one millisecond. This enhancement will facilitate real-time communication for time-sensitive operations, crucial for technologies like autonomous vehicles and telesurgery. Moreover, 6G is designed to support exponentially more connected devices (Sharma,

M. et al., 2022) per square kilometer, accommodating the burgeoning IoT (Pandey, J. K. et al., 2022) ecosystem without congestion.

Energy efficiency and sustainability are central to 6G development, with advanced network management technologies that reduce energy consumption and carbon footprints. The system will likely leverage sophisticated artificial intelligence (Pandey, B. K., & Pandey, D., 2023) to optimize network operations and traffic management dynamically. Additionally, 6G could extend the concept of ubiquitous connectivity through advanced network topologies that integrate terrestrial, aerial, and space-based networks. This integration aims to ensure consistent and reliable internet coverage to remote and previously unreachable areas, promoting global digital inclusion.

Table 2. Characteristics of 6G communication system

Feature	Description	Technology	Use cases	Challenges
Data Rates	Terabit-per-second data rates, enabling ultra-fast wireless communication.	Advanced modulation schemes, high-frequency bands	High-definition video streaming, augmented reality	Spectrum allocation, signal attenuation at higher frequencies
Latency	Sub-millisecond latency to support real-time applications like augmented reality and autonomous vehicles.	Edge computing, low-latency protocols	Autonomous vehicles, remote surgery	Network reliability, synchronization challenges
Device Density	Extremely high device (Du John, H. V. et al., 2022) connectivity with up to millions of devices per square kilometer, catering to IoT applications (Pramanik, S. et al., 2023).	Massive MIMO, advanced beamforming	Smart cities, IoT ecosystems, industrial automation (Bruntha, P. M. et al., 2023)	Interference management, energy consumption in dense deployments
Spectrum Bands	Utilization of a wide spectrum, including higher frequency bands (millimeter waves and terahertz frequencies).	Millimeter waves, terahertz communication	Increased data capacity, faster communication	Signal propagation challenges, regulatory approval for new bands
Energy Efficiency	Focus on energy-efficient technologies, including low-power components and sustainable power management.	Low-power hardware, energy-efficient protocols	Prolonged battery life for devices, reduced carbon footprint	Balancing performance with energy efficiency, trade-offs in design
AI Integration	Extensive use of artificial intelligence for network optimization, autonomous decision-making, and predictive analytics.	ML algorithms, AI-driven optimization (Pandey, D., & Pandey, B. K., 2022)	Predictive maintenance, intelligent resource allocation	Ethical considerations, transparency in AI decision-making
Quantum Communication	Incorporation of quantum communication for secure key exchange and enhanced encryption methods.	Quantum key distribution, quantum-safe cryptography	Quantum-resistant security, secure communication	Development of quantum technologies, integration challenges
Holographic Type Communication	Advanced communication methods, potentially involving holographic-type interfaces and immersive experiences.	Holographic displays, spatial audio technologies	Virtual collaboration, immersive entertainment	Technical complexity, user adoption and acceptance
Integrated Satellite Communication	Seamless integration with satellite communication for global coverage and connectivity.	Satellite networks, hybrid terrestrial-satellite links	Satellite networks, hybrid terrestrial-satellite links	Orbital debris management, cost of satellite infrastructure

continued on following page

Table 2. Continued

Feature	Description	Technology	Use cases	Challenges
Smart Surfaces	Use of smart surfaces and materials to enhance signal propagation and coverage.	Metamaterials, reconfigurable intelligent surfaces	Signal reflection and absorption optimization	Integration challenges, cost of deploying smart surfaces
Dynamic Network Slicing	Implementation of dynamic network slicing for tailored services and resource allocation based on diverse application requirements	Network virtualization, software-defined networking (Kirubasri, G. et al., (2022).)	Customized services, efficient resource utilization	Management of dynamic slices, orchestration complexities
Global Standardization	International collaboration to establish global standards for interoperability and consistency.	Standardization bodies (ITU, IEEE, 3GPP)	Global interoperability, seamless roaming	Negotiating diverse regulatory environments, alignment of standards
Security and Privacy Focus	Robust security mechanisms to address emerging threats and privacy concerns associated with massive connectivity (David, S. et al., 2023).	Advanced encryption, secure protocols	Protection against cyber threats, user data privacy	Continuous monitoring, compliance with evolving security standards

ATTACKS AND THREATS

As the development of 6G networks progresses, it is imperative to consider the various hazards and threats that may emerge within these networks. Several potential assaults and dangers that could arise in 6G networks encompass:

- **Malicious software:** The threat of malicious software looms large over 6G networks, posing a significant risk. This software can breach network security, illicitly access confidential data, disrupt network operations, or even cause physical damage to the network infrastructure.
- **DDoS assaults**: DDoS attacks are a standard method of disrupting a network's normal functioning. These assaults involve orchestrating the simultaneous use of multiple devices to overwhelm a network with traffic, effectively blocking or severely impeding the passage of legitimate traffic through the network.
- **IoT botnets**: The proliferation of IoT devices (Iyyanar, P. et al., 2023) in 6 G networks brings a significant vulnerability. If compromised, these devices can be manipulated to form botnets, which can then be used to execute DDoS attacks or other malicious activities, underscoring the need for robust security measures.
- **Unauthorized base stations**: These devices have the capacity to intercept and manipulate network traffic or direct attacks towards other devices connected to the network for malicious intent.
- **Eavesdropping:** Attacks involve clandestinely intercepting and illicitly acquiring sensitive information transmitted over a network. This nefarious activity encompasses surreptitiously monitoring data exchanges to capture confidential information, such as individual data, economic details, and other sensitive data types traversing the network.
- **Physical attacks:** Disruptions to network infrastructure, encompassing physical attacks like severing fibre optic cables or interfering with power supplies, can significantly impair network operations and result in considerable harm.

- **Man-in-the-middle attacks:** Man-in-the-middle attacks allow attackers to intercept and manipulate network traffic clandestinely, stealing Delicate data or initiating attacks on other devices connected to the network.
- **Distribution network intrusions:** supply chain attacks involve infiltrating network infrastructure by targeting vendors and suppliers responsible for providing the components utilized within the network. By compromising these supply chain entities, attackers can introduce malicious elements or vulnerabilities into the network, potentially leading to widespread security breaches or disruptions.
- **Quantum attacks:** These assaults could pose a risk to 6G networks by potentially compromising cryptography and other security measures currently considered secure, especially with the advent of quantum computers.

Ensuring the resilience of 6 G networks against a wide array of threats and attacks necessitates the early prioritization of security measures. This process is a collective effort, requiring continuous research, development of innovative security technologies and protocols, and active collaboration among telecom carriers, hardware manufacturers, and other relevant parties within the 6G network (Shaukat Dar et al., 2015). The strategic integration of two detection approaches in hybrid-based detection systems significantly enhances overall security. This approach capitalizes on the unique strengths of each method. Cyberattacks manifest in diverse forms, including Denial of Service (DoS), R2L, U2R, and Probing. Each of these attacks exploits distinct vulnerabilities, and by integrating multiple detection approaches, we can effectively neutralize them. Table 2 offers a comprehensive overview of several assault types and their respective descriptions.

Table 3. Attacks and their description

Type of Threat/Attack	Description
DDoS Attacks	DDoS attacks could target critical infrastructure components of 6G networks, overwhelming them with a flood of traffic and disrupting services.
Man-in-the-Middle Attacks	Attackers could intercept communication between devices and infrastructure in 6G networks, potentially eavesdropping on sensitive information or altering data exchanges.
Network Slicing Vulnerabilities	As 6G networks are expected to utilize network slicing, vulnerabilities in the slicing architecture could be exploited by attackers to gain unauthorized access or disrupt services within specific slices.
Insider Threats	Malicious insiders with access to network infrastructure or sensitive information could abuse their privileges to steal data, disrupt services, or sabotage the network.
Rogue Base Stations	Attackers might deploy rogue base stations to impersonate legitimate network infrastructure, enabling various attacks such as interception of communications or distribution of malware.
IoT Botnets	With the proliferation of IoT devices in 6G networks, compromised devices could be harnessed to form botnets capable of launching DDoS attacks, spreading malware, or conducting other malicious activities.
Identity Theft	Attackers may target user identities within 6G networks, stealing credentials or impersonating legitimate users to gain unauthorized access to resources or conduct fraudulent activities.
Supply Chain Attacks	Compromising the supply chain of 6G equipment could involve injecting backdoors, malicious firmware, or hardware implants into network infrastructure components, undermining the security of the entire network ecosystem.

MACHINE LEARNING TECHNIQUES IN ANAMOLY DETECTION

Recent studies highlight the essential role of artificial intelligence (AI) and machine learning (ML) within the network architecture of forthcoming 6G technologies. The 6G networking sector recognizes the significant importance of integrating AI, acknowledging its pivotal role. While incorporating AI/ML into 5G networks requires substantial training data and robust processing capabilities, their indispensability in 6G networks is evident. AI and ML algorithms are crucial in strengthening 6G's security infrastructure and enhancing autonomy, precision, and predictive capabilities. This section explores various challenges associated with integrating AI/ML in the 6G framework (Siriwardhana et al., 2021)

- *Reliability:* AI empowers network security professionals by ensuring the network's security and maintaining the accuracy of ML models and their components, making their role crucial.
- *Visibility:* The real-time security functions based on AI and ML are constantly monitored, providing a high level of control and credibility that instils confidence in the audience.
- *Legal and Ethical Implications:* AI-driven optimization methods may unintentionally exclude specific clients or applications. Furthermore, although AI-powered security solutions strive to protect all users, AI ultimately oversees security services and may encounter some shortcomings.
- *Flexibility and Adaptability:* Maintaining the confidentiality of federated learners requires secure data transfers. Nevertheless, AI and ML face difficulties when it comes to scaling the necessary computational, storage resources, and communication for this purpose.
- *Managed security duties:* Integrating AI/ML security solutions with extensive data operations can result in significant overhead. Ensuring safety and security during the learning and inference phases is crucial for maintaining the model's adaptability. The future intelligent 6G system (Revathi, T. K. et al., 2022) is anticipated to leverage advanced AI techniques to fulfill the needs of emerging scenarios, high service demands, and essential capabilities. Figure 1 depicts the structure of a secure 6G system that relies on artificial intelligence and machine learning.

Figure 1. AI/ML security architecture of 6G

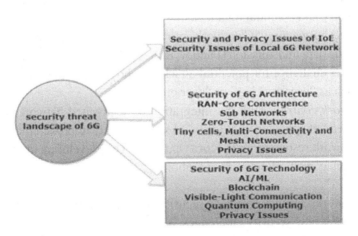

Security Improvements in 5G Networks

Although 5G networks provide significant security benefits, they also bring about fresh security concerns. Here are some important security enhancements found in 5G networks, as discussed by Giordani et al. (2021). (Plastiras et al., 2018)

- *Enhanced encryption:* Enhanced encryption (Kumar Pandey, B. et al., 2021) involves the implementation of sophisticated cryptographic methods and algorithms to safeguard data transmitted across networks. This ensures a higher level of protection against unauthorized access or decryption by potential attackers. Enhancing the security of communication channels often requires the use of more advanced encryption algorithms, longer encryption keys (Pandey, D. et al., 2021), and reliable encryption protocols.
- *Network Segmentation:* 5G networks divide the network into separate virtual networks, each with its own distinct security policies and restrictions. This segmentation allows for the separation and secure management of various types of traffic, applying different levels of protection as required.
- *SIM-based security:* This type of security is centered around Subscriber Identity Module (SIM) cards, commonly found in mobile devices. SIM cards play a crucial role in authenticating users and safeguarding their communications on cellular networks. This form of security usually entails the use of SIM cards to authenticate users, encrypt data transfers (Pandey, B. K. et al., 2022), and safeguard against unauthorized entry to mobile networks.
- *Software-controlled networking* enables the dynamic configuration of networks and the immediate adjustment of security policies to address emerging threats and attacks.
- *Network function virtualization* involves the implementation of network functions in software instead of hardware. This approach aids in reducing the vulnerability to hardware-based attacks and streamlines the process of updating and securing network operations.
- *Improved IoT protection:* 5G networks now include enhanced security protocols designed to strengthen defenses against cyber threats, specifically for IoT devices (Abdulkarim, Y. I. et al., 2024). These measures include implementing mutual authentication processes and secure communication protocols. These security measures are crucial for ensuring the integrity and confidentiality of data transfers between IoT (Vinodhini, V. et al., 2022) devices and the network, protecting against unauthorized entry and potential data leaks.
- *Enhanced privacy protection:* 5G networks provide sophisticated privacy features that conceal user credentials during data transmission (Pandey, B. K. et al., 2023). This additional level of anonymity presents a greater challenge for potential attackers who aim to monitor or identify users involved in activities such as "user tracking" or "user profiling."
- *Improving network control:* With the integration of segmented networking and software-driven network management approaches, 5G networks have greatly improved their ability to offer precise control over network resources and security policies. This development allows for a more precise and detailed management of network operations.

Although 5G networks bring significant security advantages over previous mobile network generations, they also bring new security challenges. These challenges include the increased use of edge computing and the need to implement security measures across various devices. To tackle these concerns and guarantee the safety of 5G networks, continuous efforts from telecommunication companies, device manufacturers, and other stakeholders will be necessary. This will require ongoing innovation, collaboration, and a strong focus on security (Ali et al., 2021).

ML INTEGRATION WITH 6G NETWORKS

ML technology is anticipated to have a significant impact on 6G networks by enhancing network performance, optimizing resource allocation, and enabling the introduction of innovative services and applications. Here are a few possible uses of ML technology in 6G networks:

- *Network Optimization:* ML algorithms have the ability to enhance network efficiency and minimize latency by evaluating network traffic and optimizing network resources such as capacity and energy. In addition, ML has the ability to predict network congestion in order to prevent network slowdowns and make real-time adjustments to network assets.
- *Anomaly Detection:* Machine learning algorithms can identify abnormal data flow patterns in data communication, which may indicate security breaches or other problems. This allows for a quick response to security breaches and network irregularities by detecting real-time deviations.
- *Proactive Maintenance:* ML algorithms can accurately forecast maintenance needs for network infrastructure components like base stations and antennas. By analyzing sensor data (Jayapoorani, S. et al., 2023) and other relevant information, these procedures can detect patterns that may indicate possible problems. This allows for proactive maintenance and reduces the amount of time equipment is out of service. This approach improves network efficiency by minimizing disruptions and ensuring prompt repairs.
- *Smart Edge Processing:* Configuring ML algorithms on edge devices allows for immediate data manipulation and interpretation right at the network's edge. This method minimizes the necessity of transmitting large amounts of data across long distances, resulting in enhanced network efficiency and decreased latency.
- *Intelligent Network Management:* ML algorithms can analyze network performance data to identify areas for improvement and streamline network management processes. ML has the ability to predict potential network issues and suggest proactive maintenance measures. In addition, machine learning can improve network routing to reduce congestion and improve overall efficiency.
- Machine learning algorithms are essential for enhancing network security as they quickly detect and address security threats. ML has the ability to identify abnormal devices or traffic patterns in a network and automatically trigger security protocols as a result.

In the realm of network security, machine learning techniques have the potential to greatly enhance the capabilities of 6G networks. They can also facilitate the development of new services and applications while simultaneously improving network performance (An et al., 2022). However, the successful integration of ML models into 6G networks will necessitate substantial investments in artificial intelligence and machine learning infrastructure. Furthermore, (Qiao et al. 2020), (Z. Zhang et al., 2019), and (Elfatih et al.,2021) emphasize the need for meticulous consideration of data privacy and security concerns. In order to ensure seamless connectivity and smooth functioning of the 6G ecosystem, it is crucial for network operators, device manufacturers, and software developers to work closely together. This collaboration is necessary for the successful integration of machine learning into 6G networks (Tomkos et al.,2020).

CHLLENGES AND FUTURE RESEARCH DIRECTIONS

- *Optimized Resource Allocation with Contextual Intelligence:* The current allocation techniques for random access (RA) suffer from congestion problems and excessive signaling overhead, which hampers their ability to support mass machine-type communication (mMTC) (Alhashimi et al., n.d.). In order to ensure seamless connectivity for 6G wireless networks (Devasenapathy, D. et al., 2024), it is crucial to implement context-aware smart resource allocation. Nevertheless, the allocation of resources presents difficulties for emerging 6G wireless applications because of strict response time requirements and the demand for seamless user experiences. Exploring a potential research direction could involve merging the uplink grant allocation method developed by 3GPP with machine learning techniques. This integration could be highly applicable in modern Intelligent Transport Systems (ITS) (Sahani, K. et al., 2023) to guarantee prompt and reliable Internet of Vehicles (IoV) applications while also meeting strict Quality of Experience (QoE) criteria (Alhashimi et al., n.d.). In addition, machine learning models can effectively forecast traffic patterns in vehicle-to-everything (V2X) communications, reducing random allocation behavior. This can result in a decrease in signaling overhead and the prevention of collisions, ultimately allowing IoV devices to function with low energy usage.

- *ML for Environmentally Sustainable Communication:* Despite recent research progress in AI-based services during the 6G era, there is still a requirement to implement these advancements into real-world applications (Kim & Ben-Othman, 2022; Zhu et al., 2022). Machine learning techniques have a wide range of applications in renewable energy sources such as solar, wind, tide, and other predictable but uncontrollable energy sources. In addition, machine learning techniques can also be advantageous for partially controllable energy sources such as RF energy. These methods allow for the analysis of the relationship between unpredictable yet predictable energy harvesting technologies and future power generation (Feng et al., 2021; Ryden et al., 2023). ML-driven solutions are becoming more and more efficient in analyzing the intricate relationships between network traces and future transmission policies. This trend stands out as terrestrial network transmission policies change over time (Ren et al., 2022).

- *Privacy-Focused Aerial Base Station:* The research community has shown great interest in the rise of massive machine-type communication (mMTC) devices in 6G networks. We dedicate this focus to addressing the challenges surrounding users' privacy and data security. Additionally, in areas lacking a communication network, unmanned aerial vehicles serve as aerial base stations, providing communication channels. These vehicles serve across different domains, including space, air, land, and sea, to offer 6G services. In addition, UAVs improve network coverage and connectivity (Mishra et al., 2023) (Damigos et al., 2023).

- *Fog Computing:* In the realm of modern wireless applications within 6G networks, there is an increasing need for diverse computing capabilities among devices. Traditional federated learning approaches face several challenges in this regard. Therefore, there is a strong focus on the significance of device-to-device intelligence in fog learning (Alibraheemi et al., 2023). Fog computing is a cutting-edge model that leverages the architecture of fog computing to execute ML tasks (Bani-Bakr et al., 2022). FogL expands the scope of federated learning to accommodate a wide range of computing devices within the fog network (Al Maruf et al., 2022). It enables intelligent model training by allowing device-to-device communications across various network layers.

- *Mobile Edge Learning:* The incorporation of edge computing capabilities in 6G networks enables the processing and analysis of sensory data near its origin, resulting in a substantial decrease in transmitted data volume (Abdelrahim, 2023). Mobile edge learning still encounters several technical challenges in its real-world implementation. Usually, ML training tasks involve complex computations. In addition, mobile edge learning encounters difficulties in heterogeneous environments, commonly referred to as the 'straggler's dilemma'. However, a distributed edge learning system might be able to help with optimizing computations in privacy-conscious and fast 6G apps (Du et al., 2020).

- *Extended Reality in Healthcare (Tareke, S. A. et al., 2022):* The Internet of Medical Things (IoMT) is projected to gain substantial value with the introduction of the 6G communication network (Hong et al., 2022). In addition, the improved security features will allow users to safely share sensitive information without worrying about it being intercepted or tampered with. Moreover, the utilization of Extended Reality (XR) in Teleoperation has demonstrated its potential to improve operational efficiency in intricate medical care situations (Chukhno et al., 2022). The rise of multisensory extended reality (XR) robots brings forth a novel form of traffic that requires specific Quality of Service (QoS) requirements. It is anticipated that 6G network connectivity will be accessible in the coming years to meet these demands. Investigating the potential of machine learning in enhancing cellular-connected virtual reality and unmanned aerial vehicle networks for remote surgery applications using extended reality (XR) is an intriguing area for future research and development.

- *Internet of Senses (IoS):* In the ever-evolving landscape of 6G technology, there is a growing need for advanced use cases that seamlessly connect the physical and digital worlds (Wei et al., 2023). For example, the Internet of Senses (IoS) has the potential to expand our sensory experiences beyond the limitations of our physical bodies. It will be possible to provide affordable and dependable solutions for these use cases. 6G networks will revolutionize communication by enabling immersive experiences within the Internet of Senses (IoS). With full telepresence capabilities, geographical distance will no longer be a barrier to effective communication (Demirhan & Alkhateeb, 2023). Advanced personalized immersive devices powered by AI will enable precise interaction with the human body, opening up new experiences and actions that were previously out of reach. This will greatly enhance human communication. In addition, 6G networks will enable new forms of communication with robust authentication and authorization measures.

CONCLUSION

In conclusion, the exploration of machine learning (ML) solutions for anomaly detection in 6G communication systems presents a promising approach to enhancing the security, reliability, and performance of next-generation networks. Through analysis and experimentation with diverse datasets and algorithms, this study has illuminated the potential of ML techniques in effectively identifying and mitigating anomalies in 6G networks. By using things like packet size, protocol types, network topology, and device behaviors, ML models have shown they can find strange patterns that could mean security breaches, performance issues, or system failures. Furthermore, the scalability and adaptability of ML algorithms make them well-suited for the dynamic and intricate nature of 6G networks, where numerous devices, services, and applications interact in real-time. Using advanced methods like deep learning, reinforcement learning, and ensemble methods makes anomaly detection systems even more accurate and reliable, letting you respond quickly to new threats and problems. However, challenges persist in deploying ML solutions for anomaly detection in 6G networks, including the necessity for large-scale, representative datasets specific to 6G environments and concerns regarding data privacy, model interpretability, and computational resource limitations. Ongoing research efforts are crucial for developing innovative algorithms capable of adapting to evolving network architectures, security threats, and quality of service requirements inherent to 6G communication systems. In summary, this paper underscores the significance of leveraging ML models to tackle the intricacies of anomaly detection in 6G networks. By advancing our understanding and capabilities in this field and harnessing advancements in technologies such as AI and 6G networks, we can pave the way for communication systems that are more resilient, efficient, and secure, thus meeting the demands of the future digital era.

REFERENCES

Abdelrahim, E. M. (2022, December 2). A Novel Data Offloading with Deep Learning Enabled Cyberattack Detection Model for Edge Computing in 6G Networks. *AI-Enabled 6G Networks and Applications*, 17–33. 10.1002/9781119812722.ch2

Abdulkarim, Y. I., Awl, H. N., Muhammadsharif, F. F., Saeed, S. R., Sidiq, K. R., Khasraw, S. S., Dong, J., Pandey, B. K., & Pandey, D. (2024). Metamaterial-based sensors loaded corona-shaped resonator for COVID-19 detection by using microwave techniques. *Plasmonics*, 19(2), 595–610. 10.1007/s11468-023-02007-4

Al Maruf, M., Singh, A., Azim, A., & Auluck, N. (2022, November 1). Faster Fog Computing Based Over-the-Air Vehicular Updates: A Transfer Learning Approach. *IEEE Transactions on Services Computing*, 15(6), 3245–3259. 10.1109/TSC.2021.3099897

Alhashimi, H. F., Hindia, M. N., Dimyati, K., Hanafi, E. B., Safie, N., Qamar, F., Azrin, K., & Nguyen, Q. N. (2023, January 28). A Survey on Resource Management for 6G Heterogeneous Networks: Current Research, Future Trends, and Challenges. *Electronics (Basel)*, 12(3), 647. 10.3390/electronics12030647

Ali, E. S., Hasan, M. K., Hassan, R., Saeed, R. A., Hassan, M. B., Islam, S., Nafi, N. S., & Bevinakoppa, S. (2021, March 12). Machine Learning Technologies for Secure Vehicular Communication in Internet of Vehicles: Recent Advances and Applications. *Security and Communication Networks*, 2021, 1–23. 10.1155/2021/8868355

Alibraheemi, A. M. H., Hindia, M. N., Dimyati, K., Izam, T. F. T. M. N., Yahaya, J., Qamar, F., & Abdullah, Z. H. (2023). A Survey of Resource Management in D2D Communication for B5G Networks. *IEEE Access : Practical Innovations, Open Solutions*, 11, 7892–7923. 10.1109/ACCESS.2023.3238799

Alwis, C. D., Kalla, A., Pham, Q. V., Kumar, P., Dev, K., Hwang, W. J., & Liyanage, M. (2021). Survey on 6G Frontiers: Trends, Applications, Requirements, Technologies and Future Research. *IEEE Open Journal of the Communications Society*, 2, 836–886. 10.1109/OJCOMS.2021.3071496

Anand, R., Khan, B., Nassa, V. K., Pandey, D., Dhabliya, D., Pandey, B. K., & Dadheech, P. (2023). Hybrid convolutional neural network (CNN) for kennedy space center hyperspectral image. *Aerospace Systems*, 6(1), 71–78. 10.1007/s42401-022-00168-4

Anand, R., Lakshmi, S. V., Pandey, D., & Pandey, B. K. (2024). An enhanced ResNet-50 deep learning model for arrhythmia detection using electrocardiogram biomedical indicators. *Evolving Systems*, 15(1), 83–97. 10.1007/s12530-023-09559-0

Banafaa, M., Shayea, I., Din, J., Hadri Azmi, M., Alashbi, A., Ibrahim Daradkeh, Y., & Alhammadi, A. (2023, February). 6G Mobile Communication Technology: Requirements, Targets, Applications, Challenges, Advantages, and Opportunities. *Alexandria Engineering Journal*, 64, 245–274. 10.1016/j.aej.2022.08.017

Bani-Bakr, A., Hindia, M. N., Dimyati, K., Zawawi, Z. B., & Tengku Mohmed Noor Izam, T. F. (2022). Caching and Multicasting for Fog Radio Access Networks. *IEEE Access : Practical Innovations, Open Solutions*, 10, 1823–1838. 10.1109/ACCESS.2021.3137148

Bessant, Y. A., Jency, J. G., Sagayam, K. M., Jone, A. A. A., Pandey, D., & Pandey, B. K. (2023). Improved parallel matrix multiplication using Strassen and Urdhvatiryagbhyam method. *CCF Transactions on High Performance Computing*, 5(2), 102–115. 10.1007/s42514-023-00149-9

Bilen, T., Canberk, B., Sharma, V., Fahim, M., & Duong, T. Q. (2022, May 13). AI-Driven Aeronautical Ad Hoc Networks for 6G Wireless: Challenges, Opportunities, and the Road Ahead. *Sensors (Basel)*, 22(10), 3731. 10.3390/s2210373135632140

. Bruntha, P. M., Dhanasekar, S., Hepsiba, D., Sagayam, K. M., Neebha, T. M., Pandey, D., & Pandey, B. K. (2023). Application of switching median filter with L 2 norm-based auto-tuning function for removing random valued impulse noise. *Aerospace systems, 6*(1), 53-59.

Chukhno, O., Galinina, O., Andreev, S., Molinaro, A., & Iera, A. (2022, December). Interplay of User Behavior, Communication, and Computing in Immersive Reality 6G Applications. *IEEE Communications Magazine*, 60(12), 28–34. 10.1109/MCOM.009.2200238

Damigos, G., Lindgren, T., & Nikolakopoulos, G. (2023). Toward 5G Edge Computing for Enabling Autonomous Aerial Vehicles. *IEEE Access : Practical Innovations, Open Solutions*, 11, 3926–3941. 10.1109/ACCESS.2023.3235067

David, S., Duraipandian, K., Chandrasekaran, D., Pandey, D., Sindhwani, N., & Pandey, B. K. (2023). Impact of blockchain in healthcare system. In *Unleashing the Potentials of blockchain technology for healthcare industries* (pp. 37–57). Academic Press. 10.1016/B978-0-323-99481-1.00004-3

Deepa, R., Anand, R., Pandey, D., Pandey, B. K., & Karki, B. (2022). Comprehensive performance analysis of classifiers in diagnosis of epilepsy. *Mathematical Problems in Engineering*, 2022, 2022. 10.1155/2022/1559312

Demirhan, U., & Alkhateeb, A. (2023, May). Integrated Sensing and Communication for 6G: Ten Key Machine Learning Roles. *IEEE Communications Magazine*, 61(5), 113–119. 10.1109/MCOM.006.2200480

Devasenapathy, D., Madhumathy, P., Umamaheshwari, R., Pandey, B. K., & Pandey, D. (2024). Transmission-efficient grid-based synchronized model for routing in wireless sensor networks using Bayesian compressive sensing. *SN Computer Science*, 5(1), 1–11.

Du, J., Jiang, B., Jiang, C., Shi, Y., & Han, Z. (2023, April). Gradient and Channel Aware Dynamic Scheduling for Over-the-Air Computation in Federated Edge Learning Systems. *IEEE Journal on Selected Areas in Communications*, 41(4), 1035–1050. 10.1109/JSAC.2023.3242727

Du John, H. V., Jose, T., Jone, A. A. A., Sagayam, K. M., Pandey, B. K., & Pandey, D. (2022). Polarization insensitive circular ring resonator based perfect metamaterial absorber design and simulation on a silicon substrate. *Silicon*, 14(14), 9009–9020. 10.1007/s12633-021-01645-9

Elfatih, N. M., Hasan, M. K., Kamal, Z., Gupta, D., Saeed, R. A., Ali, E. S., & Hosain, M. S. (2021, December 30). Internet of vehicle's resource management in 5G networks using AI technologies: Current status and trends. *IET Communications*, 16(5), 400–420. 10.1049/cmu2.12315

Elmoiz Alatabani, L., Sayed Ali, E., Mokhtar, R. A., Saeed, R. A., Alhumyani, H., & Kamrul Hasan, M. (2022, April 15). Deep and Reinforcement Learning Technologies on Internet of Vehicle (IoV) Applications: Current Issues and Future Trends. *Journal of Advanced Transportation*, 2022, 1–16. 10.1155/2022/1947886

Gawali, V. S., & Ranjan, N. M. (2023). Anomaly detection system in 5G networks via deep learning model. *International Journal of Wireless and Mobile Computing*, 24(3/4), 287–302. 10.1504/IJWMC.2023.131319

Giordani, M., Polese, M., Mezzavilla, M., Rangan, S., & Zorzi, M. (2020, March). Toward 6G Networks: Use Cases and Technologies. *IEEE Communications Magazine*, 58(3), 55–61. 10.1109/MCOM.001.1900411

Giordani, M., & Zorzi, M. (2021, March). Non-Terrestrial Networks in the 6G Era: Challenges and Opportunities. *IEEE Network*, 35(2), 244–251. 10.1109/MNET.011.2000493

. Govindaraj, V., Dhanasekar, S., Martinsagayam, K., Pandey, D., Pandey, B. K., & Nassa, V. K. (2023). Low-power test pattern generator using modified LFSR. *Aerospace Systems*, 1-8.

Hong, E. K., Lee, I., Shim, B., Ko, Y. C., Kim, S. H., Pack, S., Lee, K., Kim, S., Kim, J. H., Shin, Y., Kim, Y., & Jung, H. (2022, April). 6G R&D vision: Requirements and candidate technologies. *Journal of Communications and Networks (Seoul)*, 24(2), 232–245. 10.23919/JCN.2022.000015

. I, C. L. (2021, March). AI as an Essential Element of a Green 6G. *IEEE Transactions on Green Communications and Networking*, 5(1), 1–3. 10.1109/TGCN.2021.3057247

Iyyanar, P., Anand, R., Shanthi, T., Nassa, V. K., Pandey, B. K., George, A. S., & Pandey, D. (2023). A real-time smart sewage cleaning UAV assistance system using IoT. In *Handbook of Research on Data-Driven Mathematical Modeling in Smart Cities* (pp. 24–39). IGI Global.

Jayapoorani, S., Pandey, D., Sasirekha, N. S., Anand, R., & Pandey, B. K. (2023). Systolic optimized adaptive filter architecture designs for ECG noise cancellation by Vertex-5. *Aerospace Systems*, 6(1), 163–173. 10.1007/s42401-022-00177-3

Johnson, J. M., & Yadav, A. (2016). Fault Location Estimation in HVDC Transmission Line Using ANN. *Proceedings of First International Conference on Information and Communication Technology for Intelligent Systems*. Springer. 10.1007/978-3-319-30933-0_22

Kanakarajan, N. K., & Muniasamy, K. (2015, October 25). Improving the Accuracy of Intrusion Detection Using GAR-Forest with Feature Selection. *Advances in Intelligent Systems and Computing*, 539–547. 10.1007/978-81-322-2695-6_45

Khalifa, O. O., Roubleh, A., Esgiar, A., Abdelhaq, M., Alsaqour, R., Abdalla, A., Ali, E. S., & Saeed, R. (2022, October 1). An IoT-Platform-Based Deep Learning System for Human Behavior Recognition in Smart City Monitoring Using the Berkeley MHAD Datasets. *Systems*, 10(5), 177. 10.3390/systems10050177

Khan, B., Hasan, A., Pandey, D., Ventayen, R. J. M., Pandey, B. K., & Gowwrii, G. (2021). Fusion of datamining and artificial intelligence in prediction of hazardous road accidents. In *Machine learning and iot for intelligent systems and smart applications* (pp. 201–223). CRC Press. 10.1201/9781003194415-12

Kim, H., & Ben-Othman, J. (2023, January). Eco-Friendly Low Resource Security Surveillance Framework Toward Green AI Digital Twin. *IEEE Communications Letters*, 27(1), 377–380. 10.1109/LCOMM.2022.3218050

Kirubasri, G., Sankar, S., Pandey, D., Pandey, B. K., Nassa, V. K., & Dadheech, P. (2022). Software-defined networking-based Ad hoc networks routing protocols. In *Software defined networking for Ad Hoc networks* (pp. 95–123). Springer International Publishing. 10.1007/978-3-030-91149-2_5

Kirubasri, G., Sankar, S., Pandey, D., Pandey, B. K., Singh, H., & Anand, R. (2021, September). A recent survey on 6G vehicular technology, applications and challenges. In *2021 9th International Conference on Reliability, Infocom Technologies and Optimization (Trends and Future Directions)(ICRITO)* (pp. 1-5). IEEE.

Kumar Pandey, B., Pandey, D., Nassa, V. K., Ahmad, T., Singh, C., George, A. S., & Wakchaure, M. A. (2021). Encryption and steganography-based text extraction in IoT using the EWCTS optimizer. *Imaging Science Journal*, 69(1-4), 38–56. 10.1080/13682199.2022.2146885

Liang, W., Xiao, L., Zhang, K., Tang, M., He, D., & Li, K. C. (2022, August 15). Data Fusion Approach for Collaborative Anomaly Intrusion Detection in Blockchain-Based Systems. *IEEE Internet of Things Journal*, 9(16), 14741–14751. 10.1109/JIOT.2021.3053842

Loey, M., Manogaran, G., Taha, M. H. N., & Khalifa, N. E. M. (2021, January). A hybrid deep transfer learning model with machine learning methods for face mask detection in the era of the COVID-19 pandemic. *Measurement*, 167, 108288. 10.1016/j.measurement.2020.10828832834324

Mhawi, D. N., & Hashem, P. S. H. (2021, December 6). Proposed Hybrid CorrelationFeatureSelection-ForestPanalizedAttribute Approach to advance IDSs. *Karbala International Journal of Modern Science*, 7(4). 10.33640/2405-609X.3166

Mishra, R., Gupta, H. P., Kumar, R., & Dutta, T. (2023, January). Leveraging Augmented Intelligence of Things to Enhance Lifetime of UAV-Enabled Aerial Networks. *IEEE Transactions on Industrial Informatics*, 19(1), 586–593. 10.1109/TII.2022.3197410

Mohamad, M., Selamat, A., Krejcar, O., Crespo, R. G., Herrera-Viedma, E., & Fujita, H. (2021, November 30). Enhancing Big Data Feature Selection Using a Hybrid Correlation-Based Feature Selection. *Electronics (Basel)*, 10(23), 2984. 10.3390/electronics10232984

Mokhtari, S., Abbaspour, A., Yen, K. K., & Sargolzaei, A. (2021, February 8). A Machine Learning Approach for Anomaly Detection in Industrial Control Systems Based on Measurement Data. *Electronics (Basel)*, 10(4), 407. 10.3390/electronics10040407

Moon, S. H., & Kim, Y. H. (2020, August). An improved forecast of precipitation type using correlation-based feature selection and multinomial logistic regression. *Atmospheric Research*, 240, 104928. 10.1016/j.atmosres.2020.104928

Oleiwi, H. W., Mhawi, D. N., & Al-Raweshidy, H. (2022). MLTs-ADCNs: Machine Learning Techniques for Anomaly Detection in Communication Networks. *IEEE Access : Practical Innovations, Open Solutions*, 10, 91006–91017. 10.1109/ACCESS.2022.3201869

Ozpoyraz, B., Dogukan, A. T., Gevez, Y., Altun, U., & Basar, E. (2022). Deep Learning-Aided 6G Wireless Networks: A Comprehensive Survey of Revolutionary PHY Architectures. *IEEE Open Journal of the Communications Society*, 3, 1749–1809. 10.1109/OJCOMS.2022.3210648

Pajouh, H. H., Dastghaibyfard, G., & Hashemi, S. (2015, November 19). Two-tier network anomaly detection model: A machine learning approach. *Journal of Intelligent Information Systems*, 48(1), 61–74. 10.1007/s10844-015-0388-x

Pandey, B. K., & Pandey, D. (2023). Parametric optimization and prediction of enhanced thermoelectric performance in co-doped CaMnO3 using response surface methodology and neural network. *Journal of Materials Science Materials in Electronics*, 34(21), 1589. 10.1007/s10854-023-10954-1

Pandey, B. K., Pandey, D., & Agarwal, A. (2022). Encrypted information transmission by enhanced steganography and image transformation. [IJDAI]. *International Journal of Distributed Artificial Intelligence*, 14(1), 1–14. 10.4018/IJDAI.297110

Pandey, B. K., Pandey, D., Alkhafaji, M. A., Güneşer, M. T., & Şeker, C. (2023). A reliable transmission and extraction of textual information using keyless encryption, steganography, and deep algorithm with cuckoo optimization. In *Micro-Electronics and Telecommunication Engineering: Proceedings of 6th ICMETE 2022* (pp. 629–636). Springer Nature Singapore. 10.1007/978-981-19-9512-5_57

Pandey, B. K., Pandey, D., Gupta, A., Nassa, V. K., Dadheech, P., & George, A. S. (2023). Secret data transmission using advanced morphological component analysis and steganography. In *Role of data-intensive distributed computing systems in designing data solutions* (pp. 21–44). Springer International Publishing. 10.1007/978-3-031-15542-0_2

Pandey, B. K., Pandey, S. K., & Pandey, D. (2011). A survey of bioinformatics applications on parallel architectures. *International Journal of Computer Applications*, 23(4), 21–25. 10.5120/2877-3744

Pandey, D., Nassa, V. K., Jhamb, A., Mahto, D., Pandey, B. K., George, A. H., & Bandyopadhyay, S. K. (2021). An integration of keyless encryption, steganography, and artificial intelligence for the secure transmission of stego images. In *Multidisciplinary approach to modern digital steganography* (pp. 211–234). IGI Global. 10.4018/978-1-7998-7160-6.ch010

Pandey, D., & Pandey, B. K. (2022). An efficient deep neural network with adaptive galactic swarm optimization for complex image text extraction. In *Process mining techniques for pattern recognition* (pp. 121–137). CRC Press. 10.1201/9781003169550-10

Pandey, D., Pandey, B. K., & Wairya, S. (2021). Hybrid deep neural network with adaptive galactic swarm optimization for text extraction from scene images. *Soft Computing*, 25(2), 1563–1580. 10.1007/s00500-020-05245-4

Pandey, J. K., Jain, R., Dilip, R., Kumbhkar, M., Jaiswal, S., Pandey, B. K., & Pandey, D. (2022). Investigating role of iot in the development of smart application for security enhancement. In *IoT Based Smart Applications* (pp. 219–243). Springer International Publishing.

Peng, Y., Guo, Y., Hao, R., & Xu, C. (2024, April). Network traffic prediction with Attention-based Spatial–Temporal Graph Network. *Computer Networks*, 243, 110296. 10.1016/j.comnet.2024.110296

Plastiras, G., Terzi, M., Kyrkou, C., & Theocharides, T. (2018, July). Edge Intelligence: Challenges and Opportunities of Near-Sensor Machine Learning Applications. *2018 IEEE 29th International Conference on Application-Specific Systems, Architectures and Processors (ASAP)*. IEEE. 10.1109/ASAP.2018.8445118

Pramanik, S., Pandey, D., Joardar, S., Niranjanamurthy, M., Pandey, B. K., & Kaur, J. (2023, October). An overview of IoT privacy and security in smart cities. In *AIP Conference Proceedings* (*Vol. 2495*, No. 1). AIP Publishing 10.1063/5.0123511

Qiao, X., Huang, Y., Dustdar, S., & Chen, J. (2020, July 1). 6G Vision: An AI-Driven Decentralized Network and Service Architecture. *IEEE Internet Computing*, 24(4), 33–40. 10.1109/MIC.2020.2987738

Rajagopal, S., Kundapur, P. P., & Hareesha, K. S. (2020, January 24). A Stacking Ensemble for Network Intrusion Detection Using Heterogeneous Datasets. *Security and Communication Networks*, 2020, 1–9. 10.1155/2020/4586875

Ren, Y., Xie, R., Yu, F. R., Huang, T., & Liu, Y. (2022, September). Green Intelligence Networking for Connected and Autonomous Vehicles in Smart Cities. *IEEE Transactions on Green Communications and Networking*, 6(3), 1591–1603. 10.1109/TGCN.2022.3148293

Revathi, T. K., Sathiyabhama, B., Sankar, S., Pandey, D., Pandey, B. K., & Dadeech, P. (2022). An intelligent model for coronary heart disease diagnosis. In *Networking Technologies in Smart Healthcare* (pp. 309–327). CRC Press. 10.1201/9781003239888-15

Rydén, H., Farhadi, H., Palaios, A., Hévizi, L., Sandberg, D., & Kvernvik, T. (2023, October). Next Generation Mobile Networks' Enablers: Machine Learning-Assisted Mobility, Traffic, and Radio Channel Prediction. *IEEE Communications Magazine*, 61(10), 94–98. 10.1109/MCOM.001.2200592

Saad, W., Bennis, M., & Chen, M. (2020, May). A Vision of 6G Wireless Systems: Applications, Trends, Technologies, and Open Research Problems. *IEEE Network*, 34(3), 134–142. 10.1109/MNET.001.1900287

Saeed, , M., Kamrul Hasan, M., Hassan, R., Mokhtar, R., A. Saeed, R., Saeid, E., & Gupta, M. (2022). Preserving Privacy of User Identity Based on Pseudonym Variable in 5G. *Computers, Materials & Continua*, 70(3), 5551–5568. 10.32604/cmc.2022.017338

Saeed, R. A., Omri, M., Abdel-Khalek, S., Ali, E. S., & Alotaibi, M. F. (2022, April 29). Optimal path planning for drones based on swarm intelligence algorithm. *Neural Computing & Applications*, 34(12), 10133–10155. 10.1007/s00521-022-06998-9

Safaldin, M., Otair, M., & Abualigah, L. (2020, June 26). Improved binary gray wolf optimizer and SVM for intrusion detection system in wireless sensor networks. *Journal of Ambient Intelligence and Humanized Computing*, 12(2), 1559–1576. 10.1007/s12652-020-02228-z

Sahani, K., Khadka, S. S., Sahani, S. K., Pandey, B. K., & Pandey, D. (2023). A possible underground roadway for transportation facilities in Kathmandu Valley: A racking deformation of underground rectangular struct es. *Engineering Reports*, 12821. 10.1002/eng2.12821

Saxena, A., Agarwal, A., Pandey, B. K., & Pandey, D. (2024). Examination of the Criticality of Customer Segmentation Using Unsupervised Learning Methods. *Circular Economy and Sustainability*, 1–14. 10.1007/s43615-023-00336-4

Sayed Ali Ahmed, E., Mohammed, Z. T., Bakri Hassan, M., & Saeed, R. A. (2021). Algorithms Optimization for Intelligent IoV Applications. *Handbook of Research on Innovations and Applications of AI, IoT, and Cognitive Technologies*, 1–25. 10.4018/978-1-7998-6870-5.ch001

Sharma, M., Pandey, D., Khosla, D., Goyal, S., Pandey, B. K., & Gupta, A. K. (2022). Design of a GaN-based Flip Chip Light Emitting Diode (FC-LED) with au Bumps & Thermal Analysis with different sizes and adhesive materials for performance considerations. *Silicon*, 14(12), 7109–7120. 10.1007/s12633-021-01457-x

Sharma, M., Pandey, D., Palta, P., & Pandey, B. K. (2022). Design and power dissipation consideration of PFAL CMOS V/S conventional CMOS based 2: 1 multiplexer and full adder. *Silicon*, 14(8), 4401–4410. 10.1007/s12633-021-01221-1

Tareke, S. A., Lelisho, M. E., Hassen, S. S., Seid, A. A., Jemal, S. S., Teshale, B. M., & Pandey, B. K. (2022). The prevalence and predictors of depressive, anxiety, and stress symptoms among Tepi town residents during the COVID-19 pandemic lockdown in Ethiopia. *Journal of Racial and Ethnic Health Disparities*, 1–13.35028903

Tomkos, I., Klonidis, D., Pikasis, E., & Theodoridis, S. (2020, January 1). Toward the 6G Network Era: Opportunities and Challenges. *IT Professional*, 22(1), 34–38. 10.1109/MITP.2019.2963491

Tripathi, R. P., Sharma, M., Gupta, A. K., Pandey, D., Pandey, B. K., Shahul, A., & George, A. H. (2023). Timely prediction of diabetes by means of machine learning practices. *Augmented Human Research*, 8(1), 1. 10.1007/s41133-023-00062-4

Vinodhini, V., Kumar, M. S., Sankar, S., Pandey, D., Pandey, B. K., & Nassa, V. K. (2022). IoT-based early forest fire detection using MLP and AROC method. *International Journal of Global Warming*, 27(1), 55–70. 10.1504/IJGW.2022.122794

Viswanathan, H., & Mogensen, P. E. (2020). Communications in the 6G Era. *IEEE Access : Practical Innovations, Open Solutions*, 8, 57063–57074. 10.1109/ACCESS.2020.2981745

Wei, Z., Qu, H., Wang, Y., Yuan, X., Wu, H., Du, Y., Han, K., Zhang, N., & Feng, Z. (2023, July 1). Integrated Sensing and Communication Signals Toward 5G-A and 6G: A Survey. *IEEE Internet of Things Journal*, 10(13), 11068–11092. 10.1109/JIOT.2023.3235618

Zhang, S., & Zhu, D. (2020, December). Towards artificial intelligence enabled 6G: State of the art, challenges, and opportunities. *Computer Networks*, 183, 107556. 10.1016/j.comnet.2020.107556

Zhang, Z., Xiao, Y., Ma, Z., Xiao, M., Ding, Z., Lei, X., Karagiannidis, G. K., & Fan, P. (2019, September). 6G Wireless Networks: Vision, Requirements, Architecture, and Key Technologies. *IEEE Vehicular Technology Magazine*, 14(3), 28–41. 10.1109/MVT.2019.2921208

Zhu, S., Ota, K., & Dong, M. (2022, March). Green AI for IIoT: Energy Efficient Intelligent Edge Computing for Industrial Internet of Things. *IEEE Transactions on Green Communications and Networking*, 6(1), 79–88. 10.1109/TGCN.2021.3100622

Ziegler, V., Viswanathan, H., Flinck, H., Hoffmann, M., Raisanen, V., & Hatonen, K. (2020). 6G Architecture to Connect the Worlds. *IEEE Access : Practical Innovations, Open Solutions*, 8, 173508–173520. 10.1109/ACCESS.2020.3025032

Chapter 15
Advancements and Challenges in System on Chip Antennas (SoC) for 6G Communication

Manvinder Sharma
https://orcid.org/0000-0001-9158-0466
Chitkara University Institute of Engineering and Technology, Chitkara University, Punjab, India

Rajneesh Talwar
https://orcid.org/0000-0002-2109-8858
Chitkara University Institute of Engineering and Technology, Chitkara University, Punjab, India

Satyajit Anand
Chitkara University Institute of Engineering and Technology, Chitkara University, Punjab, India

Jyoti Bhola
Chitkara University Institute of Engineering and Technology, Chitkara University, Punjab, India

Digvijay Pandey
https://orcid.org/0000-0003-0353-174X
Dr. APJ Abdul Kalam, India

ABSTRACT

The upcoming 6G wireless networks aims to provide substantial improvements over existing 5G networks. These include improvements in data rate, latency, reliability, and connectivity. To enable these goals, significant advancements in all aspects of wireless communication systems are required. The major factor of communication hardware is antenna design. System on chip (SoC) antennas can be directly integrated with transceiver circuitry. These designs can meet the demanding performance requirements of 6G networks. In this chapter, recent advancements in SoC antenna technology for 6G communication systems are discussed. The key challenges that need to be addressed are also discussed. Literature related to the development of wideband millimeter wave antennas with beamforming capabilities is presented. Research directions related to modeling, design and implementation are identified to guide future work in this area.

DOI: 10.4018/979-8-3693-2931-3.ch015

INTRODUCTION

The constant growth in mobile data traffic and connectivity demands raised the generations of wireless networks from First Generation (1G) to the current 5G networks. Research and development for the next generation 6G networks has already commenced. The succesfful deployment of 6G communication plans to be implemented before 2030 (Dang, S.,et al.,2019). Compared to 5G, 6G aims to deliver up to 1000x higher data rates (1 Tbps), 10-100x lower latency (<100 µs), 10x improvement in reliability and10-100x higher connection density (Zhang, Z.,et al.,2019). The journey from 1G to 5G has not only transformed the way we connect and communicate but has also become an integral part of our socioeconomic fabric.

These extremely ambitious performance targets will require revolutionary changes in all aspects of wireless communication networks. A key hardware component that faces significant innovation needs is the antenna. Innovative antenna design will be critical to fully realize the potential of 6G networks. Figure 1 shows 6G Architecture. The architecture includes Enhanced mobile broadband which provides data rates up to 1 Tbps to individual users. This is through advanced techniques like massive MIMO, ultra dense networks, new modulation schemes and tapping of new frequency bands like terahertz. Other includes free space optical communication, Visible light communication, AI integration. It also incorporates sensing mechanisms for applications like health monitoring, industrial automation and smart cities. It provides heterogeneous access for supporting diverse access mechanisms from existing 4G/5G to WiFi, VLC and device to device connections.

Figure 1. Architecture of 6G communication

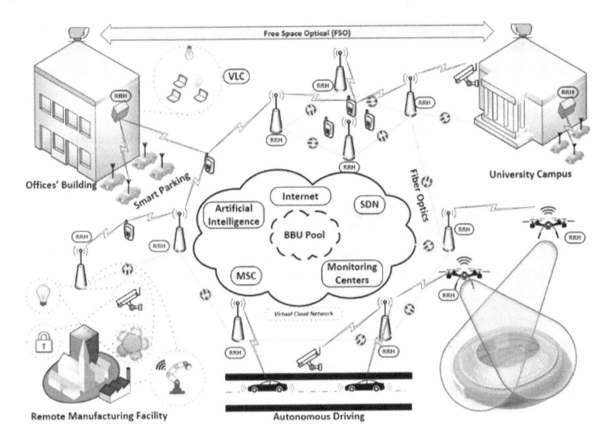

Antennas have evolved tremendously from the simple dipole antennas used in the first generation wireless networks. Current 5G smartphones already use complex antenna arrays with efficient mmWave antennas, tunable antennas and adaptive beamforming capabilities (Hong, W.,et al.,2019). However, the requirements for 6G are far more demanding. The need for wide bandwidths, high integration levels with transceiver circuitry, beam steering capabilities at higher frequencies and form factor constraints necessitate novel antenna approaches customized for 6G (Rajagopal, S.,et al.,2020).

SYSTEM ON CHIP ANTENNAS

The basic idea behind System on chip (SoC) antennas is to fabricate antenna elements on top of radio frequency (RF) integrated circuits that provide transceiver functionality. This integration can be done by patterning metal traces on chip insulation layers to create miniaturized antennas operating in millimeter wave frequencies (Pandey, D.,et al.,2021). Phased array antennas can also be constructed on chip and directly connected to RF transceiver components. Some major advantages of SoC antennas compared to conventional discrete antennas are extreme miniaturization, lower power consumptions and lower cost. However, SoC antennas also pose considerable design challenges related to efficient radiation, impedance matching, interference, bandwidth limitations and modeling complexity [6] Figure 2 shows System on chip Antenna Structure.

Figure 2. SoC antenna structure

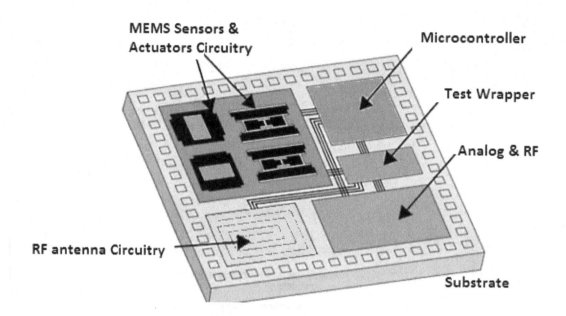

System on chip (SoC) integration of antennas has emerged as a highly promising solution to meet the needs of faster communication (Raja, D.,et al.,2024). Integrating antennas and full transceiver functionalities in a single RF SoC module enables substantial improvements(Checko, A.,et al.,2021) in overall system performance. The progress made in SoC antennas over the past decade has shown their potential to meet the technical requirements of 6G networks related to beamforming, bandwidth, form factor, cost, power efficiency and integration (Mendis, R.,et al.,2019).

RECENT ADVANCEMENTS IN SOC ANTENNAS

Several important advancements have been achieved in SoC antenna design and implementation over the past years. Table 1 shows millimeter wave (mmWave) SoC antennas operating at frequencies above 24 GHz, which are important to tap large spectrum bandwidths needed for 6G Communication data rates. The phased array antennas from 28-140 GHz using either silicon or LTCC fabrication demonstrates mmWave phased arrays are feasible on-chip.

Table 1. Summary of mmWave SoC antennas

Reference	Frequency	Antenna Type	Key Features	Findings	Impact
[9] (Chen, X. & Huang, Y. (2018))	28 GHz	4x4 phased array	Silicon-based fabrication, up to 36 dBi gain	Achieved high gain with integration on silicon	Shows feasibility of mmWave phased arrays on-chip
(Hong, W., et al., (2021))	40 GHz	16-element phased array	Analog and digital beamforming	Combined analog and digital beamforming enables flexibility	Integrated beamforming essential for 6G applications
(Björnson, E.,et al.,2019)	140 GHz	1x4 horn antenna array	LTCC fabrication, 30% bandwidth	Demonstrated mmWave array up to 140 GHz	Expands operable bandwidth for 6G
Zhang, Y. P., & Liu, D. (2009)	300 GHz	Slot antenna	CMOS fabrication, 10 dBi gain	Antenna integrated at 300 GHz	Pushes boundaries of on-chip integration
Li, Y., & Luk, K. M. (2015)	60 GHz	8x8 patch array	Phase shifters, 38% efficiency	Large scale phased array with beam steering	Highlights path for massive MiMO in 6G

Table 2 reviews work on increasing bandwidth for SoC antennas. Wider bandwidths support more frequency bands and higher data rates critical for 6G Communication. Advanced design techniques have achieved fractional bandwidths beyond 15% in compact antennas significantly improving on conventional designs.

Table 2. Summary of work on wide bandwidth

Reference	Antenna Type	Bandwidth	Key Techniques	Findings	Impact
(Nafe, M.,et al.,2021)	Stacked ME dipole	21.7%	Modified ground plane	Wide bandwidth with compact design	Enables support for more frequency bands
(Mat Amin,et al.,2021)	E-shaped patch	14.2%	Dual-band design	Achieved bandwidth >10%	Essential for high data rates in 6G
Huang, K. C. (2018)	SIW slot array	16.8%	Coupled resonators	Broad bandwidth using resonance coupling	Essential for high data rate 6G systems
(Li, Y., et al.,(2023))	Multiband monopole	18.7%	Multiple resonances	Wideband with simple design	Shows possibility for compact wideband antennas

Table 3 compares review of literature on techniques to improve antenna efficiency. It is critical for powerconstrained mobile devices. Efficiencies of 60-70% have been attained by using novel materials like graphene superstrates and high-resistivity silicon substrates

Table 3. Summary of work on efficiency

Reference	Efficiency	Techniques Used	Findings	Impact
(Rappaport, T. S.,et al.,2019)	70%	High resistivity Si substrate	Very high efficiency on silicon	Paves way for efficiency SoC antennas
(Björnson, E.,et al.,2019)	64%	Gap waveguide feeding	High efficiency with novel feeding	New feeding mechanisms can improve efficiency
(Manzillo, M., et al.,(2019))	60%	EBG structures below antenna	Efficiency enhancement using EBG	EBG structures enable highefficiency designs
(Khan, I.,et al.,2023)	57%	Metamaterial superstrate	Efficiency boost with metamaterials	Novel materials can enhance efficiency

These advancements show that SoC antenna technology has reached a level of maturity to meet many of the requirements critical for 6G networks. The capabilities demonstrated serve as a strong foundation for further development tailored to 6G applications.

CHALLENGES

Several key challenges remain to be addressed for SoC antennas to realize their full potential for 6G wireless systems:

Improving Efficiency

Efficiency in mmWave bands has been a significant challenge, with current achievements ranging between 50-70%. However, as we move towards frequencies beyond 100 GHz, maintaining high efficiency becomes even more challenging. To address this novel materials, design techniques and integration approaches are used. The use of metamaterials, (Alkhateeb, A.,et al.,2019) shows promise in manipulating electromagnetic waves at these higher frequencies. This also enhances efficiency by reducing signal losses and optimizing antenna performance. Advancements in semiconductor materials and fabrication processes are also being explored to improve the overall efficiency of the system.

The design considerations for transceiver integration and substrate effects play a crucial role in achieving high efficiency. Researchers are actively exploring innovative approaches to minimize losses associated with these factors. Integration of efficient thermal management systems has also becomes imperative to prevent overheating, especially at higher frequencies. The thermal effects at these frequencies can significantly impact performance.

Expanding Bandwidth

The demand for larger bandwidths exceeding 30%, is a key requirement for supporting new frequency bands and achieving the high data rates for 6G. Current SoC designs face limitations, typically offering only 15-20% bandwidth. Addressing this challenge involves the development of new wideband antenna topologies(Sharma, M.,et al.,2023). The unconventional antenna designs such as fractal antennas and reconfigurable antennas can achieve broader frequency coverage. The exploration of advanced signal processing techniques also contributes to achieving wider bandwidths, allowing for better utilization of the available frequency spectrum (Sengupta, R.,et al.,2020).

Beam Steering at Higher Frequencies

While beam steering has been extensively studied below 40 GHz, there is a critical need for new developments to enable efficient and wide-angle beam steering at mmWave and sub-THz frequencies. The shorter wavelengths at these frequencies pose challenges in achieving precise (Sengupta, R.,et al.,2021) and controllable beam steering. The advanced phased array antenna systems, dynamic beam forming algorithms and innovative materials can overcome these challenges.

To address mm Wave and sub THz beam steering, new materials with unique electromagnetic properties are being researched to enable faster and more accurate beam adjustments. The development of compact and efficient beam forming circuits is crucial for achieving beam steering at these high frequencies. Machine learning algorithms are also being integrated into beam forming systems to adapt and optimize beam steering in real time.

Scaling up Antenna Arrays

Enabling advanced spatial multiplexing and ultra dense networks requires the large scale integration of hundreds of antenna elements with transceivers. This integration poses challenges related to complexity, power consumption(Menon, V., et al., (2022)) and form factor. To scale up antenna arrays, innovations in materials such as flexible and low loss dielectrics are being explored to reduce the complexity and size of the antenna array. Moreover, energy efficient transceiver designs and low power circuitry are essential to mitigate the increased power consumption associated with larger antenna arrays. Advanced packaging technologies, such as 3D integration and system in package (SiP) approaches can be used to achieve the desired form factor without compromising performance.

Modeling and Analysis

Accurate modeling of System on Chip (SoC) antennas is crucial for optimizing their design accounting for substrate effects, transceiver integration, mutual coupling, bandwidth and thermal effects. The Simulation tools (Devasenapathy, D., et al., (2024).) are being refined to incorporate the intricate details of antenna structures, considering the impact of different substrates on performance. Furthermore, the modeling of mutual coupling effects between closely spaced antenna elements is essential for achieving reliable performance in complex antenna arrays(Goyal, S., et al., (2022)). The integration of thermal analysis into the modeling process is becoming increasingly important, The demand for higher data rates and frequencies leads to elevated thermal challenges (Pandey, B. K.,et al.,2011).

Further Miniaturization

Continued reduction of antenna size is imperative for integration into space constrained platforms such as smartphones and wearables. (Checko, A.,et al.,2015) Achieving further miniaturization involves exploring innovative materials and advanced fabrication techniques including 3D printing.

The use of 3D printing technologies allows for the creation of intricate and customized antenna structures that can be seamlessly integrated into various devices(Nagar, P. L., et al., (2023, August).). The exploration of unconventional antenna designs such as fractal antennas and origami spired structures contributes to the ongoing efforts to miniaturize antennas without compromising performance.

Cost Reduction

While System on Chip (SoC) antennas leverage integrated circuit (IC) fabrication processes for cost effectiveness, additional cost benefits can be realized through heterogeneous integration for mass scale production(Huang, Y. & Wu, T. (2018)).

Heterogeneous integration involves combining different materials and technologies on a single substrate, optimizing each component for its specific function. This approach (Taneja, A.,et al.,2023) allows for the incorporation of diverse functionalities, such as signal processing and power management into a single chip. This also reduce the overall production cost.

Overcoming research challenges and advancing SoC antennas through innovative design(Sharma, M.,et al.,2022), advanced materials, fabrication techniques and accurate modeling will facilitate widespread adoption in 6G networks.

CONCLUSION

System on Chip antenna technology has emerged as an essential component to unlock the potential of 6G wireless networks. This helps to deliver substantial improvements in data rate, latency and reliability and connection density. Recent research has demonstrated impressive advancements in SoC antennas in terms of mmWave design, bandwidth, efficiency, integration and beam forming capabilities. However, several important challenges related to performance, design complexity, modeling and fabrication need to be overcome through interdisciplinary research. This includes spanning materials science, antenna engineering and integrated circuit design. Addressing these research problems and advancing SoC antenna technology will enable fully integrated, high performance and cost effective antennas for next generation 6G wireless systems.

REFERENCES

Alkhateeb, A., Alex, S., Varkey, P., Li, Y., Qu, Q., & Tujkovic, D. (2019). Deep learning coordinated beamforming for highlymobile millimeter wave systems. *IEEE Access : Practical Innovations, Open Solutions*, 7, 3732837348.

Björnson, E., Sanguinetti, L., Wymeersch, H., Hoydis, J., & Marzetta, T. L. (2019). Massive MIMO is a reality—What is next?: Five promising research directions for antenna arrays. *Digital Signal Processing*, 94, 320. 10.1016/j.dsp.2019.06.007

Checko, A., Christiansen, H. L., Yan, Y., Scolari, L., De Domenico, G., & Alrabadi, O. N. (2015). Cloud RAN for Mobile Networks—A Technology Overview. *IEEE Communications Surveys and Tutorials*, 17(1), 405426. 10.1109/COMST.2014.2355255

Checko, A., Henriksen, H. L., Sacco, G., Wu, R., Pedersen, K. I., & Madsen, A. B. (2021). Optimizing data rate through antenna miniaturization: From 5G MIMO to 6G Intelligent Conjugate Beamforming. *IEEE Communications Magazine*, 59(8), 5258.

Chen, X., & Huang, Y. (2018). MillimeterWave Centralized Phased Array with Chip Integrated Antenna for 5G Wireless Networks. *IEEE Journal of Solid-State Circuits*, 53(5), 14201432.

Dang, S., Amin, O., Shihada, B., & Alouini, M. S. (2020). What should 6G be? *Nature Electronics*, 3(1), 2019. 10.1038/s41928-019-0355-6

Devasenapathy, D., Madhumathy, P., Umamaheshwari, R., Pandey, B. K., & Pandey, D. (2024). Transmission-efficient grid-based synchronized model for routing in wireless sensor networks using Bayesian compressive sensing. *SN Computer Science*, 5(1), 1–11.

Dutta, V., Sharma, S., Raizada, P., Hosseini-Bandegharaei, A., Kaushal, J., & Singh, P. (2020). Fabrication of visible light active $BiFeO_3/CuS/SiO_2$ Z-scheme photocatalyst for efficient dye degradation. *Materials Letters*, 270, 127693. 10.1016/j.matlet.2020.127693

Goyal, S., Pandey, D., Singh, H., Singh, J., Kakkar, R., & Srinivasu, P. N. (2022). Mathematical modelling for prediction of spread of corona virus and artificial intelligence/machine learning-based technique to detect COVID-19 via smartphone sensors. *International Journal of Modelling Identification and Control*, 41(1-2), 43–52. 10.1504/IJMIC.2022.127096

Hong, W., Ba, D., Winkler, V., & Zhu, Z. (2021). Integrated mmWave Phased Arrays for 5G and Beyond. *Proceedings of the IEEE*, 109(2), 168181.

Hong, W., Jiang, Z. H., Yu, C., & Zhou, Z. (2019). Multibeam Antenna Technologies for 5G Wireless Communications. *IEEE Transactions on Antennas and Propagation*, 67(6), 42814294.

Huang, K. C. (2018). *Millimeter Wave Antennas for Gigabit Wireless Communications: A Practical Guide to Design and Analysis in a System Context*. John Wiley & Sons.

Huang, Y., & Wu, T. (2018). Terahertz Graphene Optical Modulators Enhancement using DoubleLayer Graphene. *IEEE Journal of Selected Topics in Quantum Electronics*, 24(6), 17.

Khan, I., Zhang, K., Ali, L., & Wu, Q. (2023). Enhanced Quad-Port MIMO Antenna Isolation With Metamaterial Superstrate. *IEEE Antennas and Wireless Propagation Letters*.

Kumar, G., & Kumar, R. (2019). A survey on planar ultra-wideband antennas with band notch characteristics: Principle, design, and applications. *AEÜ. International Journal of Electronics and Communications*, 109, 76–98. 10.1016/j.aeue.2019.07.004

Li, Y., & Luk, K. M. (2015). 60-GHz substrate integrated waveguide fed cavity-backed aperture-coupled microstrip patch antenna arrays. *IEEE Transactions on Antennas and Propagation*, 63(3), 1075–1085. 10.1109/TAP.2015.2390228

Li, Y., Wang, L., Ge, L., & Wang, J. (2023). A Dual-Band Magneto-Electric Monopole Antenna Array Using Mode-Composite Waveguides. *IEEE Transactions on Antennas and Propagation*.

Manzillo, M., Occhiuzzi, C., & Marrocco, G. (2019). Modeling, Design and Experimentation of Wearable RFID Sensor Tag. *IEEE Transactions on Antennas and Propagation*, 67(3), 19001908.

Mat Amin, M. K., Soh, P. J., Kumar, S., & Ali, M. T. (2021). Integration Methods of Phased Array Antennas in Wireless Transceivers: A Review of Implementation Challenges and Solutions. *IEEE Access : Practical Innovations, Open Solutions*, 9, 1191711937.

Mendis, R., Randall, J. M., Higgins, M. D., & Kancleris, J. (2019). A 6G Uplink: Intelligent Surfaces Meet THz Communications. *IEEE Vehicular Technology Magazine*, 14(3), 92100.

Menon, V., Pandey, D., Khosla, D., Kaur, M., Vashishtha, H. K., George, A. S., & Pandey, B. K. (2022). A Study on COVID–19, Its Origin, Phenomenon, Variants, and IoT-Based Framework to Detect the Presence of Coronavirus. In *IoT Based Smart Applications* (pp. 1–13). Springer International Publishing.

Nafe, M., Alibakhshikenari, M., & See, C. H., AbdAlhameed, R., Limiti, E., Klemm, M. & Grani, F. (2021). A Novel Wideband Stacked Modified MagnetoElectric Dipole Antenna for 5G Mobile Handset Applications. *IEEE Access : Practical Innovations, Open Solutions*, 9, 7536275372.

Nagar, P. L., Bajpai, S., & Pandey, D. (2023, August). Design and Analysis of U-Slot Microstrip Patch Antenna for ISM Band Applications. In *International Conference on Mobile Radio Communications & 5G Networks* (pp. 439-451). Singapore: Springer Nature Singapore.

Pandey, B. K., Pandey, S. K., & Pandey, D. (2011). A survey of bioinformatics applications on parallel architectures. *International Journal of Computer Applications*, 23(4), 21–25. 10.5120/2877-3744

Pandey, D., Pandey, B. K., & Wairya, S. (2021). Hybrid deep neural network with adaptive galactic swarm optimization for text extraction from scene images. *Soft Computing*, 25(2), 1563–1580. 10.1007/s00500-020-05245-4

Raja, D., Kumar, D. R., Santhiyakumari, N., Kumarganesh, S., Sagayam, K. M., Thiyaneswaran, B., Pandey, B. K., & Pandey, D. (2024). A compact dual-feed wide-band slotted antenna for future wireless applications. *Analog Integrated Circuits and Signal Processing*, 118(2), 1–15. 10.1007/s10470-023-02233-0

Rajagopal, S., AbuSurra, S., & Pi, Z. (2020). Antenna array design for multiGbps mmWave communication. *IEEE Wireless Communications*, 27(2), 7379.

Rappaport, T. S., Xing, Y., Kanhere, O., Ju, S., Madanayake, A., Mandal, S., & Zhao, T. (2019). Wireless communications and applications above 100 GHz: Opportunities and challenges for 6G and beyond. *IEEE Access : Practical Innovations, Open Solutions*, 7, 7872978757. 10.1109/ACCESS.2019.2921522

Sengupta, R., Sengupta, D., Kamra, A. K., & Pandey, D. (2020). Artificial Intelligence and Quantum Computing for a Smarter Wireless Network. *Artificial Intelligence*, 7(19), 2020.

Sengupta, R., Sengupta, D., Pandey, D., Pandey, B. K., Nassa, V. K., & Dadeech, P. (2021). A Systematic review of 5G opportunities, architecture and challenges. Future Trends in 5G and 6G, 247-269.

Sharma, M., Gupta, A. K., Singh, J., Mittal, R., Singh, H., & Pandey, D. (2022, December). Effects of slot shape in performance of SIW based Leaky Wave Antenna. In *2022 International Conference on Computational Modelling, Simulation and Optimization (ICCMSO)* (pp. 268-273). IEEE. 10.1109/ICCMSO58359.2022.00060

Sharma, M., Saripalli, S. R., Gupta, A. K., Palta, P., & Pandey, D. (2023). Image Processing-Based Method of Evaluation of Stress from Grain Structures of Through Silicon Via (TSV). *International Journal of Image and Graphics*, 2550008. 10.1142/S0219467825500081

Sharma, M., Singh, H., Gupta, A. K., & Khosla, D. (2023). Target identification and control model of autopilot for passive homing missiles. *Multimedia Tools and Applications*, 83(20), 1–30. 10.1007/s11042-023-17804-6

Taneja, A., & Saluja, N. (2023). A transmit antenna selection based energy-harvesting mimo cooperative communication system. *Journal of the Institution of Electronics and Telecommunication Engineers*, 69(1), 368–377. 10.1080/03772063.2020.1822217

Zhang, Y. P., & Liu, D. (2009). Antenna-on-chip and antenna-in-package solutions to highly integrated millimeter-wave devices for wireless communications. *IEEE Transactions on Antennas and Propagation*, 57(10), 2830–2841. 10.1109/TAP.2009.2029295

Zhang, Z., Xiao, Y., Ma, Z., Xiao, M., Ding, Z., Lei, X., Karagiannidis, G. K., & Fan, P. (2019). 6G wireless networks: Vision, requirements, architecture andkey technologies. *IEEE Vehicular Technology Magazine*, 14(3), 2841. 10.1109/MVT.2019.2921208

Chapter 16
Teaching Tomorrow's Security:
A Curriculum for 6G Communication Challenges

Ila Dixit
Department of Management, Maharaja Agrasen International College, Raipur, India

T. Rajesh Kumar
Saveetha School of Engineering, Saveetha Institute of Medical and Technical Sciences, Chennai, India

Anslin Jegu
https://orcid.org/0000-0002-0495-2813
Department of English, Panimalar Engineering College, Poonamallee, India

Neha Munjal
Department of Physics, Lovely Professional University, India

Ashok Kumar Digal
PG Department of Education, Rama Devi Women's University, Vidya Vihar Bhubaneswar, India

Baby Shamini P.
R.M.K. Engineering College, India

ABSTRACT

The shift from 5G to 6G communication landscapes ushers in a paradigm shift that is characterized by technological advancements that have never been seen before and increased security imperatives. This chapter delves into the complex world of 6G security, shedding light on the myriad of dangers and difficulties that these networks are confronted with. Alongside the implementation of zero trust architecture, it highlights the critical importance of confidentiality, integrity, and availability (also known as the CIA Triad) by conducting an exhaustive investigation of threat actors, security principles, and encryption techniques. In order to strengthen the security of 6G transactions and data, advanced encryption techniques such as post-quantum cryptography and blockchain technology have emerged as essential tools.

INTRODUCTION

The way we interact with technology is about to change dramatically as 6G communication technology promises unheard-of improvements over 5G. Anticipated to reach full functionality by 2030, 6G signifies a noteworthy advancement in the field of telecommunications (Devasenapathy, D. et al., 2024) technology, seeking to tackle and alleviate numerous compromises and constraints found in 5G

DOI: 10.4018/979-8-3693-2931-3.ch016

networks. The key performance indicators (KPIs) for 6G, according to the International Telecommunication Union Radio communication Sector (ITU-R), highlight the technology's lofty objectives. These KPIs include ultra-low latency of 0.1 milliseconds and a peak data rate of 1 Terabyte per second (Tbps), which is an astounding 1000 times faster than 5G. In addition, 6G networks are expected to support up to 100 semiconductor (Du John, H. V. et al., 2024) devices per cubic meter of space, have a battery life of 20 years, handle 10,000 times the traffic of 5G networks, and have ten times more energy efficiency. In addition, the network aims to sustain an outage rate of no more than 1 in 1 million connections, guaranteeing outstanding dependability. 6G promises unmatched accuracy in location-based services, with 10 centimeters of indoor and 1 meter of outdoor positioning capabilities. Delivering extraordinarily high data rates per device while guaranteeing dependable connectivity and worldwide coverage is at the heart of 6G's goals. 6G aims to push the limits of network performance and efficiency by drastically lowering network latency and introducing innovations like battery-free Internet of Things (IoT) devices (Pramanik, S. et al., 2023). The incorporation of device intelligence augments the potential of 6G networks, permitting them to adjust in real time to fluctuating environmental circumstances and user demands. By taking on these obstacles head-on, 6G hopes to open up new opportunities for a variety of sectors, including manufacturing, entertainment, and healthcare (Singh, S. et al., 2023) in addition to transportation (Sahani, K. et al., 2023). 6G is positioned to spur innovation and enable the upcoming generation of connected devices and services because of its revolutionary potential.

The Imperative of Security

The potential impact and revolutionary nature of 6G, the next generation of communication technology (Raja, D. et al., 2024), make security an absolute must in this age. Despite the fact that 6G networks offer to transform connectivity (Pandey, B. K. et al., 2023a) with unmatched speed, reliability, and efficiency, they also bring a plethora of complicated security issues that need to be tackled head-on. The exponential increase in the number of connected devices and the sheer volume of data (JayaLakshmi, G. et al., 2024) transmitted across networks is one of the main reasons for the heightened focus on security in the 6G era. The attack surface for bad actors grows substantially as the number of Internet of Things (IoT) (Iyyanar, P. et al., 2023) devices increases and more industries join the digital ecosystem (Swapna, H. R. et al., 2023). Cyber dangers, such as data (Sharma, S. et al., 2023) breaches, ransomware, distributed denial-of-service (DDoS) assaults, and supply chain vulnerabilities, find more entry points due to this expanded attack surface. Cybercriminals and state-sponsored actors find 6G networks attractive targets because of the critical infrastructure and essential services they support, including healthcare (KVM, S. et al., 2024), transportation, energy, and finance. Numerous sectors, including commerce, government, and public safety, might be adversely affected by the compromise or interruption of these networks. Robust security measures and protocols are required to address the new attack vectors and vulnerabilities introduced by these technologies, despite the fact that they offer tremendous opportunities for efficiency and innovation. The increasing importance of privacy and data protection is another critical component of the 6G security imperative. There is a greater potential for personal and sensitive information to be leaked due to the enormous volumes of data created and sent through 6G networks. In order to keep users' trust and protect people's privacy, it is crucial to make sure data is available, secure, and confidential. To overcome these obstacles, the 6G ecosystem's many participants including researchers, industry leaders (George, W. K. et al., 2024), regulators, and standards organizations must work together in tight coordination to create thorough security protocols, frameworks, and best practices.

Security audits, vulnerability disclosure programs, threat intelligence sharing, and regular assessments of network readiness and resilience should all be part of this partnership. A proactive and multi-layered approach to cybersecurity is critical, especially in light of the 6G era's security imperatives. We can create a digital infrastructure (George, W. K. et al., 2023) that is both secure and resilient enough to support innovation while protecting users' privacy and security if we make security a top priority when designing, deploying, and operating 6G networks (ElMaraghy, H. et al., 2021).

THREATS AND CHALLENGES IN 6G NETWORKS

Threat Actors, Motivations

Any person or organization that threatens cybersecurity is considered a cyber threat actor. The people who carry out cyberattacks are known as threat actors, and they are frequently classified according to a number of criteria, such as the attack's type, purpose, and target industry (Sharma, M. et al., 2024). The landscape of cyber threats is arguably more dynamic than it has ever been, and threat actors are developing greater sophistication. To avoid discovery and take advantage of weaknesses in digital systems and networks, they constantly modify their tactics, techniques, and procedures (TTPs). Organizations can better defend themselves against the harm caused by threat actors by having a better understanding of these actors' intentions and how they exploit vulnerabilities, compromise user identities with elevated privileges, circumvent security measures, destroy or damage data, or alter sensitive information. Threat actors mostly target large organizations in an effort to steal money, obtain sensitive information, or disrupt operations and damage their reputation. Threat actors, however, have also made small and medium-sized businesses (SMBs) frequent targets because, due to their relative lack of resources, SMBs' security systems may be less robust than those of larger corporations. Regardless of their size or industry, the majority of organizations in today's threat landscape are probably going to be the target of a threat actor. In fact, compared to 2020, businesses encountered 50% more weekly attempts at cyberattacks in 2021. If the route is left unguarded, threat actors can and will find a direct route to the crown jewels today. Organizations need to take a proactive approach to cybersecurity as cyber threats continue to grow and change. In addition to frequently updating and patching software to fix known vulnerabilities, this entails putting in place strong security measures like firewalls, intrusion detection systems, and endpoint protection solutions.

Figure 1. Common threat actors and motivations

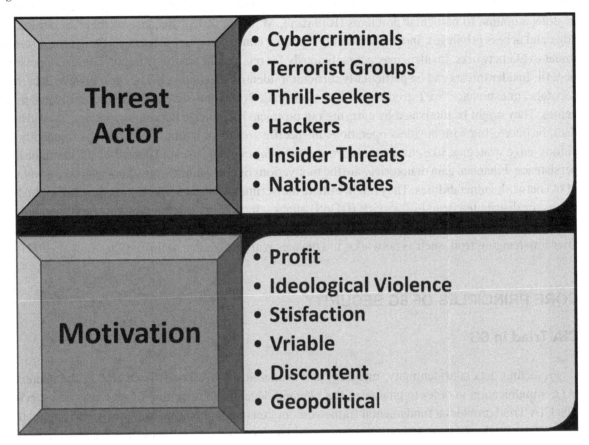

Cybercriminals looking to make money and nation-states vying for geopolitical dominance are just two examples of the wide variety of threat actors in cybersecurity are as shown in the figure 1. Risks come in different forms from insiders, hackers, and terrorist organizations. These actors' motivations range from thrill-seeking and self-satisfaction to profit, ideological violence, and geopolitical influence. For threat mitigation strategies to be effective, it is imperative to comprehend these motivations. The interaction of these elements highlights the complexity of cybersecurity environments, calling for flexible and all-encompassing defense strategies. Organizations should also set up incident response plans to lessen the impact of successful cyberattacks and invest in employee training and awareness programs to help identify and mitigate potential threats. Organizations can reduce the risk posed by cyber threat actors and safeguard their important assets and data by being watchful and well-prepared. The main incentive for cybercriminals is still money. They might target 6G networks in order to carry out ransomware attacks, steal confidential information, or commit other crimes like financial fraud or identity theft. For extortion or ideological reasons, cybercriminals may also attempt to interfere with infrastructure or services. For military, political, or economic gain, nation-states may use cyber espionage to obtain intelligence or pilfer intellectual property. Cyberattacks by state-sponsored actors are another possibility if they want to influence geopolitical developments, destroy vital infrastructure, or create chaos. Cyber operations may also be used by some states to monitor their population or quell dissent. They could be driven by a desire to promote political reform, environmental preservation, or human rights. Hacktivists might target 6G networks in an effort to interfere with normal operations, divulge private information, or draw attention to particular problems (Rüßmann, M. et al., 2015). Because of their insider knowledge and access privileges, insiders including workers, contractors, or business partners pose a serious threat to 6G networks. Insiders may act maliciously for retaliation, ideological motivations, or personal benefit. Insider threats can be particularly difficult to identify and counter. They can involve data theft, sabotage, or espionage. APT groups are often well-funded or nation-state-affiliated sophisticated threat actors. They might be motivated by carrying out protracted espionage operations, stealing confidential data, or interfering with business operations in specific sectors or nations. APT groups frequently use cutting-edge strategies, like custom malware and zero-day exploits, to avoid detection and continue their persistence. Financial gain or notoriety are the motivations of opportunistic attackers who take advantage of 6G network vulnerabilities. They might initiate indiscriminate attacks, like botnet infections, phishing scams, or distributed denial-of-service (DDoS) attacks, by taking advantage of known vulnerabilities, configuration errors, or inadequate security controls. Attackers who seek opportunities frequently go after low-hanging fruit, such as networks and devices with inadequate security (Xu, X. et al., 2021).

CORE PRINCIPLES OF 6G SECURITY

CIA Triad in 6G

Protecting data confidentiality, integrity, and availability (CIA Triad) is critical in the context of 6G communication in order to guarantee the safe and dependable operation of networks and services. The CIA Triad provides a fundamental framework for creating and putting into practice strong security measures that take these important ideas into consideration. The first pillar of the CIA Triad, confidentiality, is concerned with preventing unwanted access to or disclosure of sensitive information. Ensuring confidentiality in 6G networks entails putting in place a number of security measures and protocols to stop unauthorized users from accessing private information. This could entail giving sensitive document handlers specific training, acquainting them with risk factors, and teaching them effective risk mitigation techniques. In order to maintain confidentiality in 6G networks, multi-factor authentication (MFA), strong passwords, and encryption techniques are frequently employed. Requiring account numbers or routing numbers, for instance, helps ensure that only authorized users can access financial information during online banking transactions, protecting confidentiality. Another crucial instrument for maintaining secrecy is encryption, which renders data unintelligible into ciphertext and prevents unauthorized parties from accessing it. Additional security measures like user IDs, passwords, and two-factor authentication (two-factor) guard against unwanted access to private data on social media sites and other internet services. Extra precautions may be taken to improve confidentiality for extremely sensitive documents, such as keeping data on air-gapped computers, disconnected storage devices, or only in hard copy. Integrity, is concerned with guaranteeing the dependability and correctness of data at every stage of its existence. Using safeguards like version control, user access controls, and file permissions to stop unwanted additions or deletions is part of maintaining data integrity. Checksums and cryptographic checksums are used by organizations to confirm the integrity of data and identify any unauthorized changes. In order to further preserve data integrity in the event of data corruption or loss, backups and redundancies are essential for restoring data to its original state. Digital signatures also function as powerful non-repudiation tools since they offer unquestionable proof of actions like document signings, message transfers, and logins. Availability, is concerned with making sure that information and services are available and functional when needed. Organizations must maintain hardware reliability, make repairs on time, and keep operating systems updated to avoid disruptions in order to guarantee availability in 6G networks. In order to lessen the impact of hardware failures and avoid service interruptions, high availability clusters, redundancy, and failover mechanisms are used. In the face of unforeseen events like power (Govindaraj, V. et al., 2023) outages, natural disasters, or cyberattacks, quick and flexible disaster recovery plans which include backups kept in geographically remote locations and extra security measures like firewalls and proxy servers—help reduce downtime and guarantee continuous service availability.

Zero Trust Architecture in 6G

Traditional security models are being challenged by Zero Trust Architecture (ZTA), which eliminates the concept of implicit trust within a network. This represents a paradigm shift in the field of cybersecurity. In light of the fact that the forthcoming sixth generation (6G) network is anticipated to be more open and heterogeneous than its predecessors, traditional security architectures that are based on perimeter defense are rendered insufficient. A software-defined Zero Trust Architecture that is tailored for 6G networks emerges as a promising solution to construct a security framework that is both dynamic and scalable. This is in response to the evolving challenges that are being faced. Instead of relying on perimeter defenses to differentiate between trusted internal networks and untrusted external networks, Zero Trust Architecture operates on the principle of "never trust, always verify." This is in contrast to traditional security models, which rely on it. When this occurs, it indicates that each and every access request, irrespective of its origin or location, is subject to verification and validation prior to being granted access to network resources. In the context of 6G networks, where the lines between internal and external networks become increasingly blurry, ZTA provide an approach to access control that is both more robust and more adaptable. In order to guarantee secure access control, ZTA for 6G networks makes it possible for control domains to work together in an adaptive manner by utilizing the principles of software-defined networking (SDN). A real-time evaluation of access requests is made possible by this dynamic approach, which also makes it possible to implement granular access policies that are tailored to specific network contexts and user identities. As a consequence of this, ZTA is able to effectively prevent a variety of malicious access behaviors that are typically detected in 6G networks. These behaviors include distributed denial of service (DDoS) attacks, the propagation of malware, and zero-day exploits. Some of the most important aspects of ZTA's design for 6G networks are the implementation of identity-centric access controls, continuous monitoring and analytics, and the automation of security policies and enforcement mechanisms. ZTA improves visibility and control over network traffic by shifting the focus from network perimeters to individual users and devices. This results in a reduction in the attack surface and a reduction in the impact of security breaches to a minimum. The simulation demonstrated that ZTA for 6G networks is both effective and robust in mitigating cybersecurity threats. ZTA enables organizations to establish an elastic and scalable security regime that can evolve in tandem with the evolving threat landscape. This is accomplished by dynamically adapting to changing network conditions and access patterns based on the ZTA.

ADVANCED ENCRYPTION TECHNIQUES

Post-Quantum Cryptography in 6G

Post-Quantum Cryptography (PQC) is a cutting-edge cybersecurity technique that provides a strong barrier against the impending threat that quantum computing poses to traditional cryptographic algorithms. The integration of PQC plays a crucial role in protecting the confidentiality and integrity of data transmitted over 6G networks as the sixth generation (6G) of communication technology (Gupta, A. K. et al., 2023) draws near. Because quantum computing makes use of quantum algorithms that can quickly solve difficult mathematical problems (Sahani, S. K. et al., 2024) like discrete logarithms and integer factorization which are the foundations of popular encryption schemes like RSA and ECC it poses a serious threat to conventional cryptographic systems. With the development of quantum computers, these cryptographic algorithms could become outdated and jeopardize the security of sensitive data transferred over 6G networks. On the other hand, PQC algorithms provide long-term security guarantees against quantum threats since they are made to withstand attacks from both classical and quantum computers. These algorithms make use of mathematical puzzles like lattice-based, hash-based, code-based, and multivariate polynomial cryptography mathematical problems (Bessant, Y. A. et al., 2023) that are thought to be challenging to solve even for quantum computers. PQC is essential to 6G security because it safeguards data integrity and confidentiality over a range of network layers and communication channels (Shiroishi, Y. et al., 2018). Organizations can guarantee that sensitive data is protected against current and upcoming cryptographic threats by incorporating PQC algorithms into encryption protocols and cryptographic primitives used in 6G networks. Early integration into 6G networks is a strategic imperative as the need for reliable PQC solutions grows as quantum computers become more potent and widely available. It can also improve the security of cutting-edge applications and technologies that are anticipated to proliferate in the 6G ecosystem, including smart cities, driverless cars, and Internet of Things (IoT) devices. While reducing the possibility of potential cryptographic vulnerabilities, PQC offers a robust foundation for secure communication and data exchange, enabling the realization of creative and networked 6G-enabled services. All things considered, Post-Quantum Cryptography is extremely promising for enhancing the security posture of 6G networks and guaranteeing the availability, confidentiality, and integrity of data transferred over these cutting-edge communication infrastructures. Organizations can proactively mitigate the changing threat landscape and protect digital communication going forward from the disruptive potential of quantum computing by embracing PQC as a key component of their 6G security strategies (Ferreira, C. M., & Serpa, S., 2018).

Blockchain for 6G Security

The integration of blockchain (David, S. et al., 2023) technology into sixth-generation (6G) communication networks holds immense potential to improve security, trust, and transparency in transactions and data exchanges. Blockchain technology has emerged as a groundbreaking solution for securing data and transactions in various industries. Blockchain is a distributed, decentralized ledger technology that records transactions over a network of nodes in a safe, irreversible manner. Every transaction is cryptographically connected to every other transaction, creating a chain of blocks that cannot be changed backward without the network's members' agreement (Makridakis, S., 2017). Because of its intrinsic immutability and transparency, blockchain is a great option for protecting data and transactions in 6G networks. Blockchain technology can be used in the context of 6G to improve security in a number of important areas:

- Tamper-Proof Transactions: Blockchain uses consensus techniques and cryptographic hashing to make sure that transactions in 6G networks are secure. Every transaction is safely stored on the blockchain, creating an unchangeable series of blocks that can only be added or removed with the agreement of all network users. Because blockchain transactions are tamper-proof, there is less chance of fraud, data manipulation, and unauthorized access, which improves the integrity and dependability of smart contracts, financial transactions, and digital interactions in 6G networks.

- Data Authenticity and Integrity: A reliable method for guaranteeing the authenticity and integrity of data transferred over 6G networks is offered by blockchain technology. A distributed ledger and cryptographic signatures on every transaction allow blockchain to create an unchangeable record of data exchanges. This increases the trust and dependability of data transactions in 6G networks by preventing unwanted changes or tampering. Furthermore, users can confirm the provenance and history of data thanks to blockchain's transparency, which improves data authenticity and integrity even more.

- Decentralized Identity Management: Blockchain enables decentralized identity management systems in 6G networks, which allow users and devices to have safe methods for authorization and authentication. A distinct cryptographic identifier is given to every user or device, and this identifier is recorded on the blockchain along with pertinent identity attributes. By giving users complete control over their digital identities, this decentralized approach to identity management lowers the risk of identity theft, impersonation, and unauthorized access. Additionally, by lowering the exposure of sensitive personal data and decreasing dependency on centralized identity providers, blockchain-based identity solutions provide improved privacy protection.

- Supply Chain Traceability: By offering transparent and unchangeable records of transactions and the movements of goods, blockchain improves supply chain traceability in 6G networks. Every transaction on the blockchain, such as the transfer of goods from one party to another, is documented with all pertinent metadata, such as timestamps, location information, and product specifications. This makes it possible for interested parties to trace the origin of goods, confirm the legitimacy of products, and guarantee legal compliance. Blockchain-based supply chain solutions also make recalls and audits quicker and more effective, which lowers the possibility of fraud, fake goods, and supply chain interruptions.

- Decentralized Applications (Dapps): On top of 6G networks, decentralized applications (Dapps) can be developed thanks to blockchain platforms like Ethereum. The security, transparency, and programmability of blockchain technology are utilised by these Dapps to facilitate inventive use cases, including decentralised finance (DeFi), decentralised autonomous organisations (DAOs), and decentralised marketplaces. Dapps improve the security and resilience of 6G networks by removing single points of failure and decentralizing control, enabling users to communicate with digital services directly and without the use of middlemen.

AUTHENTICATION AND ACCESS CONTROL IN 6G

Biometric and MFA

The use of multiple authentication factors, also known as multi-factor authentication (MFA), is a complementary method to biometric authentication that provides an additional layer of security. The following are some examples of authentication factors: something you know (for example, a password), something you have (for example, a smartphone or a physical token), and something you are (for example, biometric data). The use of multi-factor authentication (MFA) significantly reduces the risk of unauthorized access that could be caused by stolen or weak passwords. This is accomplished by requiring users to provide multiple forms of verification. Multi-factor authentication, which is also referred to as two-factor authentication (2FA), is growing in popularity as a standard security practice across a variety of industries. Multi-factor authentication (MFA) can incorporate two or more authentication factors, which further enhances security. Two-factor authentication (2FA) uses two additional authentication checks. The utilization of this multi-layered approach helps to reduce the likelihood of unauthorized access and safeguards sensitive information against the possibility of it being compromised. In the highly interconnected digital landscape of today, the implementation of multi-factor authentication (MFA) is absolutely necessary for companies and organizations that deal with sensitive data, financial transactions, or confidential user information. Organizations are able to establish robust security measures that not only protect sensitive data but also enhance user trust and confidence in digital platforms when they combine multi-factor authentication (MFA) with biometric authentication. As the state of technology continues to advance, multi-factor authentication (MFA) and biometric authentication will become increasingly important in the process of ensuring the safety and authenticity of digital transactions and user identities (Van Eck, N., & Waltman, L., 2010). Types of Biometrics used for Authentication and Verification are as shown in the figure 2.

Figure 2. Types of biometrics used for authentication and verification

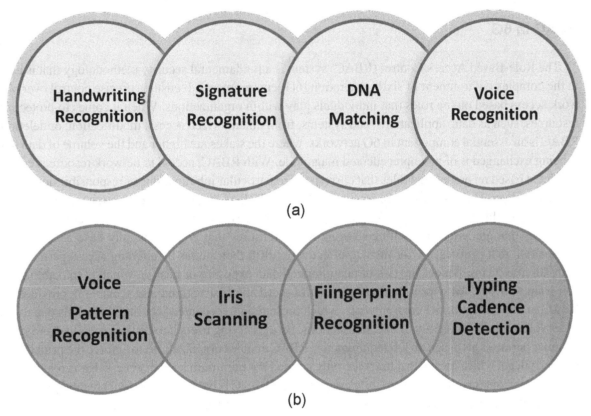

The revolutionary advancements in digital security that biometric authentication and multi-factor authentication (MFA) represent are enhancing both the level of protection and the level of user convenience across a wide range of applications. In the past, multifactor authentication frequently required users to go through the laborious process of entering a One-Time Password (OTP), which could be both time-consuming and inconvenient for the users. On the other hand, the landscape of digital security has undergone a significant transformation as a result of the introduction of biometric multifactor authentication. For the purpose of verifying user identities, biometric authentication makes use of biometric data such as face, voice, fingerprint, palm, and eye recognition. This type of authentication is exemplified by solutions offered by industry leaders such as Biocube. With this approach, not only is increased security ensured, but the process of onboarding, logging in, and gaining access is also simplified dramatically. The utilization of passive liveness technology and hardware-agnostic solutions eliminates the requirement for specialized hardware, which in turn makes biometric authentication accessible and scalable for users of digital identities in the modern era. The ability of biometric multifactor authentication to provide authentic user login and access in a matter of seconds without compromising security is one of the most significant advantages afforded by this authentication method. Using a variety of biometric modalities, such as face, voice, palm, fingerprint, and eye recognition, biometric authentication provides multimodal and contactless accessibility, even in situations where the user is located in a remote location. This not only improves the safety of the system, but it also improves the user experience by removing the need for specialized hardware (Mourtzis, D. et al., 2022).

RBAC in 6G

The Role-Based Access Control (RBAC) system is a fundamental security methodology that is used in the complex environment of sixth-generation (6G) environments. It ensures precise control over network access based on the roles that individuals play within organizations. When it comes to protecting resources, such as data, applications, and systems, from unauthorized access, modification, or deletion, RBAC is an essential component in 6G networks, where the stakes are higher and the volume of data that is being exchanged is of an unprecedented magnitude. With RBAC, access to network resources can be restricted based on predefined roles that correspond to particular job functions or responsibilities within an organization. This is the fundamental principle behind RBAC's operation. The actions and operations that a user is able to carry out within the network are specifically determined by the permissions and privileges that are associated with each role. RBAC ensures that employees only have access to the information that is necessary for them to effectively fulfill their duties by aligning access permissions with job roles. This reduces the risk of unauthorized data exposure or misuse, which is something that you want to avoid. An approach to access control that is both structured and scalable is provided by RBAC in the context of 6G environments, which present unprecedented security challenges due to the sheer volume and complexity of data transactions. By classifying users into distinct roles and assigning granular permissions based on job requirements, RBAC enables organizations to enforce the principle of least privilege, which means that users are only granted the minimum level of access that is required for them to carry out their tasks at hand. Implementation of RBAC is based on three fundamental principles: role assignment, role authorization, and role management of roles. In the process of role assignment, roles are defined and then assigned to particular users or groups on the basis of the job functions or responsibilities that they perform. In order to authorize roles, it is necessary to provide each role with the appropriate permissions and privileges, as well as to specify the actions that users in each role are authorized to carry out. Access controls are kept in accordance with the policies and regulations of the organization through the process of role management, which entails the ongoing administration and maintenance of roles and permissions. Especially in light of recent data breaches that were attributed to improper access controls, the significance of RBAC in the field of cybersecurity cannot be overstated. For example, the discovery of a massive data leak in October 2019, in which 4 terabytes of personally identifiable information (PII) belonging to over 1.2 billion individuals were exposed due to lax access controls, highlights the importance of implementing robust access management mechanisms such as response-based access control (RBAC). Putting RBAC into action in 6G environments comes with its own unique set of challenges, particularly in large organizations that have intricate hierarchies and teams that are interconnected. In order to clearly define roles, map permissions, and ensure consistency across a wide variety of systems and applications, careful planning and coordination are required. All things considered, the advantages of RBAC in terms of improving data security, reducing the likelihood of data breaches, and guaranteeing compliance with regulatory requirements far outweigh the difficulties associated with its implementation.

BUILDING SECURE 6G SYSTEMS

SDLC for 6G Apps

With the advent of 6G applications, connectivity and capabilities have advanced significantly in the quickly changing world of technology. But this progress also means that software development processes must now take security very seriously. For 6G applications, the significance of incorporating security into each stage of the software development lifecycle (SDLC) has been highlighted by the digital transformation occurring across all industry sectors. Implementing a secure SDLC approach is essential for fostering software trust while preserving the speed and agility required for market competitiveness, from architecture and design to release and maintenance. Every organization is now essentially a software business due to the blurring of boundaries between businesses and software brought about by the digital transformation. Software must be protected from threats and vulnerabilities whether it is used for internal operations or sold directly to customers (Saxena, A. et al., 2024) in order to maintain a profit. However, a lot of development teams see security as an impediment, a bottleneck that slows down development and extends the time it takes to release new features (Muniandi, B. et al., 2024). Nonetheless, the ramifications of software vulnerabilities are extensive, presenting hazards to both enterprises and clients. As a result, incorporating security into the SDLC is essential to effectively bringing safe, high-quality products to market. For companies creating 6G apps, there are many advantages to implementing a secure SDLC approach. First off, by spotting and addressing security threats early in the development cycle, it strengthens the security posture. This strengthens overall cybersecurity resilience by producing software that is less vulnerable to exploitation by bad actors. Additionally, a secure SDLC approach lowers the costs of security incidents and breaches, which include monetary losses, harm to one's reputation, and legal liabilities. Consumers are more likely to interact with businesses they believe in, which helps to maintain a brand's loyalty and reputation. Apart from these broad benefits, enterprises creating 6G applications can reap particular benefits from a secure SDLC approach. By detecting and resolving security issues early in the process, it reduces the need for rework and bug fixes, facilitating shorter development times and costs. Additionally, by lowering security risks, businesses can improve the general caliber of their software, decreasing the likelihood of errors and boosting efficiency . A secure SDLC approach also boosts employee morale because motivated staff members are more productive when they are aware that security is a top priority. Creating a culture of security within the company encourages team members to take ownership of their work and take accountability for it, which helps to promote a proactive and cooperative approach to software development.

Figure 3. Software/system development life cycle--SDLC

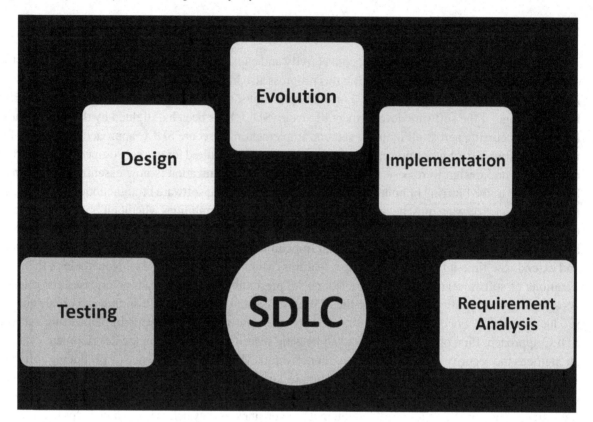

EMERGING THREATS IN 6G

AI Attacks and Defenses

The introduction of 6G networks brings with it opportunities for connectivity and innovation that have never been seen before, but it also has the potential to bring about new challenges in terms of cybersecurity. The methods and level of sophistication of cyber attacks are constantly evolving in tandem with the capabilities of artificial intelligence (AI) (Anand, R. et al., 2024), which are constantly expanding. Attacks that are driven by artificial intelligence present a significant risk in the context of 6G networks. These attacks make use of machine learning (Tripathi, R. P. et al., 2023) algorithms to take advantage of vulnerabilities and intrusion into systems. On the other hand, artificial intelligence (Pandey, B. K. et al., 2024) also provides powerful tools for defense, which enable proactive detection and defense against threats in real time. Attacks in 6G networks that are driven by artificial intelligence can refer to a wide variety of methods, ranging from targeted phishing campaigns to sophisticated malware that is designed to circumvent conventional security measures. The use of social engineering techniques that are powered by artificial intelligence (Pandey, B. K., & Pandey, D., 2023) in order to manipulate users and gain access to sensitive information is one of the primary concerns. Additionally, artificial intelligence algorithms (Anand, R. et al., 2023) can be utilized to analyze patterns of network traffic and locate vulnerabilities that can be exploited to launch large-scale attacks, such as distributed denial-of-service (DDoS) attacks.

It is necessary for organizations that are developing 6G networks to implement defense mechanisms that are driven by artificial intelligence and are able to detect and respond to attacks in real time in order to combat the evolving threats. One strategy involves the implementation of intrusion detection systems (IDS) that are powered by artificial intelligence (Pandey, B. K. et al., 2023b) and continuously monitor network traffic for any suspicious activity. The utilization of machine learning algorithms by these systems allows for the identification of abnormal patterns of behavior that are indicative of potential threats. This enables prompt action to be taken in order to mitigate risks before they become more severe. This can be accomplished by utilizing defensive mechanisms. Through the process of analyzing user behavior and device activity in real time, artificial intelligence algorithms are able to recognize deviations from normal patterns and automatically quarantine or remediate endpoints that have been compromised.

Artificial intelligence has the potential to assume a pivotal role in incident response and threat intelligence, in addition to its proactive threat detection capabilities. Security analytics platforms that are powered by artificial intelligence are able to collect and examine vast amounts of security data from all over the network in order to identify new threats and provide security teams with insights that they can put into action. Because of this, organizations are able to quickly evaluate the severity of a threat, prioritize response efforts, and effectively deploy countermeasures. There is the potential for automated incident response systems to utilize artificial intelligence algorithms in order to analyze the nature and severity of an attack and then automatically deploy countermeasures in order to contain and mitigate the impact of the attack (Friedman, B., & Hendry, D. G., 2019).

GOVERNANCE, RISK MANAGEMENT, AND COMPLIANCE (GRC) IN 6G

Regulatory Compliance for 6G Providers

Regulatory risks are associated with the possibility of unfavorable outcomes stemming from breaking or failing to comply with laws, rules, or industry standards that control a specific industry or activity. Governmental agencies, regulatory bodies, or trade associations create these rules to protect consumers, uphold the integrity of the market, encourage transparency, and guarantee adherence to the law. Regulatory risks may originate from a number of different places, such as modifications to laws, omission of required permits or licenses, violations of data privacy laws, environmental regulations, labor laws, or financial regulations. Failure to adhere to regulatory standards may result in fines, penalties, legal action, harm to one's reputation, revocation of licenses, or even the closure of one's business. Compliance risks are a subset of regulatory risks that pertain to the possible negative outcomes that may arise from an organization's inability to follow internal policies, procedures, or standards that are put in place to ensure compliance with industry best practices and external regulations. By outlining precise standards and guidelines for worker behavior, operational procedures, risk management, and governance processes, these internal policies and procedures aim to reduce regulatory risks. Internal control weaknesses or gaps, insufficient supervision or training, ignorance of legal requirements, or a failure to keep an eye on and enforce policy compliance are all potential sources of compliance risks. Failure to adhere to internal policies and procedures may result in financial losses, reputational harm, operational disruptions, and regulatory penalties. Compliance risks are more strongly linked to internal elements like corporate culture, governance frameworks, and risk management procedures than regulatory risks, which mostly originate from external sources like laws and industry standards. But regulatory risks and compliance risks also present serious obstacles for businesses, necessitating early detection, evaluation, and mitigation of possible effects. Comprehensive approaches to address regulatory and compliance risks should be included in effective risk management strategies. These approaches should include strong compliance programs, frequent audits and assessments, employee education and awareness initiatives, and ongoing reporting and monitoring systems to guarantee continued adherence to internal standards and regulatory requirements. In an increasingly complex and regulated business environment, organizations can strengthen their resilience, safeguard their reputation, and uphold stakeholder trust by taking proactive measures to manage regulatory and compliance risks.

IMPLEMENTING AND TESTING

Security Assessment in Real-World Scenarios

The evaluation of the security posture of systems, networks, applications, or environments is an essential part of security assessment in real-world scenarios. The purpose of this evaluation is to identify vulnerabilities, weaknesses, and potential risks that could be exploited by malicious actors. With the help of these evaluations, the effectiveness of the security controls, policies, and procedures that are currently in place will be evaluated, and appropriate measures will be recommended in order to improve security and reduce risks. When applied to situations that occur in the real world, security assessments can take on a variety of forms, depending on the particular context and the goals of the assessor. Different kinds of security evaluations include the following: Vulnerability Assessment: Using both automated tools and manual techniques, vulnerability assessment entails performing systematic scans of systems, networks, and applications. The purpose of these assessments is to find known vulnerabilities, configuration errors, and lax security measures that an attacker could exploit. Organizations can create efficient remediation plans to take care of the most important problems first by ranking vulnerabilities according to their severity and possible effects. Maintaining a proactive security posture and lowering the probability of successful cyberattacks require regular vulnerability assessments. Penetration testing: Also referred to as ethical hacking, penetration testing entails mimicking actual cyberattacks in order to assess how well a company's security measures are working. Expert security specialists try to take advantage of holes in networks, apps, and systems in order to obtain unwanted access or steal private information. Organizations can strengthen incident response capabilities, validate the efficacy of security controls, and find gaps in their security posture with the aid of penetration testing. Organizations can guarantee thorough coverage and improve their overall security posture by regularly conducting penetration tests with various attack scenarios. The process of locating, examining, and assessing possible dangers, weaknesses, and risks to the resources and operations of an organization is known as risk assessment. This entails estimating the probability and possible effects on crucial business operations of different threats, including cyberattacks, natural disasters, and human error. Risks can be prioritized according to severity and likelihood, which helps organizations allocate resources efficiently to address the biggest threats. Risk assessment is a continuous process that needs to be reviewed and updated frequently in order to keep up with changing threats and the business environment. Security Architecture Review:

This process entails assessing how security controls are designed and implemented in the systems, applications, and infrastructure of an organization. This entails evaluating the segmentation, intrusion detection/prevention systems, encryption methods, access controls, and network architecture. Ensuring that security controls are in line with industry best practices, legal requirements, and the organization's risk management goals is made easier with a comprehensive security architecture review. Organizations can strengthen their overall security posture by implementing corrective measures to address potential weaknesses and gaps in their security architecture. Code review is the process of looking through an application's source code to find coding mistakes, security holes, and poor design. In order to do this, the code must be examined for common security problems like buffer overflows, injection holes, weak authentication procedures, and incorrect error handling. Expert security professionals must manually review code in order to find complex security flaws that automated scanning tools might miss. As part of the software development lifecycle, organizations can enhance the security and dependability of their applications by performing comprehensive code reviews.

CONCLUSION

The future of 6G communication will be very different from the ones that came before it. This is because it will combine new technologies and keep looking for ways to make connections better. The move from 5G to 6G is a natural one, driven by the need for security and new technologies like AI-powered wireless communication, personalized network experiences, and a focus on operations that are machine-centered. The in-depth look at possible problems and threats shows how important security is in the 6G era, calling for strong encryption methods, authentication systems, and access controls. Adopting principles like Zero Trust Architecture and adding advanced encryption methods like Post-Quantum Cryptography and Blockchain Technology also show that 6G networks are dedicated to protecting privacy, security, and availability. People are excited about 6G's capabilities going beyond just faster data speeds. They want it to be able to connect all kinds of devices without any problems, be used for a wide range of things from the Internet of Things (IoT) to augmented reality and virtual reality (AR/VR), and create new experiences like haptic communication and tactile internetworking. In addition, 6G promises to change how people interact with each other, how efficiently things are done, and how we care for the environment by bringing the digital, physical, and social worlds together. The road to 6G shows how constantly changing technological progress is, where new ideas and flexibility are very important. As the industry moves forward, it is important to put security, standardization, and alignment with real-world needs at the top of the list in order to get the most out of 6G networks. With people working together and thinking ahead about how to improve technology, the era of 6G has the potential to completely change how we connect with each other and create a future where our senses are enhanced and the digital and physical worlds are seamlessly blended.

REFERENCES

Anand, R., Khan, B., Nassa, V. K., Pandey, D., Dhabliya, D., Pandey, B. K., & Dadheech, P. (2023). Hybrid convolutional neural network (CNN) for kennedy space center hyperspectral image. *Aerospace Systems*, 6(1), 71–78. 10.1007/s42401-022-00168-4

Anand, R., Lakshmi, S. V., Pandey, D., & Pandey, B. K. (2024). An enhanced ResNet-50 deep learning model for arrhythmia detection using electrocardiogram biomedical indicators. *Evolving Systems*, 15(1), 83–97. 10.1007/s12530-023-09559-0

Bessant, Y. A., Jency, J. G., Sagayam, K. M., Jone, A. A. A., Pandey, D., & Pandey, B. K. (2023). Improved parallel matrix multiplication using Strassen and Urdhvatiryagbhyam method. *CCF Transactions on High Performance Computing*, 5(2), 102–115. 10.1007/s42514-023-00149-9

David, S., Duraipandian, K., Chandrasekaran, D., Pandey, D., Sindhwani, N., & Pandey, B. K. (2023). Impact of blockchain in healthcare system. In *Unleashing the Potentials of blockchain technology for healthcare industries* (pp. 37–57). Academic Press. 10.1016/B978-0-323-99481-1.00004-3

Devasenapathy, D., Madhumathy, P., Umamaheshwari, R., Pandey, B. K., & Pandey, D. (2024). Transmission-efficient grid-based synchronized model for routing in wireless sensor networks using Bayesian compressive sensing. *SN Computer Science*, 5(1), 1–11.

Du John, H. V., Jose, T., Sagayam, K. M., Pandey, B. K., & Pandey, D. (2024). Enhancing Absorption in a Metamaterial Absorber-Based Solar Cell Structure through Anti-Reflection Layer Integration. *Silicon*, 1-11.

ElMaraghy, H., Monostori, L., Schuh, G., & ElMaraghy, W. (2021). Evolution and future of manufacturing systems. *CIRP Annals*, 70(2), 635–658. 10.1016/j.cirp.2021.05.008

Ferreira, C. M., & Serpa, S. (2018). Society 5.0 and social development: Contributions to a discussion. *Management and Organizational Studies*, 5(4), 26–31. 10.5430/mos.v5n4p26

Friedman, B., & Hendry, D. G. (2019). *Value sensitive design: Shaping technology with moral imagination*. Mit Press. 10.7551/mitpress/7585.001.0001

George, W. K., Ekong, M. O., Pandey, D., & Pandey, B. K. (2023). Pedagogy for Implementation of TVET Curriculum for the Digital World. In *Applications of Neuromarketing in the Metaverse* (pp. 117-136). IGI Global. 10.4018/978-1-6684-8150-9.ch009

George, W. K., Silas, E. I., Pandey, D., & Pandey, B. K. (2024). Utilization of Industry 4.0 Technologies in Nigerian Technical and Vocational Education: A Conundrum for Educators. In *Examining the Rapid Advance of Digital Technology in Africa* (pp. 270–293). IGI Global. 10.4018/978-1-6684-9962-7.ch014

. Govindaraj, V., Dhanasekar, S., Martinsagayam, K., Pandey, D., Pandey, B. K., & Nassa, V. K. (2023). Low-power test pattern generator using modified LFSR. *Aerospace Systems*, 1-8.

Gupta, A. K., Sharma, R., Pandey, D., Nassa, V. K., Pandey, B. K., George, A. S., & Dadheech, P. (2023). Performance analysis of eight-channel WDM optical network with different optical amplifiers for industry 4.0. In *Innovation and Competitiveness in Industry 4.0 Based on Intelligent Systems* (pp. 197–212). Springer International Publishing. 10.1007/978-3-031-29775-5_9

Iyyanar, P., Anand, R., Shanthi, T., Nassa, V. K., Pandey, B. K., George, A. S., & Pandey, D. (2023). A real-time smart sewage cleaning UAV assistance system using IoT. In *Handbook of Research on Data-Driven Mathematical Modeling in Smart Cities* (pp. 24–39). IGI Global.

JayaLakshmi. G., Pandey, D., Pandey, B. K., Kaur, P., Mahajan, D. A., & Dari, S. S. (2024). Smart Big Data Collection for Intelligent Supply Chain Improvement. In *AI and Machine Learning Impacts in Intelligent Supply Chain* (pp. 180-195). IGI Global.

KVM, S., Pandey, B. K., & Pandey, D. (2024). Design of Surface Plasmon Resonance (SPR) Sensors for Highly Sensitive Biomolecular Detection in Cancer Diagnostics. *Plasmonics*, 1–13.

Makridakis, S. (2017). The forthcoming Artificial Intelligence (AI) revolution: Its impact on society and firms. *Futures*, 90, 46–60. 10.1016/j.futures.2017.03.006

Mourtzis, D., Angelopoulos, J., & Panopoulos, N. (2022). Digital Manufacturing: The evolution of traditional manufacturing toward an automated and interoperable Smart Manufacturing Ecosystem. In *The digital supply chain* (pp. 27-45). Elsevier.

Muniandi, B., Nassa, V. K., Pandey, D., Pandey, B. K., Dadheech, P., & George, A. S. (2024). Pattern Analysis for Feature Extraction in Complex Images. In *Using Machine Learning to Detect Emotions and Predict Human Psychology* (pp. 145-167). IGI Global. 10.4018/979-8-3693-1910-9.ch007

Pandey, B. K., & Pandey, D. (2023). Parametric optimization and prediction of enhanced thermoelectric performance in co-doped CaMnO3 using response surface methodology and neural network. *Journal of Materials Science Materials in Electronics*, 34(21), 1589. 10.1007/s10854-023-10954-1

Pandey, B. K., Pandey, D., Alkhafaji, M. A., Güneşer, M. T., & Şeker, C. (2023a). A reliable transmission and extraction of textual information using keyless encryption, steganography, and deep algorithm with cuckoo optimization. In *Micro-Electronics and Telecommunication Engineering: Proceedings of 6th ICMETE 2022* (pp. 629–636). Springer Nature Singapore. 10.1007/978-981-19-9512-5_57

Pandey, B. K., Pandey, D., Dadheech, P., Mahajan, D. A., George, A. S., & Hameed, A. S. (2023b). Review on Smart Sewage Cleaning UAV Assistance for Sustainable Development. In *Handbook of Research on Safe Disposal Methods of Municipal Solid Wastes for a Sustainable Environment* (pp. 69–79). IGI Global. 10.4018/978-1-6684-8117-2.ch005

Pandey, B. K., Pandey, D., & Sahani, S. K. (2024). Autopilot control unmanned aerial vehicle system for sewage defect detection using deep learning. *Engineering Reports*, 12852. 10.1002/eng2.12852

Pramanik, S., Pandey, D., Joardar, S., Niranjanamurthy, M., Pandey, B. K., & Kaur, J. (2023, October). An overview of IoT privacy and security in smart cities. In *AIP Conference Proceedings* (*Vol. 2495*, No. 1). AIP Publishing. 10.1063/5.0123511

Raja, D., Kumar, D. R., Santhiyakumari, N., Kumarganesh, S., Sagayam, K. M., Thiyaneswaran, B., Pandey, B. K., & Pandey, D. (2024). A compact dual-feed wide-band slotted antenna for future wireless applications. *Analog Integrated Circuits and Signal Processing*, 118(2), 1–15. 10.1007/s10470-023-02233-0

. Rüßmann, M., Lorenz, M., Gerbert, P., Waldner, M., Justus, J., Engel, P., & Harnisch, M. (2015). Industry 4.0: The future of productivity and growth in manufacturing industries. *Boston consulting group, 9*(1), 54-89.

Sahani, K., Khadka, S. S., Sahani, S. K., Pandey, B. K., & Pandey, D. (2023). A possible underground roadway for transportation facilities in Kathmandu Valley: A racking deformation of underground rectangular structures. *Engineering Reports*, 12821. 10.1002/eng2.12821

Sahani, S. K., Pandey, B. K., & Pandey, D. (2024). *Single-valued Signals, Multi-valued Signals and Fixed-Point of Contractive Signals*. Mathematics Open. 10.1142/S2811007224500020

Saxena, A., Agarwal, A., Pandey, B. K., & Pandey, D. (2024). Examination of the Criticality of Customer Segmentation Using Unsupervised Learning Methods. *Circular Economy and Sustainability*, 1–14. 10.1007/s43615-023-00336-4

Sharma, M., Talwar, R., Pandey, D., Nassa, V. K., Pandey, B. K., & Dadheech, P. (2024). A Review of Dielectric Resonator Antennas (DRA)-Based RFID Technology for Industry 4.0. *Robotics and Automation in Industry*, 4(0), 303–324.

Sharma, S., Pandey, B. K., Pandey, D., Anand, R., Sharma, A., & Saini, S. (2023, March). Character Recognition Technique Implementation for Complicated Deteriorated Scene. In *2023 6th International Conference on Information Systems and Computer Networks (ISCON)* (pp. 1-4). IEEE. 10.1109/IS-CON57294.2023.10112185

Shiroishi, Y., Uchiyama, K., & Suzuki, N. (2018). Society 5.0: For human security and well-being. *Computer*, 51(7), 91–95. 10.1109/MC.2018.3011041

Singh, S., Madaan, G., Kaur, J., Swapna, H. R., Pandey, D., Singh, A., & Pandey, B. K. (2023). *Bibliometric Review on Healthcare Sustainability. Handbook of Research on Safe Disposal Methods of Municipal Solid Wastes for a Sustainable Environment*, (pp. 142-161). IGI Global. 10.4018/978-1-6684-8117-2.ch011

Swapna, H. R., Bigirimana, E., Madaan, G., Hasan, A., Pandey, B. K., & Pandey, D. (2023). Impact of neuromarketing on consumer psychology in digitally connected networks. In *Applications of Neuromarketing in the Metaverse* (pp. 193–205). IGI Global. 10.4018/978-1-6684-8150-9.ch015

Tripathi, R. P., Sharma, M., Gupta, A. K., Pandey, D., Pandey, B. K., Shahul, A., & George, A. H. (2023). Timely prediction of diabetes by means of machine learning practices. *Augmented Human Research*, 8(1), 1. 10.1007/s41133-023-00062-4

. Van Eck, N., & Waltman, L. (2010). Software survey: VOSviewer, a computer program for bibliometric mapping. *scientometrics, 84*(2), 523-538.

Xu, X., Lu, Y., Vogel-Heuser, B., & Wang, L. (2021). Industry 4.0 and Industry 5.0—Inception, conception and perception. *Journal of Manufacturing Systems*, 61, 530–535. 10.1016/j.jmsy.2021.10.006

Compilation of References

. Bruntha, P. M., Dhanasekar, S., Hepsiba, D., Sagayam, K. M., Neebha, T. M., Pandey, D., & Pandey, B. K. (2023). Application of switching median filter with L 2 norm-based auto-tuning function for removing random valued impulse noise. *Aerospace systems, 6*(1), 53-59.

. Bruntha, P. M., Dhanasekar, S., Hepsiba, D., Sagayam, K. M., Neebha, T. M., Pandey, D., & Pandey, B.K. (2023). Application of switching median filter with L 2 norm-based auto-tuning function for removing random valued impulse noise. *Aerospace systems, 6*(1), 53-59.

. David, S., Duraipandian, K., Chandrasekaran, D., Pandey, D., Sindhwani, N., & Pandey, B. K. (2023).Impact of blockchain in healthcare system. In *Unleashing the Potentials of blockchain technology forhealthcare industries* (pp. 37-57). Academic Press..

. Du John, H. V., Jose, T., Sagayam, K. M., Pandey, B. K., & Pandey, D. (2024). Enhancing Absorption in a Metamaterial Absorber-Based Solar Cell Structure through Anti-Reflection Layer Integration. *Silicon*, 1-11.

. Du John, H. V., Moni, D. J., Ponraj, D. N., Sagayam, K. M., Pandey, D., & Pandey, B. K. (2021). Design of Si based nano strip resonator with polarization-insensitive metamaterial (MTM) absorber on a glass substrate. *Silicon*, 1-10.

. Elmeadawy, S., & Shubair, R. M. (2019, November). 6G wireless communications: Future technologies and research challenges. In *2019 international conference on electrical and computing technologies and applications (ICECTA)* (pp. 1-5). IEEE.

. Govindaraj, V., Dhanasekar, S., Martinsagayam, K., Pandey, D., Pandey, B. K., & Nassa, V. K. (2023). Low-power test pattern generator using modified LFSR. *Aerospace Systems*, 1-8.

. Govindaraj, V., Dhanasekar, S., Martinsagayam, K., Pandey, D., Pandey, B. K., & Nassa, V. K. (2023).Low-power test pattern generator using modified LFSR. *Aerospace Systems*, 1-8.

. I, C. L. (2021, March). AI as an Essential Element of a Green 6G. *IEEE Transactions on Green Communications and Networking, 5*(1), 1–3. 10.1109/TGCN.2021.3057247

. Iyyanar, P., Anand, R., Shanthi, T., Nassa, V. K., Pandey, B. K., George, A. S., & Pandey, D. (2023). Areal-time smart sewage cleaning UAV assistance system using IoT. In *Handbook of Research on Data-Driven Mathematical Modeling in Smart Cities* (pp. 24-39). IGI Global.

. Kirubasri, G., Sankar, S., Pandey, D., Pandey, B. K., Nassa, V. K., & Dadheech, P. (2022). Software-defined networking-based Ad hoc networks routing protocols. In *Software defined networking for AdHoc networks* (pp. 95-123). Cham: Springer International Publishing.

. Kirubasri, G., Sankar, S., Pandey, D., Pandey, B. K., Singh, H., and Anand, R. (2021, September). A recentsurvey on 6G vehicular technology, applications and challenges. In *2021 9th InternationalConference on Reliability, Infocom Technologies and Optimization (Trends and FutureDirections)(ICRITO)* (pp. 1-5). IEEE.

. Pandey, D., Nassa, V. K., Jhamb, A., Mahto, D., Pandey, B. K., George, A. H., & Bandyopadhyay, S. K.(2021). An integration of keyless encryption, steganography, and artificial intelligence for the securetransmission of stego images. In *Multidisciplinary approach to modern digital steganography* (pp.211-234). IGI Global.

. Pandey, D., Ogunmola, G. A., Enbeyle, W., Abdullahi, M., Pandey, B. K., & Pramanik, S. (2021b). *COVID-19: A framework for effective delivering of online classes during lockdown.* Human Arenas, 1-15.

. Pandey, D., Ogunmola, G. A., Enbeyle, W., Abdullahi, M., Pandey, B. K., & Pramanik, S. (2021). COVID-19: A framework for effective delivering of online classes during lockdown. *Human Arenas*, 1-15.

. Pandey, J. K., Jain, R., Dilip, R., Kumbhkar, M., Jaiswal, S., Pandey, B. K., & Pandey, D. (2022). Investigating role of iot in the development of smart application for security enhancement. In *IoT Based Smart Applications* (pp. 219-243). Cham: Springer International Publishing.

. Pramanik, S., Pandey, D., Joardar, S., Niranjanamurthy, M., Pandey, B. K., & Kaur, J. (2023, October). An overview of IoT privacy and security in smart cities. In *AIP Conference Proceedings* (*Vol. 2495*, No.1). AIP Publishing .

. Revathi, T. K., Sathiyabhama, B., Sankar, S., Pandey, D., Pandey, B. K., & Dadeech, P. (2022). An intelligent model for coronary heart disease diagnosis. In *Networking Technologies in Smart Healthcare* (pp. 309-327). CRC Press.

. Rüßmann, M., Lorenz, M., Gerbert, P., Waldner, M., Justus, J., Engel, P., & Harnisch, M. (2015). Industry 4.0: The future of productivity and growth in manufacturing industries. *Boston consulting group, 9*(1), 54-89.

. Sengupta, R., Sengupta, D., Pandey, D., Pandey, B. K., Nassa, V. K., & Dadeech, P. (2021). A Systematic review of 5G opportunities, architecture and challenges. *Future Trends in 5G and 6G*, 247-269.

. Sengupta, R., Sengupta, D., Pandey, D., Pandey, B. K., Nassa, V. K., & Dadeech, P. (2021). A Systematic review of 5G opportunities, architecture and challenges. *Future Trends in 5G and 6G*, 247-269. Research Gate.

. Talwar, R., & Sharma, M. (2024). A Comprehensive Review on Artificial Intelligence-Driven Radiomics for Early Cancer Detection and Intelligent Medical Supply Chain. AI and Machine Learning Impacts in Intelligent Supply Chain, 226-254.

. Van Eck, N., & Waltman, L. (2010). Software survey: VOSviewer, a computer program for bibliometric mapping. *scientometrics, 84*(2), 523-538.

Abdelrahim, E. M. (2022, December 2). A Novel Data Offloading with Deep Learning Enabled Cyberattack Detection Model for Edge Computing in 6G Networks. *AI-Enabled 6G Networks and Applications*, 17–33. 10.1002/9781119812722.ch2

Abdulkarim, Y. I., & Awl, H. N., Sharif, F. F., Saeed, S. R., Sidiq, K. R., Khasraw, S. S., & Pandey, D. (2023). Metamaterial-based sensors loaded corona-shaped resonator for COVID-19 detection by using microwave techniques. *Plasmonics*, 1–16.

Adel, A. (2022). Future of Industry 5.0 in society: Human-centric solutions, challenges and prospective research areas. *Journal of Cloud Computing (Heidelberg, Germany)*, 11(1), 40. 10.1186/s13677-022-00314-536101900

Ahmed, N., & Rahman, A. ur, Malik, H., Kaleem, Z., Mumtaz, S., Huang, J., & Rodrigues, J. J. P. C. (2020). A survey on socially-aware device-to-device communications. *IEEE Access : Practical Innovations, Open Solutions*, 8, 14857–14868.

Ai, Y., Peng, M., & Zhang, K. (2018). Edge computing technologies for Internet of Things: A primer. *Digital Communications and Networks*, 4(2), 77–86. 10.1016/j.dcan.2017.07.001

Akpakwu, G. A., Silva, B. J., Hancke, G. P., & Abu-Mahfouz, A. M. (2017). A survey on 5G networks for the Internet of Things: Communication technologies and challenges. *IEEE Access : Practical Innovations, Open Solutions*, 6, 3619–3647. 10.1109/ACCESS.2017.2779844

Akundi, A., Euresti, D., Luna, S., Ankobiah, W., Lopes, A., & Edinbarough, I. (2022). State of Industry 5.0—Analysis and identification of current research trends. *Applied System Innovation*, 5(1), 27. 10.3390/asi5010027

Al Maruf, M., Singh, A., Azim, A., & Auluck, N. (2022, November 1). Faster Fog Computing Based Over-the-Air Vehicular Updates: A Transfer Learning Approach. *IEEE Transactions on Services Computing*, 15(6), 3245–3259. 10.1109/TSC.2021.3099897

Alhashimi, H. F., Hindia, M. N., Dimyati, K., Hanafi, E. B., Safie, N., Qamar, F., Azrin, K., & Nguyen, Q. N. (2023, January 28). A Survey on Resource Management for 6G Heterogeneous Networks: Current Research, Future Trends, and Challenges. *Electronics (Basel)*, 12(3), 647. 10.3390/electronics12030647

Alibraheemi, A. M. H., Hindia, M. N., Dimyati, K., Izam, T. F. T. M. N., Yahaya, J., Qamar, F., & Abdullah, Z. H. (2023). A Survey of Resource Management in D2D Communication for B5G Networks. *IEEE Access : Practical Innovations, Open Solutions*, 11, 7892–7923. 10.1109/ACCESS.2023.3238799

Ali, E. S., Hasan, M. K., Hassan, R., Saeed, R. A., Hassan, M. B., Islam, S., Nafi, N. S., & Bevinakoppa, S. (2021, March 12). Machine Learning Technologies for Secure Vehicular Communication in Internet of Vehicles: Recent Advances and Applications. *Security and Communication Networks*, 2021, 1–23. 10.1155/2021/8868355

Alkhateeb, A., Alex, S., Varkey, P., Li, Y., Qu, Q., & Tujkovic, D. (2019). Deep learning coordinated beamforming for highlymobile millimeter wave systems. *IEEE Access : Practical Innovations, Open Solutions*, 7, 3732837348.

Anand, R., Khan, B., Nassa, V. K., Pandey, D., Dhabliya, D., Pandey, B. K., & Dadheech, P. (2023). Hybrid convolutional neural network (CNN) for kennedy space center hyperspectral image. *Aerospace Systems*, 6(1), 71–78. 10.1007/s42401-022-00168-4

Anand, R., Lakshmi, S. V., Pandey, D., & Pandey, B. K. (2024). An enhanced ResNet-50 deep learning model for arrhythmia detection using electrocardiogram biomedical indicators. *Evolving Systems*, 15(1), 83–97. 10.1007/s12530-023-09559-0

Apriliyanti, M. (2022). Challenges of The Industrial Revolution Era 1.0 to 5.0: University Digital Library In Indoensia. *Library Philosophy and Practice*, 1-17.

Attanasio, B., Mazayev, A., du Plessis, S., & Correia, N. (2021). Cognitive load balancing approach for 6G MEC serving iot mashups. *Mathematics*, 10(1), 101. 10.3390/math10010101

Banafaa, M., Shayea, I., Din, J., Azmi, M. H., Alashbi, A., Daradkeh, Y. I., & Alhammadi, A. (2023). 6G mobile communication technology: Requirements, targets, applications, challenges, advantages, and opportunities. *Alexandria Engineering Journal*, 64, 245–274. 10.1016/j.aej.2022.08.017

Bangerter, B., Talwar, S., Arefi, R., & Stewart, K. (2014). Networks and devices for the 5G era. *IEEE Communications Magazine*, 52(2), 90–96. 10.1109/MCOM.2014.6736748

Bani-Bakr, A., Hindia, M. N., Dimyati, K., Zawawi, Z. B., & Tengku Mohmed Noor Izam, T. F. (2022). Caching and Multicasting for Fog Radio Access Networks. *IEEE Access : Practical Innovations, Open Solutions*, 10, 1823–1838. 10.1109/ACCESS.2021.3137148

Bessant, Y. A., Jency, J. G., Sagayam, K. M., Jone, A. A. A., Pandey, D., & Pandey, B. K. (2023). Improved parallel matrix multiplication using Strassen and Urdhvatiryagbhyam method. *CCF Transactions on High Performance Computing*, 5(2), 102–115. 10.1007/s42514-023-00149-9

Bhattarai, S., Park, J. M. J., Gao, B., Bian, K., & Lehr, W. (2016). An overview of dynamic spectrum sharing: Ongoing initiatives, challenges, and a roadmap for future research. *IEEE Transactions on Cognitive Communications and Networking*, 2(2), 110–128. 10.1109/TCCN.2016.2592921

Bikos, A. N., & Sklavos, N. (2012). LTE/SAE security issues on 4G wireless networks. *IEEE Security and Privacy*, 11(2), 55–62. 10.1109/MSP.2012.136

Bilen, T., Canberk, B., Sharma, V., Fahim, M., & Duong, T. Q. (2022, May 13). AI-Driven Aeronautical Ad Hoc Networks for 6G Wireless: Challenges, Opportunities, and the Road Ahead. *Sensors (Basel)*, 22(10), 3731. 10.3390/s2210373135632140

Bishoyi, P. K., & Misra, S. (2021). Enabling green mobile-edge computing for 5G-based healthcare applications. *IEEE Transactions on Green Communications and Networking*, 5(3), 1623–1631. 10.1109/TGCN.2021.3075903

Björnson, E., Sanguinetti, L., Wymeersch, H., Hoydis, J., & Marzetta, T. L. (2019). Massive MIMO is a reality—What is next?: Five promising research directions for antenna arrays. *Digital Signal Processing*, 94, 320. 10.1016/j.dsp.2019.06.007

Bruntha, P. M., Dhanasekar, S., Hepsiba, D., Sagayam, K. M., Neebha, T. M., Pandey, D., & Pandey, B. K. (2023). Application of switching median filter with L 2 norm-based auto-tuning function for removing random valued impulse noise. *Aerospace systems, 6*(1), 53-59.

Burbank, J. L., Andrusenko, J., Everett, J. S., & Kasch, W. T. (2013). Second-generation (2G) cellular communications. In *Wireless Networking: Understanding Internetworking Challenges* (pp. 250–365). IEEE. 10.1002/9781118590775.ch6

Castelo-Branco, I., Oliveira, T., Simões-Coelho, P., Portugal, J., & Filipe, I. (2022). Measuring the fourth industrial revolution through the Industry 4.0 lens: The relevance of resources, capabilities and the value chain. *Computers in Industry*, 138, 103639. 10.1016/j.compind.2022.103639

Chataut, R., & Akl, R. (2020). Massive MIMO systems for 5G and beyond networks—Overview, recent trends, challenges, and future research direction. *Sensors (Basel)*, 20(10), 2753. 10.3390/s2010275332408531

Checko, A., Christiansen, H. L., Yan, Y., Scolari, L., De Domenico, G., & Alrabadi, O. N. (2015). Cloud RAN for Mobile Networks—A Technology Overview. *IEEE Communications Surveys and Tutorials*, 17(1), 405426. 10.1109/COMST.2014.2355255

Checko, A., Henriksen, H. L., Sacco, G., Wu, R., Pedersen, K. I., & Madsen, A. B. (2021). Optimizing data rate through antenna miniaturization: From 5G MIMO to 6G Intelligent Conjugate Beamforming. *IEEE Communications Magazine*, 59(8), 5258.

Chen, J., Wang, C. X., Zhong, Z., Ng, D. W. K., Hanzo, L., Müller, A., & Zhang, R. (2021, June). Reconfigurable intelligent surface assisted 6G wireless networks: Challenges and opportunities. *IEEE Network*, 35(4), 215–223.

Chen, M., Gündüz, D., Huang, K., Saad, W., Bennis, M., Feljan, A. V., & Poor, H. V. (2021). Distributed learning in wireless networks: Recent progress and future challenges. *IEEE Journal on Selected Areas in Communications*, 39(12), 3579–3605. 10.1109/JSAC.2021.3118346

Chen, X., & Huang, Y. (2018). MillimeterWave Centralized Phased Array with Chip Integrated Antenna for 5G Wireless Networks. *IEEE Journal of Solid-State Circuits*, 53(5), 14201432.

Chen, Y., Weng, Q., Tang, L., Wang, L., Xing, H., & Liu, Q. (2023). Developing an intelligent cloud attention network to support global urban green spaces mapping. *ISPRS Journal of Photogrammetry and Remote Sensing*, 198, 197–209. 10.1016/j.isprsjprs.2023.03.005

Chowdhury, M. Z., Shahjalal, M., Ahmed, S., & Jang, Y. M. (2020). 6G wireless communication systems: Applications, requirements, technologies, challenges, and research directions. *IEEE Open Journal of the Communications Society*, 1, 957–975. 10.1109/OJCOMS.2020.3010270

Chukhno, O., Galinina, O., Andreev, S., Molinaro, A., & Iera, A. (2022, December). Interplay of User Behavior, Communication, and Computing in Immersive Reality 6G Applications. *IEEE Communications Magazine*, 60(12), 28–34. 10.1109/MCOM.009.2200238

Dahlman, E., Parkvall, S., & Skold, J. (2016). *4G, LTE-advanced Pro and the Road to 5G*. Academic Press.

Damigos, G., Lindgren, T., & Nikolakopoulos, G. (2023). Toward 5G Edge Computing for Enabling Autonomous Aerial Vehicles. *IEEE Access : Practical Innovations, Open Solutions*, 11, 3926–3941. 10.1109/ACCESS.2023.3235067

Dang, S., Amin, O., Shihada, B., & Alouini, M. S. (2020). What should 6G be? *Nature Electronics*, 3(1), 2019. 10.1038/s41928-019-0355-6

Darsena, D., Sciancalepore, V., & Trifiletti, D. (2021). On the authentication of reconfigurable intelligent surfaces in the context of 6G systems. *2022 IEEE Wireless Communications and Networking Conference (WCNC)*, (pp. 1-6). IEEE.

David, S., Duraipandian, K., Chandrasekaran, D., Pandey, D., Sindhwani, N., & Pandey, B. K. (2023). Impact of blockchain in healthcare system. In *Unleashing the Potentials of blockchain technology for healthcare industries* (pp. 37–57). Academic Press. 10.1016/B978-0-323-99481-1.00004-3

De Alwis, C., Kalla, A., Pham, Q. V., Kumar, P., Dev, K., Hwang, W. J., & Liyanage, M. (2021). Survey on 6G frontiers: Trends, applications, requirements, technologies and future research. *IEEE Open Journal of the Communications Society*, 2, 836–886. 10.1109/OJCOMS.2021.3071496

Deepa, R., Anand, R., Pandey, D., Pandey, B. K., & Karki, B. (2022). Comprehensive performance analysis of classifiers in diagnosis of epilepsy. *Mathematical Problems in Engineering*, 2022, 2022. 10.1155/2022/1559312

Demirhan, U., & Alkhateeb, A. (2023, May). Integrated Sensing and Communication for 6G: Ten Key Machine Learning Roles. *IEEE Communications Magazine*, 61(5), 113–119. 10.1109/MCOM.006.2200480

Deng, S., Zhao, H., Fang, W., Yin, J., Dustdar, S., & Zomaya, A. Y. (2020). Edge intelligence: The confluence of edge computing and artificial intelligence. *IEEE Internet of Things Journal*, 7(8), 7457–7469. 10.1109/JIOT.2020.2984887

Devasenapathy, D., Madhumathy, P., Umamaheshwari, R., Pandey, B. K., & Pandey, D. (2024). Transmission-efficient grid-based synchronized model for routing in wireless sensor networks using Bayesian compressive sensing. *SN Computer Science*, 5(1), 1–11.

Dhanasekar, S., Martin Sagayam, K., Pandey, B. K., & Pandey, D. (2023). Refractive Index Sensing Using Metamaterial Absorbing Augmentation in Elliptical Graphene Arrays. *Plasmonics*, 1–11. 10.1007/s11468-023-02152-w

Dogra, A., Jha, R. K., & Jain, S. (2020). A survey on beyond 5G network with the advent of 6G: Architecture and emerging technologies. *IEEE Access : Practical Innovations, Open Solutions*, 9, 67512–67547. 10.1109/ACCESS.2020.3031234

Dong, R., She, C., Hardjawana, W., Li, Y., & Vucetic, B. (2019). Deep learning for hybrid 5G services in mobile edge computing systems: Learn from a digital twin. *IEEE Transactions on Wireless Communications*, 18(10), 4692–4707. 10.1109/TWC.2019.2927312

Dong, Y., Liao, F., Pang, T., Su, H., Zhu, J., Hu, X., & Li, J. (2018). Boosting adversarial attacks with momentum. *Proceedings of the IEEE conference on computer vision and pattern recognition*. IEEE.

Du John, H. V., Jose, T., Sagayam, K. M., Pandey, B. K., & Pandey, D. (2024). Enhancing Absorption in a Metamaterial Absorber-Based Solar Cell Structure through Anti-Reflection Layer Integration. *Silicon*, 1-11.

Du John, H. V., Moni, D. J., Ponraj, D. N., Sagayam, K. M., Pandey, D., & Pandey, B. K. (2021). Design of Si based nano strip resonator with polarization-insensitive metamaterial (MTM) absorber on a glass substrate. *Silicon*, 1-10.

Du John, H. V., Ajay, T., Reddy, G. M. K., Ganesh, M. N. S., Hembram, A., Pandey, B. K., & Pandey, D. (2023). Design and simulation of SRR-based tungsten metamaterial absorber for biomedical sensing applications. *Plasmonics*, 18(5), 1903–1912. 10.1007/s11468-023-01910-0

Du John, H. V., Jose, T., Jone, A. A. A., Sagayam, K. M., Pandey, B. K., & Pandey, D. (2022). Polarization insensitive circular ring resonator based perfect metamaterial absorber design and simulation on a silicon substrate. *Silicon*, 14(14), 9009–9020. 10.1007/s12633-021-01645-9

Du John, H. V., Sagayam, K. M., Jose, T., Pandey, D., Pandey, B. K., Kotti, J., & Kaur, P. (2023a). Design simulation and parametric investigation of a metamaterial light absorber with tungsten resonator for solar cell applications using silicon as dielectric layer. *Silicon*, 15(9), 4065–4079. 10.1007/s12633-023-02321-w

Du, J., Jiang, B., Jiang, C., Shi, Y., & Han, Z. (2023, April). Gradient and Channel Aware Dynamic Scheduling for Over-the-Air Computation in Federated Edge Learning Systems. *IEEE Journal on Selected Areas in Communications*, 41(4), 1035–1050. 10.1109/JSAC.2023.3242727

Dutta, V., Sharma, S., Raizada, P., Hosseini-Bandegharaei, A., Kaushal, J., & Singh, P. (2020). Fabrication of visible light active BiFeO3/CuS/SiO2 Z-scheme photocatalyst for efficient dye degradation. *Materials Letters*, 270, 127693. 10.1016/j.matlet.2020.127693

Dwork, C. (2008). Differential privacy: A survey of results. *International conference on theory and applications of models of computation*. Springer, Berlin, Heidelberg. 10.1007/978-3-540-79228-4_1

Ekong, M. O., George, W. K., Pandey, B. K., & Pandey, D. (2023). Enhancing the Fundamentals of Industrial Safety Management in TVET for Metaverse Realities. In *Applications of Neuromarketing in the Metaverse* (pp. 19-41). IGI Global. 10.4018/978-1-6684-8150-9.ch002

Elfatih, N. M., Hasan, M. K., Kamal, Z., Gupta, D., Saeed, R. A., Ali, E. S., & Hosain, M. S. (2021, December 30). Internet of vehicle's resource management in 5G networks using AI technologies: Current status and trends. *IET Communications*, 16(5), 400–420. 10.1049/cmu2.12315

ElMaraghy, H., Monostori, L., Schuh, G., & ElMaraghy, W. (2021). Evolution and future of manufacturing systems. *CIRP Annals*, 70(2), 635–658. 10.1016/j.cirp.2021.05.008

Elmoiz Alatabani, L., Sayed Ali, E., Mokhtar, R. A., Saeed, R. A., Alhumyani, H., & Kamrul Hasan, M. (2022, April 15). Deep and Reinforcement Learning Technologies on Internet of Vehicle (IoV) Applications: Current Issues and Future Trends. *Journal of Advanced Transportation*, 2022, 1–16. 10.1155/2022/1947886

Emmert-Streib, F., Yang, Z., Feng, H., Tripathi, S., & Dehmer, M. (2020). An introductory review of deep learning for prediction models with big data. *Frontiers in Artificial Intelligence*, 3, 4. 10.3389/frai.2020.0000433733124

Ferreira, C. M., & Serpa, S. (2018). Society 5.0 and social development: Contributions to a discussion. *Management and Organizational Studies*, 5(4), 26–31. 10.5430/mos.v5n4p26

Friedman, B., & Hendry, D. G. (2019). *Value sensitive design: Shaping technology with moral imagination*. Mit Press. 10.7551/mitpress/7585.001.0001

Garg, V. K., Halpern, S., & Smolik, K. F. (1999, February). Third generation (3G) mobile communications systems. In *1999 IEEE International Conference on Personal Wireless Communications (Cat. No. 99TH8366)* (pp. 39-43). IEEE.

Gawali, V. S., & Ranjan, N. M. (2023). Anomaly detection system in 5G networks via deep learning model. *International Journal of Wireless and Mobile Computing*, 24(3/4), 287–302. 10.1504/IJWMC.2023.131319

George, W. K., Ekong, M. O., Pandey, D., & Pandey, B. K. (2023). Pedagogy for Implementation of TVET Curriculum for the Digital World. In *Applications of Neuromarketing in the Metaverse* (pp. 117-136). IGI Global. 10.4018/978-1-6684-8150-9.ch009

George, W. K., Silas, E. I., Pandey, D., & Pandey, B. K. (2024). Utilization of Industry 4.0 Technologies in Nigerian Technical and Vocational Education: A Conundrum for Educators. In *Examining the Rapid Advance of Digital Technology in Africa* (pp. 270–293). IGI Global. 10.4018/978-1-6684-9962-7.ch014

Ghobakhloo, M., Iranmanesh, M., Mubarak, M. F., Mubarik, M., Rejeb, A., & Nilashi, M. (2022). Identifying industry 5.0 contributions to sustainable development: A strategy roadmap for delivering sustainability values. *Sustainable Production and Consumption*, 33, 716–737. 10.1016/j.spc.2022.08.003

Giordani, M., Polese, M., Mezzavilla, M., Rangan, S., & Zorzi, M. (2020). Toward 6G networks: Use cases and technologies. *IEEE Communications Magazine*, 58(3), 55–61. 10.1109/MCOM.001.1900411

Giordani, M., & Zorzi, M. (2021, March). Non-Terrestrial Networks in the 6G Era: Challenges and Opportunities. *IEEE Network*, 35(2), 244–251. 10.1109/MNET.011.2000493

Gladysz, B., Tran, T. A., Romero, D., van Erp, T., Abonyi, J., & Ruppert, T. (2023). Current development on the Operator 4.0 and transition towards the Operator 5.0: A systematic literature review in light of Industry 5.0. *Journal of Manufacturing Systems*, 70, 160–185. 10.1016/j.jmsy.2023.07.008

Gohar, A., & Nencioni, G. (2021). The role of 5G technologies in a smart city: The case for intelligent transportation system. *Sustainability (Basel)*, 13(9), 5188. 10.3390/su13095188

Goyal, J., Singla, K., & Singh, S. (2019). A Survey of Wireless Communication Technologies from 1G to 5G. In *Seond International Conference on Computer Networks and Inventive Communication Technologies*. Springer: Berlin/Heidelberg, Germany.

Goyal, S., Pandey, D., Singh, H., Singh, J., Kakkar, R., & Srinivasu, P. N. (2022). Mathematical modelling for prediction of spread of corona virus and artificial intelligence/machine learning-based technique to detect COVID-19 via smartphone sensors. *International Journal of Modelling Identification and Control*, 41(1-2), 43–52. 10.1504/IJMIC.2022.127096

Grieves, M., & Vickers, J. (2016). *Origins of the digital twin concept. Florida Institute of Technology, 8*, 3-20.

Groshev, M., Guimarães, C., Martín-Pérez, J., & de la Oliva, A. (2021). Toward intelligent cyber-physical systems: Digital twin meets artificial intelligence. *IEEE Communications Magazine*, 59(8), 14–20. 10.1109/MCOM.001.2001237

Guevara, L., & Auat Cheein, F. (2020). The role of 5G technologies: Challenges in smart cities and intelligent transportation systems. *Sustainability (Basel)*, 12(16), 6469. 10.3390/su12166469

Gui, G., Liu, M., Tang, F., Kato, N., & Adachi, F. (2020). 6G: Opening new horizons for integration of comfort, security, and intelligence. *IEEE Wireless Communications*, 27(5), 126–132. 10.1109/MWC.001.1900516

Guo, H., Liu, C., Li, Q., Kang, S., & Nallanathan, A. (2021). Deep reinforcement learning aided intelligent reflecting surface for secure wireless communications. *IEEE Wireless Communications Letters*, 10(7), 1469–1473.

Guo, Q., Tang, F., & Kato, N. (2022). Federated reinforcement learning-based resource allocation for D2D-aided digital twin edge networks in 6G industrial IoT. *IEEE Transactions on Industrial Informatics*.

Guo, Y., Zhao, R., Lai, S., Fan, L., Lei, X., & Karagiannidis, G. K. (2022). Distributed machine learning for multiuser mobile edge computing systems. *IEEE Journal of Selected Topics in Signal Processing*, 16(3), 460–473. 10.1109/JSTSP.2022.3140660

Gupta, A. K., Sharma, R., Pandey, D., Nassa, V. K., Pandey, B. K., George, A. S., & Dadheech, P. (2023). Performance analysis of eight-channel WDM optical network with different optical amplifiers for industry 4.0. In *Innovation and Competitiveness in Industry 4.0 Based on Intelligent Systems* (pp. 197–212). Springer International Publishing. 10.1007/978-3-031-29775-5_9

Habibi, M. A., Han, B., Fellan, A., Jiang, W., Sánchez, A. G., Pavón, I. L., Boubendir, A., & Schotten, H. D. (2023). Towards an open, intelligent, and end-to-end architectural framework for network slicing in 6G communication systems. *IEEE Open Journal of the Communications Society*, 4, 1615–1658. 10.1109/OJCOMS.2023.3294445

Hassan, W. U., Yaqoob, I., Imran, M., Shoaib, M., & Hossain, M. S. (2022). AI-enabled next generation (6G and beyond) wireless networks: A comprehensive survey. *IEEE Access : Practical Innovations, Open Solutions*, 10, 40690–40728.

Hong, E. K., Lee, I., Shim, B., Ko, Y. C., Kim, S. H., Pack, S., Lee, K., Kim, S., Kim, J. H., Shin, Y., Kim, Y., & Jung, H. (2022, April). 6G R&D vision: Requirements and candidate technologies. *Journal of Communications and Networks (Seoul)*, 24(2), 232–245. 10.23919/JCN.2022.000015

Hong, W., Ba, D., Winkler, V., & Zhu, Z. (2021). Integrated mmWave Phased Arrays for 5G and Beyond. *Proceedings of the IEEE*, 109(2), 168181.

Hong, W., Jiang, Z. H., Yu, C., & Zhou, Z. (2019). Multibeam Antenna Technologies for 5G Wireless Communications. *IEEE Transactions on Antennas and Propagation*, 67(6), 42814294.

Huang, K. C. (2018). *Millimeter Wave Antennas for Gigabit Wireless Communications: A Practical Guide to Design and Analysis in a System Context*. John Wiley & Sons.

Huang, T., Yang, W., Wu, J., Ma, J., Zhang, X., & Zhang, D. (2019). A survey on green 6G network: Architecture and technologies. *IEEE Access : Practical Innovations, Open Solutions*, 7, 175758–175768. 10.1109/ACCESS.2019.2957648

Huang, Y., & Wu, T. (2018). Terahertz Graphene Optical Modulators Enhancement using DoubleLayer Graphene. *IEEE Journal of Selected Topics in Quantum Electronics*, 24(6), 17.

Huseien, G. F., & Shah, K. W. (2022). A review on 5G technology for smart energy management and smart buildings in Singapore. *Energy and AI*, 7, 100116. 10.1016/j.egyai.2021.100116

Iyyanar, P., Anand, R., Shanthi, T., Nassa, V. K., Pandey, B. K., George, A. S., & Pandey, D. (2023). A real-time smart sewage cleaning UAV assistance system using IoT. In *Handbook of Research on Data-Driven Mathematical Modeling in Smart Cities* (pp. 24–39). IGI Global.

Javaid, A., Niyaz, Q., Sun, W., & Alam, M. (2016). A deep learning approach for network intrusion detection system. *Proceedings of the 9th EAI International Conference on Bio-inspired Information and Communications Technologies*. IEEE. 10.4108/eai.3-12-2015.2262516

Jayalakshmi. G., Pandey, D., Pandey, B. K., Kaur, P., Mahajan, D. A., & Dari, S. S. (2024). Smart Big Data Collection for Intelligent Supply Chain Improvement. In *AI and Machine Learning Impacts in Intelligent Supply Chain* (pp. 180-195). IGI Global.

JayaLakshmi. G., Pandey, D., Pandey, B. K., Kaur, P., Mahajan, D. A., & Dari, S. S. (2024). Smart Big Data Collection for Intelligent Supply Chain Improvement. In *AI and Machine Learning Impacts in Intelligent Supply Chain* (pp. 180-195). IGI Global.

Jayapoorani, S., Pandey, D., Sasirekha, N. S., Anand, R., & Pandey, B. K. (2023). Systolic optimized adaptive filter architecture designs for ECG noise cancellation by Vertex-5. *Aerospace Systems*, 6(1), 163–173. 10.1007/s42401-022-00177-3

Jefferies, N. (1995, February). Security in third-generation mobile systems. In *IEE Colloquium on Security in Networks (Digest No. 1995/024)* (pp. 8-1). IET. 10.1049/ic:19950136

Jiang, W. (2021). Toward AI-enabled 6G: State of the art, challenges, and opportunities. *IEEE Internet of Things Journal*.

Johnson, J. M., & Yadav, A. (2016). Fault Location Estimation in HVDC Transmission Line Using ANN. *Proceedings of First International Conference on Information and Communication Technology for Intelligent Systems*. Springer. 10.1007/978-3-319-30933-0_22

Kanakarajan, N. K., & Muniasamy, K. (2015, October 25). Improving the Accuracy of Intrusion Detection Using GAR-Forest with Feature Selection. *Advances in Intelligent Systems and Computing*, 539–547. 10.1007/978-81-322-2695-6_45

Karjaluoto, H. (2006). An investigation of third generation (3G) mobile technologies and services. *Contemporary Management Research*, 2(2), 91–91. 10.7903/cmr.653

Kato, N., Mao, B., Tang, F., Kawamoto, Y., & Liu, J. (2020). Ten challenges in advancing machine learning technologies toward 6G. *IEEE Wireless Communications*, 27(3), 96–103. 10.1109/MWC.001.1900476

Kaur, S. P., & Sharma, M. (2015). Radially optimized zone-divided energy-aware wireless sensor networks (WSN) protocol using BA (bat algorithm). *Journal of the Institution of Electronics and Telecommunication Engineers*, 61(2), 170–179. 10.1080/03772063.2014.999833

Khalifa, O. O., Roubleh, A., Esgiar, A., Abdelhaq, M., Alsaqour, R., Abdalla, A., Ali, E. S., & Saeed, R. (2022, October 1). An IoT-Platform-Based Deep Learning System for Human Behavior Recognition in Smart City Monitoring Using the Berkeley MHAD Datasets. *Systems*, 10(5), 177. 10.3390/systems10050177

Khan, B., Hasan, A., Pandey, D., Ventayen, R. J. M., Pandey, B. K., & Gowwrii, G. (2021). Fusion of datamining and artificial intelligence in prediction of hazardous road accidents. In *Machine learning and iot for intelligent systems and smart applications* (pp. 201–223). CRC Press. 10.1201/9781003194415-12

Khan, I., Zhang, K., Ali, L., & Wu, Q. (2023). Enhanced Quad-Port MIMO Antenna Isolation With Metamaterial Superstrate. *IEEE Antennas and Wireless Propagation Letters*.

Khan, L. U., & Portmann, M. (2020). Security challenges of 6G wireless networks. *IEEE Internet of Things Magazine*, 3(2), 10–15. 10.1109/IOTM.0001.1900110

Khanna, T., Kaur, A., Dubey, R., & Sharma, I. (2023, November). SecuGrid: Artificial Intelligent Enabled Framework for Securing Power Grid Communication using Honeypot. In *2023 International Conference on Sustainable Communication Networks and Application (ICSCNA)* (pp. 1103-1107). IEEE. 10.1109/ICSCNA58489.2023.10370153

Khan, R., Kumar, P., Jayakody, D. N. K., & Liyanage, M. (2019). A survey on security and privacy of 5G technologies: Potential solutions, recent advancements, and future directions. *IEEE Communications Surveys and Tutorials*, 22(1), 196–248. 10.1109/COMST.2019.2933899

Kim, H., & Ben-Othman, J. (2023, January). Eco-Friendly Low Resource Security Surveillance Framework Toward Green AI Digital Twin. *IEEE Communications Letters*, 27(1), 377–380. 10.1109/LCOMM.2022.3218050

Kirubasri, G., Sankar, S., Pandey, D., Pandey, B. K., Singh, H., & Anand, R. (2021, September). A recent survey on 6G vehicular technology, applications and challenges. In *2021 9th International Conference on Reliability, Infocom Technologies and Optimization (Trends and Future Directions)(ICRITO)* (pp. 1-5). IEEE.

Kirubasri, G., Sankar, S., Pandey, D., Pandey, B. K., Nassa, V. K., & Dadheech, P. (2022). Software-defined networking-based Ad hoc networks routing protocols. In *Software defined networking for Ad Hoc networks* (pp. 95–123). Springer International Publishing. 10.1007/978-3-030-91149-2_5

Kouachi, S. R. (2006). Privacy-preserving machine learning. In *Proceedings of the Second International Workshop on Security in Machine Learning*. IEEE.

Kukliński, S., Tomaszewski, L., Kołakowski, R., & Chemouil, P. (2021). 6G-LEGO: A framework for 6G network slices. *Journal of Communications and Networks (Seoul)*, 23(6), 442–453. 10.23919/JCN.2021.000025

Kumar Pandey, B., Pandey, D., Nassa, V. K., Ahmad, T., Singh, C., George, A. S., & Wakchaure, M. A. (2021). Encryption and steganography-based text extraction in IoT using the EWCTS optimizer. *Imaging Science Journal*, 69(1-4), 38–56. 10.1080/13682199.2022.2146885

Kumar, G., & Kumar, R. (2019). A survey on planar ultra-wideband antennas with band notch characteristics: Principle, design, and applications. *AEÜ. International Journal of Electronics and Communications*, 109, 76–98. 10.1016/j.aeue.2019.07.004

KVM, S., Pandey, B. K., & Pandey, D. (2024). Design of Surface Plasmon Resonance (SPR) Sensors for Highly Sensitive Biomolecular Detection in Cancer Diagnostics. *Plasmonics*, 1-13.

Kwatra, C. V., Jain, A., Royappa, A., & Bagchi, S. (2022). Artificial Intelligence Application for Security Issues and Challenges in IoT. In *2022 5th International Conference on Contemporary Computing and Informatics (IC3I)*. IEEE.

Letaief, K. B., Chen, W., Shi, Y., Zhang, J., & Zhang, Y. A. (2019). The roadmap to 6G: AI empowered wireless networks. *IEEE Communications Magazine*, 57(8), 84–90. 10.1109/MCOM.2019.1900271

Liang, W., Xiao, L., Zhang, K., Tang, M., He, D., & Li, K. C. (2021). Data fusion approach for collaborative anomaly intrusion detection in blockchain-based systems. *IEEE Internet of Things Journal*, 9(16), 14741–14751. 10.1109/JIOT.2021.3053842

Li, L., Xu, J., Li, C., Zhang, J., & Zhang, R. (2021). Security vulnerabilities and countermeasures for 6G network intelligent surfaces. *IEEE Internet of Things Journal*, 8(24), 17552–17567.

Li, M. (2021). FLaaS: Federated learning as a service for privacy-preserving IoT applications. *IEEE Internet of Things Journal*.

Lim, W. Y. B., Huang, J., Sarwat, A. I., & Xiong, N. X. (2021). Privacy-preserving human mobility data analytics in cellular networks: A federated federated-learning approach. *IEEE Transactions on Vehicular Technology*, 70(6), 5723–5737.

Liu, Y., Chen, H., & Xing, C. G. (2021). Detecting 6G wireless insider attacks using blockchain and federated learning. *IEEE Transactions on Vehicular Technology*, 70(10), 10983–10993.

Li, Y., & Luk, K. M. (2015). 60-GHz substrate integrated waveguide fed cavity-backed aperture-coupled microstrip patch antenna arrays. *IEEE Transactions on Antennas and Propagation*, 63(3), 1075–1085. 10.1109/TAP.2015.2390228

Li, Y., Wang, L., Ge, L., & Wang, J. (2023). A Dual-Band Magneto-Electric Monopole Antenna Array Using Mode-Composite Waveguides. *IEEE Transactions on Antennas and Propagation*.

Li, Y., Yu, Y., Susilo, W., Hong, Z., & Guizani, M. (2021). Security and privacy for edge intelligence in 5G and beyond networks: Challenges and solutions. *IEEE Wireless Communications*, 28(2), 63–69. 10.1109/MWC.001.2000318

Loey, M., Manogaran, G., Taha, M. H. N., & Khalifa, N. E. M. (2021, January). A hybrid deep transfer learning model with machine learning methods for face mask detection in the era of the COVID-19 pandemic. *Measurement*, 167, 108288. 10.1016/j.measurement.2020.10828832834324

Longo, F., Padovano, A., & Umbrello, S. (2020). Value-oriented and ethical technology engineering in industry 5.0: A human-centric perspective for the design of the factory of the future. *Applied Sciences (Basel, Switzerland)*, 10(12), 4182. 10.3390/app10124182

Lu, Y., Maharjan, S., & Zhang, Y. (2021). Adaptive edge association for wireless digital twin networks in 6G. *IEEE Internet of Things Journal*, 8(22), 16219–16230. 10.1109/JIOT.2021.3098508

Lu, Y., & Zheng, X. (2020). 6G: A survey on technologies, scenarios, challenges, and the related issues. *Journal of Industrial Information Integration*, 19, 100158. 10.1016/j.jii.2020.100158

Mahmoud, H. H. H., Amer, A. A., & Ismail, T. (2021). 6G: A comprehensive survey on technologies, applications, challenges, and research problems. *Transactions on Emerging Telecommunications Technologies*, 32(4), e4233. 10.1002/ett.4233

Makridakis, S. (2017). The forthcoming Artificial Intelligence (AI) revolution: Its impact on society and firms. *Futures*, 90, 46–60. 10.1016/j.futures.2017.03.006

Malhotra, P., Pandey, D., Pandey, B. K., & Patra, P. M. (2021). Managing agricultural supply chains in COVID-19 lockdown. *International Journal of Quality and Innovation*, 5(2), 109–118. 10.1504/IJQI.2021.117181

Manzillo, M., Occhiuzzi, C., & Marrocco, G. (2019). Modeling, Design and Experimentation of Wearable RFID Sensor Tag. *IEEE Transactions on Antennas and Propagation*, 67(3), 19001908.

Mat Amin, M. K., Soh, P. J., Kumar, S., & Ali, M. T. (2021). Integration Methods of Phased Array Antennas in Wireless Transceivers: A Review of Implementation Challenges and Solutions. *IEEE Access : Practical Innovations, Open Solutions*, 9, 1191711937.

Mendis, R., Randall, J. M., Higgins, M. D., & Kancleris, J. (2019). A 6G Uplink: Intelligent Surfaces Meet THz Communications. *IEEE Vehicular Technology Magazine*, 14(3), 92100.

Meslie, Y., Enbeyle, W., Pandey, B. K., Pramanik, S., Pandey, D., Dadeech, P., & Saini, A. (2021). Machine intelligence-based trend analysis of COVID-19 for total daily confirmed cases in Asia and Africa. In *Methodologies and Applications of Computational Statistics for Machine Intelligence* (pp. 164–185). IGI Global. 10.4018/978-1-7998-7701-1.ch009

Mhawi, D. N., & Hashem, P. S. H. (2021, December 6). Proposed Hybrid CorrelationFeatureSelectionForestPanalizedAttribute Approach to advance IDSs. *Karbala International Journal of Modern Science*, 7(4). 10.33640/2405-609X.3166

Mishra, R., Gupta, H. P., Kumar, R., & Dutta, T. (2023, January). Leveraging Augmented Intelligence of Things to Enhance Lifetime of UAV-Enabled Aerial Networks. *IEEE Transactions on Industrial Informatics*, 19(1), 586–593. 10.1109/TII.2022.3197410

Mohamad, M., Selamat, A., Krejcar, O., Crespo, R. G., Herrera-Viedma, E., & Fujita, H. (2021, November 30). Enhancing Big Data Feature Selection Using a Hybrid Correlation-Based Feature Selection. *Electronics (Basel)*, 10(23), 2984. 10.3390/electronics10232984

Mohapatra, S. K., Swain, B. R., & Das, P. (2015). Comprehensive survey of possible security issues on 4G networks. *International Journal of Network Security & its Applications*, 7(2), 61–69. 10.5121/ijnsa.2015.7205

Mokhtari, S., Abbaspour, A., Yen, K. K., & Sargolzaei, A. (2021). A machine learning approach for anomaly detection in industrial control systems based on measurement data. *Electronics (Basel)*, 10(4), 407. 10.3390/electronics10040407

Moon, S. H., & Kim, Y. H. (2020, August). An improved forecast of precipitation type using correlation-based feature selection and multinomial logistic regression. *Atmospheric Research*, 240, 104928. 10.1016/j.atmosres.2020.104928

Moro, S. R., Cauchick-Miguel, P. A., de Sousa-Zomer, T. T., & de Sousa Mendes, G. H. (2023). Design of a sustainable electric vehicle sharing business model in the Brazilian context. *International Journal of Industrial Engineering and Management*, 14(2), 147–161. 10.24867/IJIEM-2023-2-330

Mourtzis, D., Angelopoulos, J., & Panopoulos, N. (2022). Digital Manufacturing: The evolution of traditional manufacturing toward an automated and interoperable Smart Manufacturing Ecosystem. In *The digital supply chain* (pp. 27-45). Elsevier.

Muniandi, B., Nassa, V. K., Pandey, D., Pandey, B. K., Dadheech, P., & George, A. S. (2024). Pattern Analysis for Feature Extraction in Complex Images. In *Using Machine Learning to Detect Emotions and Predict Human Psychology* (pp. 145–167). IGI Global. 10.4018/979-8-3693-1910-9.ch007

Murroni, M., Anedda, M., Fadda, M., Ruiu, P., Popescu, V., Zaharia, C., & Giusto, D. (2023). 6G—Enabling the New Smart City: A Survey. *Sensors (Basel)*, 23(17), 7528. 10.3390/s2317752837687986

Nafe, M., Alibakhshikenari, M., & See, C. H., AbdAlhameed, R., Limiti, E., Klemm, M. & Grani, F. (2021). A Novel Wideband Stacked Modified MagnetoElectric Dipole Antenna for 5G Mobile Handset Applications. *IEEE Access : Practical Innovations, Open Solutions*, 9, 7536275372.

Nagar, P. L., Bajpai, S., & Pandey, D. (2023, August). Design and Analysis of U-Slot Microstrip Patch Antenna for ISM Band Applications. In *International Conference on Mobile Radio Communications & 5G Networks* (pp. 439-451). Singapore: Springer Nature Singapore.

Ngo, H. Q., Jiang, C., Lozano, A., Rahul, A. A., & Nieman, K. F. (2021). Cell-free two-tier massive MIMO networks with multi-antenna user association. *IEEE Transactions on Wireless Communications*, 20(11), 7480–7494.

Nguyen, H. X., Trestian, R., To, D., & Tatipamula, M. (2021). Digital twin for 5G and beyond. *IEEE Communications Magazine*, 59(2), 10–15. 10.1109/MCOM.001.2000343

Oleiwi, H. W., Mhawi, D. N., & Al-Raweshidy, H. (2022). MLTs-ADCNs: Machine Learning Techniques for Anomaly Detection in Communication Networks. *IEEE Access : Practical Innovations, Open Solutions*, 10, 91006–91017. 10.1109/ACCESS.2022.3201869

Ozpoyraz, B., Dogukan, A. T., Gevez, Y., Altun, U., & Basar, E. (2022). Deep Learning-Aided 6G Wireless Networks: A Comprehensive Survey of Revolutionary PHY Architectures. *IEEE Open Journal of the Communications Society*, 3, 1749–1809. 10.1109/OJCOMS.2022.3210648

Pajouh, H. H., Dastghaibyfard, G., & Hashemi, S. (2015, November 19). Two-tier network anomaly detection model: A machine learning approach. *Journal of Intelligent Information Systems*, 48(1), 61–74. 10.1007/s10844-015-0388-x

Pandey, B. K., Pandey, D., Gupta, A., Nassa, V. K., Dadheech, P., & George, A. S. (2023). Secret data transmission using advanced morphological component analysis and steganography. In *Role of data-intensive distributed computing systems in designing data solutions* (pp. 21-44). Cham: Springer International Publishing.

Pandey, D. (2022). An efficient deep neural network with adaptive galactic swarm optimization for complex image text extraction. In *Process mining techniques for pattern recognition* (pp. 121-137). CRC Press.

Pandey, B. K., Mane, D., Nassa, V. K. K., Pandey, D., Dutta, S., Ventayen, R. J. M., & Rastogi, R. (2021). Secure text extraction from complex degraded images by applying steganography and deep learning. In *Multidisciplinary approach to modern digital steganography* (pp. 146–163). IGI Global. 10.4018/978-1-7998-7160-6.ch007

Pandey, B. K., & Pandey, D. (2023). Parametric optimization and prediction of enhanced thermoelectric performance in co-doped CaMnO3 using response surface methodology and neural network. *Journal of Materials Science Materials in Electronics*, 34(21), 1589. 10.1007/s10854-023-10954-1

Pandey, B. K., Pandey, D., & Agarwal, A. (2022). Encrypted information transmission by enhanced steganography and image transformation. [IJDAI]. *International Journal of Distributed Artificial Intelligence*, 14(1), 1–14. 10.4018/IJDAI.297110

Pandey, B. K., Pandey, D., Alkhafaji, M. A., Güneşer, M. T., & Şeker, C. (2023). A reliable transmission and extraction of textual information using keyless encryption, steganography, and deep algorithm with cuckoo optimization. In *Micro-Electronics and Telecommunication Engineering: Proceedings of 6th ICMETE 2022* (pp. 629–636). Springer Nature Singapore. 10.1007/978-981-19-9512-5_57

Pandey, B. K., Pandey, D., Dadheech, P., Mahajan, D. A., George, A. S., & Hameed, A. S. (2023b). Review on Smart Sewage Cleaning UAV Assistance for Sustainable Development. In *Handbook of Research on Safe Disposal Methods of Municipal Solid Wastes for a Sustainable Environment* (pp. 69–79). IGI Global. 10.4018/978-1-6684-8117-2.ch005

Pandey, B. K., Pandey, D., Gupta, A., Nassa, V. K., Dadheech, P., & George, A. S. (2023). Secret data transmission using advanced morphological component analysis and steganography. In *Role of data-intensive distributed computing systems in designing data solutions* (pp. 21–44). Springer International Publishing. 10.1007/978-3-031-15542-0_2

Pandey, B. K., Pandey, D., & Sahani, S. K. (2024). Autopilot control unmanned aerial vehicle system for sewage defect detection using deep learning. *Engineering Reports*, 12852. 10.1002/eng2.12852

Pandey, B. K., Pandey, D., Wairya, S., & Agarwal, G. (2021). An advanced morphological component analysis, steganography, and deep learning-based system to transmit secure textual data. [IJDAI]. *International Journal of Distributed Artificial Intelligence*, 13(2), 40–62. 10.4018/IJDAI.2021070104

Pandey, B. K., Pandey, D., Wariya, S., & Agarwal, G. (2021). A deep neural network-based approach for extracting textual images from deteriorate images. *EAI Endorsed Transactions on Industrial Networks and Intelligent Systems*, 8(28), e3–e3. 10.4108/eai.17-9-2021.170961

Pandey, B. K., Pandey, S. K., & Pandey, D. (2011). A survey of bioinformatics applications on parallel architectures. *International Journal of Computer Applications*, 23(4), 21–25. 10.5120/2877-3744

Pandey, B. K., Paramashivan, M. A., Kanike, U. K., Mahajan, D. A., Mahajan, R., George, A. S., & Hameed, A. S. H. (2024). Impacts of Artificial Intelligence and Machine Learning on Intelligent Supply Chains. In *AI and Machine Learning Impacts in Intelligent Supply Chain* (pp. 57–73). IGI Global. 10.4018/979-8-3693-1347-3.ch005

Pandey, D., Hasan, A., Pandey, B. K., Lelisho, M. E., George, A. H., & Shahul, A. (2023). COVID-19 epidemic anxiety, mental stress, and sleep disorders in developing country university students. *CSI Transactions on ICT*, 11(2), 119–127. 10.1007/s40012-023-00383-0

Pandey, D., Nassa, V. K., Pandey, B. K., Thankachan, B., Dadheech, P., Mahajan, D. A., & George, A. S. (2024). Artificial Intelligence and Machine Learning and Its Application in the Field of Computational Visual Analysis. In El Kacimi, Y., & Alaoui, K. (Eds.), *Emerging Engineering Technologies and Industrial Applications* (pp. 36–57). IGI Global. 10.4018/979-8-3693-1335-0.ch003

Pandey, D., & Pandey, B. K. (2022). An efficient deep neural network with adaptive galactic swarm optimization for complex image text extraction. In *Process mining techniques for pattern recognition* (pp. 121–137). CRC Press. 10.1201/9781003169550-10

Pandey, D., Pandey, B. K., & Wairya, S. (2021). Hybrid deep neural network with adaptive galactic swarm optimization for text extraction from scene images. *Soft Computing*, 25(2), 1563–1580. 10.1007/s00500-020-05245-4

Pandey, D., Wairya, S., Sharma, M., Gupta, A. K., Kakkar, R., & Pandey, B. K. (2022). An approach for object tracking, categorization, and autopilot guidance for passive homing missiles. *Aerospace Systems*, 5(4), 553–566. 10.1007/s42401-022-00150-0

Pandey, J. K., Jain, R., Dilip, R., Kumbhkar, M., Jaiswal, S., Pandey, B. K., & Pandey, D. (2022). Investigating role of iot in the development of smart application for security enhancement. In *IoT Based Smart Applications* (pp. 219–243). Springer International Publishing.

Panwar, N., Sharma, S., & Singh, A. K. (2016). A survey on 5G: The next generation of mobile communication. *Physical Communication*, 18, 64–84. 10.1016/j.phycom.2015.10.006

Parthiban, K., Pandey, D., & Pandey, B. K. (2021). Impact of SARS-CoV-2 in online education, predicting and contrasting mental stress of young students: A machine learning approach. *Augmented Human Research*, 6(1), 10. 10.1007/s41133-021-00048-0

Paschek, D., Mocan, A., & Draghici, A. (2019, May). *Industry 5.0—The expected impact of next industrial revolution. In Thriving on future education, industry, business, and Society*. Proceedings of the MakeLearn and TIIM International Conference, Piran, Slovenia.

Peng, Y., Guo, Y., Hao, R., & Xu, C. (2024, April). Network traffic prediction with Attention-based Spatial–Temporal Graph Network. *Computer Networks*, 243, 110296. 10.1016/j.comnet.2024.110296

Pereira, V., & Sousa, T. (2004). *Evolution of Mobile Communications: from 1G to 4G*. Department of Informatics Engineering of the University of Coimbra, Portugal.

Petrov, V., Kurner, T., & Hosako, I. (2020). IEEE 802.15. 3d: First standardization efforts for sub-terahertz band communications toward 6G. *IEEE Communications Magazine*, 58(11), 28–33. 10.1109/MCOM.001.2000273

Pindoria, N. M. (2021). Wireless network security: Threats, challenges and countermeasures using machine learning and deep learning techniques. *Wireless Networks*, 1–29.

Plastiras, G., Terzi, M., Kyrkou, C., & Theocharides, T. (2018, July). Edge Intelligence: Challenges and Opportunities of Near-Sensor Machine Learning Applications. *2018 IEEE 29th International Conference on Application-Specific Systems, Architectures and Processors (ASAP)*. IEEE. 10.1109/ASAP.2018.8445118

Polese, M., Giordani, M., Zang, M., Santhi, N., Lagen, S., Aminikashani, M., & Rost, P. (2021). *6G Security: A Holistic Perspective*.

Pramanik, S., Pandey, D., Joardar, S., Niranjanamurthy, M., Pandey, B. K., & Kaur, J. (2023, October). An overview of IoT privacy and security in smart cities. In *AIP Conference Proceedings* (Vol. 2495, No. 1). AIP Publishing. 10.1063/5.0123511

Qiao, X., Huang, Y., Dustdar, S., & Chen, J. (2020, July 1). 6G Vision: An AI-Driven Decentralized Network and Service Architecture. *IEEE Internet Computing*, 24(4), 33–40. 10.1109/MIC.2020.2987738

Qin, S., Xu, J., Lin, J., & Zhong, C. (2021). 6G Security for Reconfigurable Intelligent Surface Empowered Communication Systems: Challenges and Open Issues. *IEEE Access : Practical Innovations, Open Solutions*, 9, 100503–100518.

Quy, V. K., Chehri, A., Quy, N. M., Han, N. D., & Ban, N. T. (2023). Innovative trends in the 6G era: A comprehensive survey of architecture, applications, technologies, and challenges. *IEEE Access : Practical Innovations, Open Solutions*, 11, 39824–39844. 10.1109/ACCESS.2023.3269297

Raja, D., Kumar, D. R., Santhiyakumari, N., Kumarganesh, S., Sagayam, K. M., Thiyaneswaran, B., Pandey, B. K., & Pandey, D. (2024). A compact dual-feed wide-band slotted antenna for future wireless applications. *Analog Integrated Circuits and Signal Processing*, 118(2), 1–15. 10.1007/s10470-023-02233-0

Rajagopal, S., AbuSurra, S., & Pi, Z. (2020). Antenna array design for multiGbps mmWave communication. *IEEE Wireless Communications*, 27(2), 7379.

Rajagopal, S., Kundapur, P. P., & Hareesha, K. S. (2020). A stacking ensemble for network intrusion detection using heterogeneous datasets. *Security and Communication Networks*, 2020, 1–9. 10.1155/2020/4586875

Rappaport, T. S., Xing, Y., Kanhere, O., Ju, S., Madanayake, A., Mandal, S., & Zhao, T. (2019). Wireless communications and applications above 100 GHz: Opportunities and challenges for 6G and beyond. *IEEE Access : Practical Innovations, Open Solutions*, 7, 7872978757. 10.1109/ACCESS.2019.2921522

Rathore, M. M., Shah, S. A., Shukla, D., Bentafat, E., & Bakiras, S. (2021). The role of ai, machine learning, and big data in digital twinning: A systematic literature review, challenges, and opportunities. *IEEE Access : Practical Innovations, Open Solutions*, 9, 32030–32052. 10.1109/ACCESS.2021.3060863

Ray, P. P., Kumar, N., & Guizani, M. (2021). A vision on 6G-enabled NIB: Requirements, technologies, deployments, and prospects. *IEEE Wireless Communications*, 28(4), 120–127. 10.1109/MWC.001.2000384

Ren, Y., Xie, R., Yu, F. R., Huang, T., & Liu, Y. (2022, September). Green Intelligence Networking for Connected and Autonomous Vehicles in Smart Cities. *IEEE Transactions on Green Communications and Networking*, 6(3), 1591–1603. 10.1109/TGCN.2022.3148293

Revathi, T. K., Sathiyabhama, B., Sankar, S., Pandey, D., Pandey, B. K., & Dadeech, P. (2022). An intelligent model for coronary heart disease diagnosis. In *Networking Technologies in Smart Healthcare* (pp. 309–327). CRC Press. 10.1201/9781003239888-15

Rupprecht, D., Dabrowski, A., Holz, T., Weippl, E., & Pöpper, C. (2018). On security research towards future mobile network generations. *IEEE Communications Surveys and Tutorials*, 20(3), 2518–2542. 10.1109/COMST.2018.2820728

Rydén, H., Farhadi, H., Palaios, A., Hévizi, L., Sandberg, D., & Kvernvik, T. (2023, October). Next Generation Mobile Networks' Enablers: Machine Learning-Assisted Mobility, Traffic, and Radio Channel Prediction. *IEEE Communications Magazine*, 61(10), 94–98. 10.1109/MCOM.001.2200592

Saad, W., Bennis, M., & Chen, M. (2019). A vision of 6G wireless systems: Applications, trends, technologies, and open research problems. *IEEE Network*, 34(3), 134–142. 10.1109/MNET.001.1900287

Saeed, , M., Kamrul Hasan, M., Hassan, R., Mokhtar, R., A. Saeed, R., Saeid, E., & Gupta, M. (2022). Preserving Privacy of User Identity Based on Pseudonym Variable in 5G. *Computers, Materials & Continua*, 70(3), 5551–5568. 10.32604/cmc.2022.017338

Saeed, R. A., Omri, M., Abdel-Khalek, S., Ali, E. S., & Alotaibi, M. F. (2022, April 29). Optimal path planning for drones based on swarm intelligence algorithm. *Neural Computing & Applications*, 34(12), 10133–10155. 10.1007/s00521-022-06998-9

Safaldin, M., Otair, M., & Abualigah, L. (2020, June 26). Improved binary gray wolf optimizer and SVM for intrusion detection system in wireless sensor networks. *Journal of Ambient Intelligence and Humanized Computing*, 12(2), 1559–1576. 10.1007/s12652-020-02228-z

Sahani, K., Khadka, S. S., Sahani, S. K., Pandey, B. K., & Pandey, D. (2023). A possible underground roadway for transportation facilities in Kathmandu Valley: A racking deformation of underground rectangular structures. *Engineering Reports*, 12821. 10.1002/eng2.12821

Sahani, S. K., Pandey, B. K., & Pandey, D. (2024). *Single-valued Signals, Multi-valued Signals and Fixed-Point of Contractive Signals*. Mathematics Open. 10.1142/S2811007224500020

Sasidevi, S., Kumarganesh, S., Saranya, S., Thiyaneswaran, B., Shree, K. V. M., & Martin Sagayam, K. (2024, May 15). Design of Surface Plasmon Resonance (SPR) Sensors for Highly Sensitive Biomolecular Detection in Cancer Diagnostics. *Plasmonics*. 10.1007/s11468-024-02343-z

Saxena, A., Agarwal, A., Pandey, B. K., & Pandey, D. (2024). Examination of the Criticality of Customer Segmentation Using Unsupervised Learning Methods. *Circular Economy and Sustainability*, 1–14. 10.1007/s43615-023-00336-4

Saxena, A., Agarwal, A., Pandey, B. K., & Pandey, D. (2024). Examination of the Criticality of Customer Segmentation Using Unsupervised Learning Methods. Circular Economy and Sustainability, 1-14. Saxena.

Saxena, A., Sharma, N. K., Pandey, D., & Pandey, B. K. (2021). Influence of tourists satisfaction on future behavioral intentions with special reference to desert triangle of Rajasthan. *Augmented Human Research*, 6(1), 13. 10.1007/s41133-021-00052-4

Saxena, N., & Chaudhari, N. S. (2014). Secure-aka: An efficient aka protocol for umts networks. *Wireless Personal Communications*, 78(2), 1345–1373. 10.1007/s11277-014-1821-0

Sayed Ali Ahmed, E., Mohammed, Z. T., Bakri Hassan, M., & Saeed, R. A. (2021). Algorithms Optimization for Intelligent IoV Applications. *Handbook of Research on Innovations and Applications of AI, IoT, and Cognitive Technologies*, 1–25. 10.4018/978-1-7998-6870-5.ch001

Schmidhuber, J. (2015). Deep learning in neural networks: An overview. *Neural Networks*, 61, 85–117. 10.1016/j.neunet.2014.09.00325462637

Sengupta, R., Sengupta, D., Pandey, D., Pandey, B. K., Nassa, V. K., & Dadeech, P. (2021). A Systematic review of 5G opportunities, architecture and challenges. Future Trends in 5G and 6G, 247-269.

Sengupta, R., Sengupta, D., Kamra, A. K., & Pandey, D. (2020). Artificial Intelligence and Quantum Computing for a Smarter Wireless Network. *Artificial Intelligence*, 7(19), 2020.

Sennan, S., Alotaibi, Y., Pandey, D., & Alghamdi, S. (2022). EACR-LEACH: Energy-Aware Cluster-based Routing Protocol for WSN Based IoT. *Computers, Materials & Continua*, 72(2), 2159–2174. 10.32604/cmc.2022.025773

Serôdio, C., Cunha, J., Candela, G., Rodriguez, S., Sousa, X. R., & Branco, F. (2023). The 6G Ecosystem as Support for IoE and Private Networks: Vision, Requirements, and Challenges. *Future Internet*, 15(11), 348. 10.3390/fi15110348

Sharma, M., Gupta, A. K., Arora, T., Pandey, D., & Vats, S. (2023, March). Comprehensive Analysis of Multiband Microstrip Patch Antennas used in IoT-based Networks. In *2023 10th International Conference on Computing for Sustainable Global Development (INDIACom)* (pp. 1424-1429). IEEE.

Sharma, M., Saripalli, S. R., Gupta, A. K., Talwar, R., Dadheech, P., & Kanike, U. K. (2023). Real-Time Pothole Detection During Rainy Weather Using Dashboard Cameras for Driverless Cars. In *Handbook of Research on Thrust Technologies' Effect on Image Processing* (pp. 384-394). IGI Global. 10.4018/978-1-6684-8618-4.ch023

Sharma, S., Pandey, B. K., Pandey, D., Anand, R., Sharma, A., & Saini, S. (2023, March). Character Recognition Technique Implementation for Complicated Deteriorated Scene. In *2023 6th International Conference on Information Systems and Computer Networks (ISCON)* (pp. 1-4). IEEE. 10.1109/ISCON57294.2023.10112185

Sharma, M., Gupta, A. K., Singh, J., Mittal, R., Singh, H., & Pandey, D. (2022, December). Effects of slot shape in performance of SIW based Leaky Wave Antenna. In *2022 International Conference on Computational Modelling, Simulation and Optimization (ICCMSO)* (pp. 268-273). IEEE. 10.1109/ICCMSO58359.2022.00060

Sharma, M., Pandey, D., Khosla, D., Goyal, S., Pandey, B. K., & Gupta, A. K. (2022). Design of a GaN-based Flip Chip Light Emitting Diode (FC-LED) with au Bumps & Thermal Analysis with different sizes and adhesive materials for performance considerations. *Silicon*, 14(12), 7109–7120. 10.1007/s12633-021-01457-x

Sharma, M., Pandey, D., Palta, P., & Pandey, B. K. (2022). Design and power dissipation consideration of PFAL CMOS V/S conventional CMOS based 2: 1 multiplexer and full adder. *Silicon*, 14(8), 4401–4410. 10.1007/s12633-021-01221-1

Sharma, M., Saripalli, S. R., Gupta, A. K., Palta, P., & Pandey, D. (2023). Image Processing-Based Method of Evaluation of Stress from Grain Structures of Through Silicon Via (TSV). *International Journal of Image and Graphics*, 2550008. 10.1142/S0219467825500081

Sharma, M., Sharma, B., Gupta, A. K., & Pandey, D. (2023). Recent developments of image processing to improve explosive detection methodologies and spectroscopic imaging techniques for explosive and drug detection. *Multimedia Tools and Applications*, 82(5), 6849–6865. 10.1007/s11042-022-13578-5

Sharma, M., & Singh, H. (2021). Substrate integrated waveguide based leaky wave antenna for high frequency applications and IoT. *International Journal of Sensors, Wireless Communications and Control*, 11(1), 5–13. 10.2174/2210327909666190401210659

Sharma, M., & Singh, H. (2022). Contactless methods for respiration monitoring and design of SIW-LWA for real-time respiratory rate monitoring. *Journal of the Institution of Electronics and Telecommunication Engineers*, 1–11.

Sharma, M., Singh, H., Gupta, A. K., & Khosla, D. (2023). Target identification and control model of autopilot for passive homing missiles. *Multimedia Tools and Applications*, 83(20), 1–30. 10.1007/s11042-023-17804-6

Sharma, M., Talwar, R., Pandey, D., Nassa, V. K., Pandey, B. K., & Dadheech, P. (2024). A Review of Dielectric Resonator Antennas (DRA)-Based RFID Technology for Industry 4.0. *Robotics and Automation in Industry*, 4(0), 303–324.

Shen, F., Shi, H., & Yang, Y. (2021, August). A comprehensive study of 5G and 6G networks. In *2021 international conference on wireless communications and smart grid (ICWCSG)* (pp. 321-326). IEEE.

Sheth, K., Patel, K., Shah, H., Tanwar, S., Gupta, R., & Kumar, N. (2020). A taxonomy of AI techniques for 6G communication networks. *Computer Communications*, 161, 279–303. 10.1016/j.comcom.2020.07.035

Shinde, S. S., Marabissi, D., & Tarchi, D. (2021). A network operator-biased approach for multi-service network function placement in a 5G network slicing architecture. *Computer Networks*, 201, 108598. 10.1016/j.comnet.2021.108598

Shiroishi, Y., Uchiyama, K., & Suzuki, N. (2018). Society 5.0: For human security and well-being. *Computer*, 51(7), 91–95. 10.1109/MC.2018.3011041

Shone, N., Ngoc, T. N., Phai, V. D., & Shi, Q. (2018). A deep reinforcement learning framework for the dynamic control of data exposure in smart cities. *IEEE Transactions on Industrial Informatics*, 14(12), 5366–5375.

Shrivastava, U., & Verma, J. K. (2021, December). A Study on 5G Technology and Its Applications in Telecommunications. In *2021 International Conference on Computational Performance Evaluation (ComPE)* (pp. 365-371). IEEE.

Sinclair, M., Maadi, S., Zhao, Q., Hong, J., Ghermandi, A., & Bailey, N. (2023). Assessing the socio-demographic representativeness of mobile phone application data. *Applied Geography (Sevenoaks, England)*, 158, 102997. 10.1016/j.apgeog.2023.102997

Singh, S., Madaan, G., Kaur, J., Swapna, H. R., Pandey, D., Singh, A., & Pandey, B. K. (2023). *Bibliometric Review on Healthcare Sustainability*. Handbook of Research on Safe Disposal Methods of Municipal Solid Wastes for a Sustainable Environment, (pp. 142-161). IGI Global. 10.4018/978-1-6684-8117-2.ch011

Singh, S., Madaan, G., Kaur, J., Swapna, H. R., Pandey, D., Singh, A., & Pandey, B. K. (2023). Bibliometric Review on Healthcare Sustainability. *Handbook of Research on Safe Disposal Methods of Municipal Solid Wastes for a Sustainable Environment*, 142-161.

Singh, M., Kumar, R., Tandon, D., Sood, P., & Sharma, M. (2020, December). Artificial intelligence and iot based monitoring of poultry health: A review. In *2020 IEEE International Conference on Communication, Networks and Satellite (Comnetsat)* (pp. 50-54). IEEE. 10.1109/Comnetsat50391.2020.9328930

Singh, P. R., Singh, V. K., Yadav, R., & Chaurasia, S. N. (2023). 6G networks for artificial intelligence-enabled smart cities applications: A scoping review. *Telematics and Informatics Reports*, 9, 100044. 10.1016/j.teler.2023.100044

Sun, Y., Su, L., Wang, B. H., & Yang, W. H. (2021). Federated machine learning-based authentication in Fog RAN intelligent edge for space-air-ground IoT networks. *IEEE Internet of Things Journal*.

Swapna, H. R., Bigirimana, E., Madaan, G., Hasan, A., Pandey, B. K., & Pandey, D. (2023). Impact of neuromarketing on consumer psychology in digitally connected networks. In *Applications of Neuromarketing in the Metaverse* (pp. 193–205). IGI Global. 10.4018/978-1-6684-8150-9.ch015

Taneja, A., & Saluja, N. (2023). A transmit antenna selection based energy-harvesting mimo cooperative communication system. *Journal of the Institution of Electronics and Telecommunication Engineers*, 69(1), 368–377. 10.1080/03772063.2020.1822217

Tao, Y., Wu, J., Lin, X., & Yang, W. (2023). *DRL-Driven Digital Twin Function Virtualization for Adaptive Service Response in 6G Networks*. IEEE Networking Letters.

Tareke, S. A., Lelisho, M. E., Hassen, S. S., Seid, A. A., Jemal, S. S., & Teshale, B. M., & Pandey, B. K. (2022). The prevalence and predictors of depressive, anxiety, and stress symptoms among Tepi townresidents during the COVID-19 pandemic lockdown in Ethiopia. *Journal of Racial and Ethnic Health Disparities*, 1–13.35028903

Tareke, S. A., Lelisho, M. E., Hassen, S. S., Seid, A. A., Jemal, S. S., Teshale, B. M., & Pandey, B. K. (2022). The prevalence and predictors of depressive, anxiety, and stress symptoms among Tepi town residents during the COVID-19 pandemic lockdown in Ethiopia. *Journal of Racial and Ethnic Health Disparities*, 1–13.35028903

Tataria, H., Shafi, M., Molisch, A. F., Dohler, M., Sjöland, H., & Tufvesson, F. (2021). 6G wireless systems: Vision, requirements, challenges, insights, and opportunities. *Proceedings of the IEEE*, 109(7), 1166–1199. 10.1109/JPROC.2021.3061701

Tomkos, I., Klonidis, D., Pikasis, E., & Theodoridis, S. (2020, January 1). Toward the 6G Network Era: Opportunities and Challenges. *IT Professional*, 22(1), 34–38. 10.1109/MITP.2019.2963491

Tragos, E. Z., Maglogiannis, V., Mukherjee, M., Dagiuklas, T., & Ranganathan, P. (2020). 6G Security Requirements: The Holistic Landscape. *IEEE Vehicular Technology Magazine*, 15(4), 70–77.

Tripathi, R. P., Sharma, M., Gupta, A. K., Pandey, D., Pandey, B. K., Shahul, A., & George, A. H. (2023).*Timely prediction of diabetes by means of machine learning practices*. Augmented Human Research.

Tripathi, R. P., Sharma, M., Gupta, A. K., Pandey, D., Pandey, B. K., Shahul, A., & George, A. H. (2023). Timely prediction of diabetes by means of machine learning practices. *Augmented Human Research*, 8(1), 1. 10.1007/s41133-023-00062-4

Tuegel, E. J., Ingraffea, A. R., Eason, T. G., & Spottswood, S. M. (2011). Reengineering aircraft structural life prediction using a digital twin. *International Journal of Aerospace Engineering*, 2011, 2011. 10.1155/2011/154798

Ullah, Y., Roslee, M. B., Mitani, S. M., Khan, S. A., & Jusoh, M. H. (2023). a survey on handover and mobility management in 5G HetNets: Current state, challenges, and future directions. *Sensors (Basel)*, 23(11), 5081. 10.3390/s2311508137299808

Vieira, J. (2017). Deep convolutional neural networks for massive MIMO fingerprint-based positioning. *Proceedings of the 2017 IEEE 28th Annual International Symposium on Personal, Indoor, and Mobile Radio Communications*. IEEE. 10.1109/PIMRC.2017.8292280

Vinodhini, V., Kumar, M. S., Sankar, S., Pandey, D., Pandey, B. K., & Nassa, V. K. (2022). IoT-based early forest fire detection using MLP and AROC method. *International Journal of Global Warming*, 27(1), 55–70. 10.1504/IJGW.2022.122794

Viswanathan, H., & Mogensen, P. E. (2020). Communications in the 6G era. *IEEE Access : Practical Innovations, Open Solutions*, 8, 57063–57074. 10.1109/ACCESS.2020.2981745

Wang, B., Zhou, H., Li, X., Yang, G., Zheng, P., Song, C., Yuan, Y., Wuest, T., Yang, H., & Wang, L. (2024). Human Digital Twin in the context of Industry 5.0. *Robotics and Computer-integrated Manufacturing*, 85, 102626. 10.1016/j.rcim.2023.102626

Wei, Z., Qu, H., Wang, Y., Yuan, X., Wu, H., Du, Y., Han, K., Zhang, N., & Feng, Z. (2023, July 1). Integrated Sensing and Communication Signals Toward 5G-A and 6G: A Survey. *IEEE Internet of Things Journal*, 10(13), 11068–11092. 10.1109/JIOT.2023.3235618

Wu, W., Zhou, C., Li, M., Wu, H., Zhou, H., Zhang, N., Shen, X. S., & Zhuang, W. (2022). AI-native network slicing for 6G networks. *IEEE Wireless Communications*, 29(1), 96–103. 10.1109/MWC.001.2100338

Wu, Y., Zhang, K., & Zhang, Y. (2021). Digital twin networks: A survey. *IEEE Internet of Things Journal*, 8(18), 13789–13804. 10.1109/JIOT.2021.3079510

Xu, X., Lu, Y., Vogel-Heuser, B., & Wang, L. (2021). Industry 4.0 and Industry 5.0—Inception, conception and perception. *Journal of Manufacturing Systems*, 61, 530–535. 10.1016/j.jmsy.2021.10.006

Yang, H., Alphones, A., Xiong, Z., Niyato, D., Zhao, J., & Wu, K. (2020). Artificial-intelligence-enabled intelligent 6G networks. *IEEE Network*, 34(6), 272–280. 10.1109/MNET.011.2000195

Yang, P., Xiao, Y., Xiao, M., & Li, S. (2019). 6G wireless communications: Vision and potential techniques. *IEEE Network*, 33(4), 70–75. 10.1109/MNET.2019.1800418

Yang, Q., Liu, Y., Chen, T., & Tong, Y. (2019). Federated machine learning: Concept and applications. *ACM Transactions on Intelligent Systems and Technology*, 10(2), 1–19. 10.1145/3298981

Yang, Z. (2021). AI-enabled intelligent 6G networks. *IEEE Network*, 35(2), 126–132.

Yazar, A., Doğan Tusha, S., & Arslan, H. (2020). 6G vision: An ultra-flexible perspective. *ITU Journal : ICT Discoveries*, 1(1), 121–140. 10.52953/IKVY9186

Zhang, C., Patras, P., & Haddadi, H. (2019). Deep learning in mobile and wireless networking: A survey. *IEEE Communications Surveys and Tutorials*, 21(3), 2224–2287. 10.1109/COMST.2019.2904897

Zhang, S., & Zhu, D. (2020, December). Towards artificial intelligence enabled 6G: State of the art, challenges, and opportunities. *Computer Networks*, 183, 107556. 10.1016/j.comnet.2020.107556

Zhang, Y. P., & Liu, D. (2009). Antenna-on-chip and antenna-in-package solutions to highly integrated millimeter-wave devices for wireless communications. *IEEE Transactions on Antennas and Propagation*, 57(10), 2830–2841. 10.1109/TAP.2009.2029295

Zhang, Z., Xiao, Y., Ma, Z., Xiao, M., Ding, Z., Lei, X., Karagiannidis, G. K., & Fan, P. (2019, September). 6G wireless networks: Vision, requirements, architecture, and key technologies. *IEEE Vehicular Technology Magazine*, 14(3), 28–41. 10.1109/MVT.2019.2921208

Zhao, L., Han, G., Li, Z., & Shu, L. (2020). Intelligent digital twin-based software-defined vehicular networks. *IEEE Network*, 34(5), 178–184. 10.1109/MNET.011.1900587

Zhou, C., Yang, H., Duan, X., Lopez, D., Pastor, A., Wu, Q., & Jacquenet, C. (2021). *Digital twin network: Concepts and reference architecture*. Internet Engineering Task Force.

Zhu, S., Ota, K., & Dong, M. (2022, March). Green AI for IIoT: Energy Efficient Intelligent Edge Computing for Industrial Internet of Things. *IEEE Transactions on Green Communications and Networking*, 6(1), 79–88. 10.1109/TGCN.2021.3100622

Ziegler, V., Viswanathan, H., Flinck, H., Hoffmann, M., Raisanen, V., & Hatonen, K. (2020). 6G Architecture to Connect the Worlds. *IEEE Access, 8*.

Ziegler, V., Viswanathan, H., Flinck, H., Hoffmann, M., Räisänen, V., & Hätönen, K. (2020). 6G architecture to connect the worlds. *IEEE Access : Practical Innovations, Open Solutions*, 8, 173508–173520. 10.1109/ACCESS.2020.3025032

About the Contributors

Tanveer Ahmed has spent the past 19 years working at TCS London as a Senior Project Manager. During this time, he has participated in a wide variety of projects across the globe, including those located in the United Kingdom, the United States of America, and Europe. 2023 will be the year that he defends his doctoral thesis in Cloud Communication and Security.

Binay Kumar Pandey currently working as an Assistant Professor in Department of Information Technology of Govind Ballabh Pant University of Agriculture and Technology Pantnagar Uttrakhand, India. He obtained his M. Tech with Specialization in Bioinformatics from Maulana Azad National Institute of Technology Bhopal M. P. India, in 2008 . He obtained his First Degree B. Tech at the IET Lucknow (Uttar Pradesh Technical University, Uttar Pradesh and Lucknow) India, in 2005. In 2010, he joined Department of Information Technology of College of Technology in Govind Ballabh Pant University of Agriculture and Technology Pantnagar as an Assistant Professor and worked for various UG and PG projects till date. He has more than ten years of experience in the field of teaching and research.He has more than 40 publications in reputed peer journal reputed journal Springer,Inderscience,(sci and socopus indexed journal and others) and 3 patent.He has many awards such PM Scholarship etc . He session chair in IEEE International Conference on Advent Trends in Multidisciplinary Research and Innovation (ICATMRI-2020) on December 30, 2020 organized by Pankaj Laddhad Institute of Technology and Management Studies; Buldhana, Maharashtra, India

Joel Alanya-Beltran has a PhD. in Education and Scientific Training, Doctor in Education, Master in Management and Communication of Social and Solidarity Entities at Universitat Abat Oliba CEU in Spain, Master in University Teaching, Electronic Engineer at Pontificia Universidad Católica del Perú. With diploma studies in: Use of TICS in education for teaching; Business Intelligence with Power BI; Training for research with itinerancy in scientific production; Soft Skills and Management Skills and in Fundamentals of the Digital Educational Community. With MOS Microsoft Certified Specialist in Excel 2016; Specialization in: Pedagogical use of Information and Communication Technologies for teaching; and in Business Intelligence and Analysis with Excel and Power BI. Researcher teacher, with more than 8 years of experience in university teaching with participation in different presentations in International Congresses since 2020.

Joshuva Arockia Dhanraj is currently working as an Associate Professor in Computer Science Cluster (AI&ML), School of Engineering at Dayanada Sagar University. He is also associated as Adjunct Faculty in University Centre for Research & Development (UCRD), Chandigarh University, Mohali, Punjab, India. He has totally 5.5 years of teaching experience, 3 years of research experience and 1 year of post-doc experience. He has published 160+ papers in peer-reviewed/SCI/Scopus indexed international journals and he filed nearly 10 National and 5 International Patents. His research area is in the field of Machine Learning, Machine Fault Diagnosis and Prognosis, Structural Health and Condition Monitoring, Renewable Energy Applications towards SDG Development.

Joshuva Arockia Dhanraj is currently working as an Associate Professor in Computer Science Cluster (AI&ML), School of Engineering at Dayanada Sagar University. He is also associated as Adjunct Faculty in University Centre for Research & Development (UCRD), Chandigarh University, Mohali, Punjab, India. He has totally 5.5 years of teaching experience, 3 years of research experience and 1 year of post-doc experience. He has published 160+ papers in peer-reviewed/SCI/Scopus indexed international journals and he filed nearly 10 National and 5 International Patents. His research area is in the field of Machine Learning, Machine Fault Diagnosis and Prognosis, Structural Health and Condition Monitoring, Renewable Energy Applications towards SDG Development.

B. Kiruthiga graduated B.E. in Instrumentation and Control Engineering from Madurai Kamaraj University, Madurai and M.E. in Power Electronics and Drives from Anna University, Chennai. She received her Ph.D. degree in Electrical Engineering from Anna University, Chennai. Since 2006, she is in the field of teaching and currently, She is working as an Assistant Professor in Department of Electrical and Electronics Engineering in Velammal College of Engineering and Technology, Madurai. She has received R &D funding worth Rs. 77.55 lakhs from DST. Her main field of interest is Renewable Energy System, Power Quality and Power Electronics.

Kallol Bhaumik completed his B.Tech in Electrical Engineering from the West Bengal University of Technology, West Bengal, India, in 2011. He further pursued M.Tech in Power Systems from the West Bengal University of Technology, West Bengal, India, in 2013. In 2023, he successfully earned his Ph.D. from IIT (ISM), Dhanbad, India. Dr. Bhaumik currently serves as an Associate Professor in the Department of Electrical and Electronics Engineering at Malla Reddy Engineering College and Management Sciences, located in Hyderabad, Telangana, India. With a decade of combined experience in both teaching and industry, he has obtained three patents and contributed to numerous publications in national and international journals and conferences. His research areas of interest encompass Wireless Power Transfer, EV, Power Electronics Applications, Application of High-Frequency Converter, Resonant power conversion, mainly applied to contactless energy transfer, high-frequency induction heating systems, and Multi-zone Multi-output Inverters.

Jyoti Bhola has done her B. Tech. in Electronics and Communication Engineering from Kurukshetra University. She did her M. Tech. in Electronics and Communication Engineering, from Punjab Technical University, Jalandhar. She did her PhD from National Institute of Technology, Hamirpur in Electronics and Communication Engineering. She is currently working as Assistant Professor at Chitkara University, Punjab, India. Her area of interest in research is Wireless Sensors Networks and Adhoc Networks. She has published more than 40 research articles in reputed SCI / SCOPUS journals and conferences.

Pankaj Dadheech is currently working as a Professor & Deputy Head in the Department of Computer Science & Engineering (NBA Accredited), Swami Keshvanand Institute of Technology, Management & Gramothan (SKIT), Jaipur, Rajasthan, India (Accredited by NAAC A++ Grade). He has more than 18 years of experience in teaching. He is currently working a Professor & Dy. HOD in the Department of Computer Science & Engineering (NBA Accredited), Swami Keshvanand Institute of Technology, Management & Gramothan (SKIT), Jaipur, Rajasthan, India. He has published 25 Patents at Intellectual Property India, Office of the Controller General of Patents, Design and Trade Marks, Department of Industrial Policy and Promotion, Ministry of Commerce and Industry, Government of India. He has published 8 International Patents (USA, South African, Australian, Germany) & 2 Copyrights. He has 80 publications in various International & National Journals, 63 papers in various National & International conferences. He has published 9 Books & 40 Book Chapters.

Yagya Dutta Dwivedi is a faculty member in the Department of Aeronautical Engineering at the Institute of Aeronautical Engineering, located in Hyderabad, Telangana 500043, India.

A. Shaji George is a highly respected and accomplished figure in the fields of education and the ICT industry. With over 30 years of experience in designing and deploying large-scale projects, Dr. George has made significant contributions to industry. He has a passion for working with clients to help them overcome complex business challenges, drive innovation, and deliver results. Dr. George is recognized as an expert in the field of ICT and has published over 250 research papers in national and international journals and conferences. He is also the author of 30 books that have been published by reputed international publishers. His extensive contributions to the field have earned him 20 awards, recognizing his outstanding achievements and contributions to the industry. Dr. George is the founder of PU Publications, PUIRJ, PUIIJ and PUIRP International Journal. He has a keen interest in wireless networking, cloud computing, big data analytics, the Internet of Things, and industrial automation systems. He works with clients from a wide range of industries to develop strategies and implement next-generation solutions that transform their businesses.

Vishal Ashok Ingole, working as an Assistant Professor in the Department of Management, P.R.Pote Patil College of Engg. & Management, Amravati, Maharashtra..He has pursing Ph.D in subject Digital Marketing. He has more than 13 years of teaching and 17 years of as Industrialist Corporate experience. He has published 2 book and more than 12 papers Research papers in various National and International Journals and Proceedings in National Conferences.

C.R. Komala is in the Department of Information Science and Engineering, HKBK College of Engineering, Bengaluru, Karnataka 560045, India.

T. Rajesh Kumar received the Bachelor of Engineering degree in Electronics and Communication Engineering from the Madras University, Madras in 1996, the Master of Engineering degree in Computer Science and Engineering from Manonmaniam Sundaranar University, Tirunelveli, in 2004 and the Ph.D degree in Information and Communication Engineering from Anna University, Chennai during January 2020. He is having more than 23 years of teaching experiences in Electronics and Computer Science Engineering Courses. He is currently working as Associate Professor in Computer Science and Engineering department at Saveetha School of Engineering, Saveetha Institute of Medical and Technical Sciences, Chennai, Tamil Nadu, India. T.Rajesh Kumar has published 4 SCI journals, 22 Scopus and WOS papers, 2 Indian Patents, 2 book chapters, one Text book and 6 peer review research papers in National, International journals and conferences. He chaired in various conferences as panel member and also reviewed various journal papers in Scopus/SCI journals. His research area includes Speech/Image Signal Processing, Machine learning, Data mining, Embedded System and Knowledge Engineering. He is having membership in ISTE, IAEngg., IAET, SDIWC, RED and IEEE.

T. Madhavi hails from Guntur, Andhra Pradesh, India. She obtained her B.Sc. from Nagarjuna University in 1991 and her B.Ed. from Sri Padmavathi Mahila Visvavidyalayam in 1993. In 2012, she earned an MBA in Human Resource Management and Marketing from Birla Institute of Technology (BIT), Mesra, Ranchi. Her academic journey continued with a Ph.D. in Human Resource Management from Jaipur National University, Jaipur, Rajasthan, and a PG Diploma from IIM Indore. Dr. Madhavi has held positions as an Assistant Professor at Jaipur National University and Vignan's Nirula Institute of Women's Technology and Science, among others. She has also served as an Associate Professor at K.R. Mangalam University and Vivekananda Global University. Currently, since November 2023, she has been working as an Associate Professor at the School of Commerce and Management, Mohan Babu University. With 17 and a half years of teaching and 7 years of research experience, she has contributed significantly to the fields of Human Resource Management, Talent Management, and Customer Satisfaction, among others, publishing over 25 research articles in esteemed international journals and conference proceedings.

P. Baby Shamini is presently working as Assistant Professor in the department of Computer Science and Engineering at RMK Engineering College, Chennai, India. She has 8 years of teaching experience. She completed her bachelor's degree in computer science and engineering at Bethlahem Institute of Technology, Nagercoil in the year 2012 and obtained her master's degree in Computer Science and Engineering at Java Engineering College, Chennai, in the year 2014. She has over 20 Research Articles published in reputed journal including IEEE, Springer and Scopus Indexed Journal. She is pursuing Ph.D in faculty of Information and communication in Anna University, Chennai, India. Her research interest includes Data Analytics and Deep Learning.

Sabyasachi Pramanik is a professional IEEE member. He obtained a PhD in Computer Science and Engineering from Sri Satya Sai University of Technology and Medical Sciences, Bhopal, India. Presently, he is an Associate Professor, Department of Computer Science and Engineering, Haldia Institute of Technology, India. He has many publications in various reputed international conferences, journals, and book chapters (Indexed by SCIE, Scopus, ESCI, etc). He is doing research in the fields of Artificial Intelligence, Data Privacy, Cybersecurity, Network Security, and Machine Learning. He also serves on the editorial boards of several international journals. He is a reviewer of journal articles from IEEE, Springer, Elsevier, Inderscience, IET and IGI Global. He has reviewed many conference papers, has been a keynote speaker, session chair, and technical program committee member at many international conferences. He has authored a book on Wireless Sensor Network. He has edited 8 books from IGI Global, CRC Press, Springer and Wiley Publications.

Ardly Melba Reena B has received her B. Tech.,(IT) degree from the Anna University, Chennai, in 2005, M.E.,(CSE) degree from Anna University, Chennai, in 2008, and pursuing Ph.D., degree in SRM University, Chennai, India, from 2021. For the past 17 years since 2005, she has worked in both teaching and in software. She has 8 years teaching experience and held different positions like Head of the Department in various Engineering Colleges in Tamil Nadu. She has nearly 8 years experience in software industry as a software engineer. She is currently working as Assistant Professor SG in the Institute of Computer Science and Engineering at Saveetha School of Engineering, Chennai. Her research interests include Deep learning, Cloud Systems, Machine Learning, Artificial Intelligence, Network Security etc.

Bathrinath Sankaranarayanan's areas of interests are Industrial Engineering, Production Scheduling and Optimization, Supply Chain Management, Industrial Safety. He obtained his Post Doctoral Fellowship in University of Southern Denmark. He has several publications in the area of Industrial Engineering.

Manvinder Sharma is working as Associate Professor in Department of Interdisciplinary Courses in Engineering (DICE) at Chitkara University. He has published more than 50 International Research papers including 8 SCI, 7 ESCI and 26 Scopus. He received his PhD from Punjabi University. He has guided 1 Ph.D and 5 M.tech students. His area of interest is Antenna, Biomedical, IoT and Digital Image Processing. He also received "MASTER" award from IIT Bombay Spoken tutorial programme.

Index

Symbols

6G 1, 2, 3, 4, 5, 6, 7, 8, 9, 10, 11, 12, 13, 14, 15, 16, 17, 18, 19, 20, 21, 22, 23, 25, 26, 27, 28, 29, 32, 36, 39, 40, 41, 42, 43, 44, 45, 46, 47, 48, 49, 50, 52, 53, 54, 55, 56, 57, 58, 59, 60, 61, 62, 63, 64, 65, 67, 68, 69, 74, 75, 77, 78, 79, 80, 81, 82, 83, 84, 85, 86, 87, 88, 89, 90, 91, 92, 93, 94, 96, 97, 98, 99, 100, 101, 103, 104, 114, 116, 117, 118, 120, 121, 122, 123, 124, 126, 129, 134, 135, 136, 137, 138, 139, 140, 141, 142, 154, 156, 158, 160, 161, 163, 164, 165, 166, 167, 168, 172, 174, 176, 177, 178, 179, 180, 184, 190, 191, 193, 197, 198, 199, 200, 202, 203, 204, 205, 206, 207, 209, 210, 211, 212, 213, 214, 216, 217, 218, 219, 220, 221, 223, 224, 226, 227, 230, 231, 232, 233, 234, 235, 236, 237, 239, 240, 241, 242, 243, 244, 245, 246, 247, 248, 249, 250, 251, 252, 253, 254, 255, 256, 258, 259, 260, 261, 262, 263, 264, 265, 266, 267, 268, 269, 270, 272, 273, 274, 275, 277

6G communication 60, 80, 81, 84, 94, 137, 163, 165, 176, 197, 198, 212, 214, 221, 230, 231, 233, 234, 241, 242, 251, 252, 254, 255, 262, 266, 277

6G Networks 1, 2, 3, 5, 6, 7, 12, 13, 15, 16, 17, 19, 20, 21, 22, 25, 26, 27, 28, 29, 42, 46, 47, 48, 49, 50, 52, 53, 54, 55, 56, 60, 61, 62, 63, 64, 65, 67, 68, 69, 74, 75, 80, 81, 82, 83, 84, 85, 86, 87, 88, 89, 90, 91, 92, 93, 94, 98, 99, 100, 101, 103, 104, 114, 118, 121, 122, 123, 124, 137, 140, 142, 154, 160, 163, 164, 165, 166, 168, 172, 176, 179, 180, 190, 200, 205, 211, 216, 218, 221, 223, 230, 231, 235, 236, 237, 239, 240, 241, 242, 243, 245, 251, 252, 254, 256, 258, 263, 264, 265, 266, 267, 268, 269, 272, 274, 275, 277

6G technology 49, 53, 63, 69, 103, 104, 121, 122, 141, 142, 154, 172, 178, 179, 190, 191, 241

A

ai 3, 5, 6, 7, 9, 12, 13, 14, 15, 16, 17, 18, 19, 20, 21, 22, 23, 25, 26, 27, 28, 29, 31, 32, 33, 36, 37, 38, 39, 40, 42, 47, 60, 61, 62, 63, 64, 65, 68, 69, 70, 71, 72, 73, 74, 75, 77, 80, 81, 82, 83, 84, 85, 86, 87, 88, 89, 90, 91, 92, 93, 94, 96, 100, 101, 102, 104, 122, 123, 140, 142, 164, 165, 174, 176, 178, 183, 190, 192, 194, 200, 205, 206, 207, 209, 210, 212, 213, 214, 215, 216, 221, 222, 224, 225, 226, 234, 237, 240, 241, 242, 243, 244, 245, 246, 248, 249, 252, 274, 277, 279

AI 3, 5, 6, 7, 9, 12, 13, 14, 15, 16, 17, 18, 19, 20, 21, 22, 23, 25, 26, 27, 28, 29, 31, 32, 33, 36, 37, 38, 39, 40, 42, 47, 60, 61, 62, 63, 64, 65, 68, 69, 70, 71, 72, 73, 74, 75, 77, 80, 81, 82, 83, 84, 85, 86, 87, 88, 89, 90, 91, 92, 93, 94, 96, 100, 101, 102, 104, 122, 123, 140, 142, 164, 165, 174, 176, 178, 183, 190, 192, 194, 200, 205, 206, 207, 209, 210, 212, 213, 214, 215, 216, 221, 222, 224, 225, 226, 234, 237, 240, 241, 242, 243, 244, 245, 246, 248, 249, 252, 274, 277, 279

Anomaly Detection 28, 29, 36, 37, 39, 42, 50, 52, 53, 54, 55, 64, 71, 72, 83, 85, 93, 147, 171, 172, 186, 191, 230, 231, 233, 239, 241, 242, 245, 246, 247

antenna 10, 11, 13, 14, 25, 43, 59, 78, 117, 122, 139, 142, 158, 176, 210, 217, 251, 252, 253, 254, 255, 256, 257, 258, 259, 260, 261, 279

Artificial Intelligence 3, 10, 12, 15, 23, 24, 25, 26, 27, 29, 31, 32, 33, 34, 36, 37, 38, 39, 43, 47, 50, 53, 58, 61, 62, 64, 68, 75, 76, 78, 80, 82, 83, 84, 85, 86, 88, 89, 90, 93, 94, 96, 97, 98, 99, 101, 104, 105, 114, 116, 117, 122, 123, 126, 138, 142, 157, 165, 175, 178, 190, 193, 194, 204, 209, 210, 212, 214, 224, 227, 228, 234, 237, 239, 245, 247, 249, 259, 261, 274, 275, 279

authentication 1, 2, 3, 4, 5, 6, 7, 8, 17, 18, 20, 21, 25, 28, 29, 30, 33, 34, 48, 49, 62, 64, 68, 72, 74, 81, 89, 92, 103, 122, 130, 131, 146, 152, 153, 154, 161, 168, 170, 171, 172, 179, 199, 205, 213, 215, 216, 218, 219, 220, 221, 222, 223, 238, 241, 266, 269, 270, 271, 277

B

biometric authentication 28, 30, 33, 34, 220, 270, 271

Blockchain 1, 2, 3, 4, 5, 6, 7, 8, 10, 15, 28, 41, 42, 76, 92, 95, 100, 107, 115, 126, 136, 144, 155, 161, 162, 165, 166, 168, 169, 170, 173, 192, 198, 202, 203, 205, 206, 208, 212, 213, 216, 218, 219, 221, 222, 227, 244, 246, 262, 268, 269, 277, 278

blockchain technology 8, 41, 76, 92, 95, 100, 115, 126, 136, 155, 161, 162, 165, 166, 173, 192, 208, 212, 216, 221, 227, 244, 262, 268, 269, 277, 278

C

challenges 1, 2, 5, 7, 8, 9, 10, 12, 13, 15, 16, 17, 21, 23, 24, 29, 35, 36, 41, 42, 46, 47, 48, 49, 55, 58, 60, 61, 62, 64, 67, 68, 69, 71, 75, 76, 77, 81, 82, 84, 86, 87, 88, 90, 91, 92, 94, 95, 96, 97, 98, 100, 101,

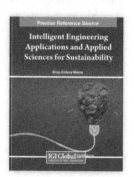
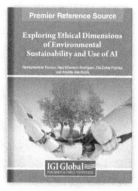

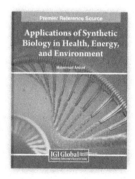

Ensure Quality Research is Introduced to the Academic Community

Become a Reviewer for IGI Global Authored Book Projects

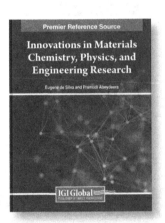
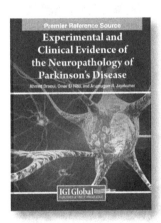
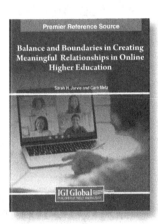

The overall success of an authored book project is dependent on quality and timely manuscript evaluations.

Applications and Inquiries may be sent to:
development@igi-global.com

Applicants must have a doctorate (or equivalent degree) as well as publishing, research, and reviewing experience. Authored Book Evaluators are appointed for one-year terms and are expected to complete at least three evaluations per term. Upon successful completion of this term, evaluators can be considered for an additional term.

If you have a colleague that may be interested in this opportunity, we encourage you to share this information with them.

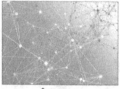

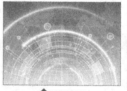

Printed in the United States
by Baker & Taylor Publisher Services